Artificial Intelligence for Disease Diagnosis and Prognosis in Smart Healthcare

Artificial Intelligence (AI) in general and machine learning (ML) and deep learning (DL) in particular and related digital technologies are a couple of fledging paradigms that next-generation healthcare services are sprinting towards. These digital technologies can transform various aspects of healthcare, leveraging advances in computing and communication power. With a new spectrum of business opportunities, AI-powered healthcare services will improve the lives of patients, their families, and societies. However, the application of AI in the healthcare field requires special attention given the direct implication with human life and well-being. Rapid progress in AI leads to the possibility of exploiting healthcare data for designing practical tools for automated diagnosis of chronic diseases such as dementia and diabetes. This book highlights the current research trends in applying AI models in various disease diagnoses and prognoses to provide enhanced healthcare solutions. The primary audience of the book are postgraduate students and researchers in the broad domain of healthcare technologies.

Artificial Intelligence for Disease Diagnosis and Prognosis in Smart Healthcare

Edited by
Ghita Kouadri Mostefaoui
S. M. Riazul Islam
Faisal Tariq

CRC Press
Taylor & Francis Group
Boca Raton London New York

CRC Press is an imprint of the
Taylor & Francis Group, an **informa** business

First edition published 2023
by CRC Press
6000 Broken Sound Parkway NW, Suite 300, Boca Raton, FL 33487-2742

and by CRC Press
4 Park Square, Milton Park, Abingdon, Oxon, OX14 4RN

CRC Press is an imprint of Taylor & Francis Group, LLC

Library of Congress Cataloging-in-Publication Data
Names: Mostefaoui, Ghita K., editor. | Islam, S. M. Riazul, editor. | Tariq, Faisal, editor.
Title: Artificial intelligence for disease diagnosis and prognosis in smart healthcare / edited by Ghita K. Mostefaoui, S.M. Riazul Islam, Faisal Tariq.
Description: Boca Raton : CRC Press, 2022. | Includes bibliographical references and index. | Summary: "Artificial Intelligence (AI) in general and machine learning (ML) and deep learning (DL) in particular and related digital technologies are a couple of fledging paradigms that the next generation healthcare services are sprouting towards. These digital technologies can transform various aspects of healthcare, leveraging advances in computing and communication power. With a new spectrum of business opportunities, AI-powered healthcare services would improve the lives of patients, their families, and societies. However, the application of AI in the healthcare field requires special attention given the direct implication with human life and well-being. Rapid progress in AI leads to the possibility of exploiting healthcare data for designing practical tools for automated diagnosis of chronic diseases such as dementia and diabetes. This book highlights the current research trends in applying AI models in various disease diagnoses and prognoses to provide enhanced healthcare solutions. The primary audience of the book will be postgraduate students and researchers in the broad domain of healthcare technologies"-- Provided by publisher.
Identifiers: LCCN 2022034534 | ISBN 9781032168302 (hardback) | ISBN 9781032171265 (paperback) | ISBN 9781003251903 (ebook)
Subjects: MESH: Artificial Intelligence | Diagnosis, Computer-Assisted | Precision Medicine | Medical Informatics Applications
Classification: LCC R855.3 | NLM W 26.55.A7 | DDC 610.285--dc23/eng/20221206
LC record available at https://lccn.loc.gov/2022034534

ISBN: 978-1-032-16830-2 (hbk)
ISBN: 978-1-032-17126-5 (pbk)
ISBN: 978-1-003-25190-3 (ebk)

DOI: 10.1201/9781003251903

Typeset in Latin Roman font
by KnowledgeWorks Global Ltd.

Contents

Preface

Artificial Intelligence (AI) in general and machine learning and deep learning, in particular, are becoming increasingly sophisticated and continuously pushing the boundary of machine capabilities to be more efficient, more rapid, and low cost. In line with this, the potential for AI in healthcare is enormous. Indeed, AI and related digital technologies are a couple of fledging paradigms that next-generation healthcare services are sprinting towards. These digital technologies can transform various aspects of healthcare, leveraging advances in computing and communication power. With a new spectrum of business opportunities, AI-powered healthcare services will improve the lives of patients, their families and societies. Because of the AI-driven paradigm shifts in healthcare approaches, researchers across the globe have started to explore various technological solutions. This book provides comprehensive coverage of AI models across multiple disease diagnoses and prognoses highlighting the recent advances, working principles, and relevant challenges.

The book introduces a broad perspective of how AI has become an integral part of medical research and has contributed to producing new drugs. It helps us to understand how AI can better analyze the evolving nature of diseases. In clinical practice, the use of multiple medications, called polypharmacy, is a well-known problem, which can occur in many situations. For example, new medication administration where a proper diagnosis is not noted can result in the unnecessary use of drugs. Therefore, drug sensitivity prediction that helps to prevent polypharmacy is crucial for quality treatment. The book provides a thorough discussion on how AI approaches can be helpful in drug response prediction. Precision medicine is an emerging approach to disease treatment and prevention. To personalize treatment and medical services, it can exploit people's unique characteristics, such as their genome sequence, microbiome composition, and lifestyle. The book pays adequate attention to how better-quality patient care can be ensured using precision medicine. Apart from generic AI approaches irrespective of diseases, the book introduces particular AI methods and systems for a set of specific diseases, including breast cancer, glioblastoma, COVID-19, skin lesions, and Parkinson's disease. The book also provides good tutorial content on data-driven cancer analysis using web-based online tools in the context of a specific gene and cancer type.

Currently, there are few books that cover AI applications in healthcare in general. Unlike those, this book is more focused on the diagnosis and prognosis of human diseases. Moreover, this book covers a few additional topics, such as privacy and security in AI-based healthcare, which are becoming vitally important but largely ignored by existing books. As this is a rapidly developing field, we captured the latest developments in the area during the last couple of years, making this book distinct from the others.

Since we have presented most of the book's content in a tutorial manner, it will be well-perceived by the general audience with no solid technical background. However, in particular, the book will be a good fit for postgraduate students and researchers in the broad domain of healthcare technologies. Also, the book can be helpful as a reference for BSc/MSc project/thesis works. Overall, we expect that this edited book will help readers understand how to achieve enhanced healthcare solutions in the era of AI.

Editors

Ghita Kouadri Mostefaoui, PhD, is currently Associate Professor at the Department of Computer Science, University College London (UCL). Her current teaching includes programming, computer architecture and software engineering. Before joining UCL, Dr. Mostefaoui worked as Software Engineer at the UK Synchrotron where she developed the beamlines archiving application. Prior to that, she worked as Research Officer at Oxford University, where she contributed to several e-health projects. She received her PhD in computer science from both the University of Fribourg, Switzerland and Université Pierre et Marie Curie (Paris 6). Dr. Mostefaoui is also a Fellow of the Higher Education Academy.

S. M. Riazul Islam, PhD, is currently a Senior Lecturer in Computer Science at the University of Huddersfield, United Kingdom. Before, moving to the UK, he was an Assistant Professor at the Department of Computer Science and Engineering, Sejong University, South Korea, from 2017 to 2022. His prior affiliations were at Inha University as a Postdoctoral Fellow, at Samsung R&D Institute as a Senior Engineer, and at the University of Dhaka as an Assistant Professor in EEE. He received his PhD in information and communication engineering from Inha University, South Korea, in 2012. His research interests include applied AI, machine learning, data science, and IoT.

Faisal Tariq, PhD, is currently Senior Lecturer at the James Watt School of Engineering, University of Glasgow, United Kingdom. Prior to his current role, he worked at Queen Mary University of London as a lecturer. He received his MSc in engineering from Chalmers University of Technology, Sweden and PhD from the Open University, UK. His research interests include applications of AI and machine learning (ML) in various domains including smart wireless communications, healthcare technology, cyber security, and intelligent Internet of Things (IIoT). He received the best paper award in IEEE WPMC 2013. He is Associate Editor for *IEEE Wireless Communications Letters* and Editor for the *Journal of Networks and Computer Applications*. He is a senior member of IEEE and Fellow of the Higher Education Academy.

Contributors

S. M. Riazul Islam
Department of Computer Science
University of Huddersfield
Huddersfield, UK

Faisal Tariq
James Watt School of Engineering
University of Glasgow
Glasgow, UK

Ghita Kouadri Mostefaoui
Department of Computer Science
University College London
London, UK

Sujay Saha
Department of Electrical and Electronic
    Engineering
University of Dhaka
Dhaka, Bangladesh

Shekh Mahmudul Islam
Department of Electrical and Electronic
    Engineering
University of Dhaka
Dhaka, Bangladesh

Muhammad Afzal
Department of Software
Sejong University
Sejong, South Korea

Maqbool Hussain
Department of Computer Science
University of Derby
Derby, UK

Julhash U. Kazi
Division of Translational Cancer Research
Department of Laboratory Medicine and
    Lund Stem Cell Center
Lund University
Lund, Sweden

Masum Shah Junayed
Department of Computer Engineering
Bahcesehir University
Istanbul, Turkey

Baharul Islam
Department of Computer Engineering
Bahcesehir University
Istanbul, Turkey

Arezoo Sadeghzadeh
Department of Computer Engineering
Bahcesehir University
Istanbul, Turkey

Esraa Hassan
Department of Machine Learning and
    Information Retrieval Faculty of
    Artificial Intelligence
Kafrelsheikh University
Kafrelsheikh, Egypt
    and
Department of Computer Science Faculty
    of Computers and Information
Mansoura University
Mansoura, Egypt

Mahmoud Y. Shams
Department of Machine Learning and
    Information Retrieval Faculty of
    Artificial Intelligence
Kafrelsheikh University
Kafrelsheikh, Egypt

Noha A. Hikal
Department of Information Technology
    Faculty of Computers and Information
Mansoura University
Mansoura, Egypt

Samir Elmougy
Department of Computer Science Faculty
    of Computers and Information
Mansoura University
Mansoura, Egypt

Hossain Shakhawat
Memorial Sloan Kettering Cancer Center
New York, New York
and
Tokyo Institute of Technology
Yokohama, Japan

Matthew Hanna
Memorial Sloan Kettering Cancer Center
New York, New York

Kareem Ibrahim
Memorial Sloan Kettering Cancer Center
New York, New York

Rene Serrette
Memorial Sloan Kettering Cancer Center
New York, New York

Peter Ntiamoah
Memorial Sloan Kettering Cancer Center
New York, New York

Marcia Edelweiss
Memorial Sloan Kettering Cancer Center
New York, New York

Edi Brogi
Memorial Sloan Kettering Cancer Center
New York, New York

Meera Hameed
Memorial Sloan Kettering Cancer Center
New York, New York

Masahiro Yamaguchi
Tokyo Institute of Technology
Yokohama, Japan

Dara Ross
Memorial Sloan Kettering Cancer Center
New York, New York

Yukako Yagi
Memorial Sloan Kettering Cancer Center
New York, New York

Subbroto Kumar Saha
Department of Biochemistry and Molecular
    Medicine
University of California
Davis, California

Afsana Nishat
Microbiology and Cell Science
University of Florida
Gainesville, Florida

Shaker El-Sappagh
Faculty of Computer Science and
    Engineering
Galala University
Suez, Egypt

Moradul SiddiqueYeasir
Department of Computer Science and
    Engineering
Jashore University of Science and
    Technology
Jashore, Bangladesh

Arefin Tusher
Department of Computer Science and
    Engineering
Jashore University of Science and
    Technology
Jashore, Bangladesh

Humaun Kabir
Department of Computer Science and
  Engineering
Jashore University of Science and
  Technology
Jashore, Bangladesh

Mohammad Farhad Bulbul
Department of Mathematics
Jashore University of Science and
  Technology
Jashore, Bangladesh
  and
Department of Computer Science and
  Engineering
Pohang University of Science and
  Technology
Pohang, South Korea

Syed Galib
Department of Computer Science and
  Engineering
Jashore University of Science and
  Technology
Jashore, Bangladesh

Fatma M. Talaat
Faculty of Artificial Intelligence
Kafrelsheikh University
Kafrelsheikh, Egypt

Zeinab Hassan
Faculty of Artificial Intelligence
Kafrelsheikh University
Kafrelsheikh, Egypt

Nora El-Rashidy
Faculty of Artificial Intelligence
Kafrelsheikh University
Kafrelsheikh, Egypt

Sakir Hossain
Memorial Sloan Kettering Cancer Center
New York, New York

Alamgir Kabir
Memorial Sloan Kettering Cancer Center
New York, New York

S. M. Hasan Mahmud
Memorial Sloan Kettering Cancer Center
New York, New York

Shafiqul Islam
School of Computing Science
University of Glasgow
Glasgow, United Kingdom

Khairul Islam
Department of Information Communication
  Technology
Islamic University
Kushtia, Bangladesh

Al Amin
Department of Computer Science and
  Engineering
Prime University
Dhaka, Bangladesh

Mojibur Rahman Redoy Akanda
Department of Computer Science and
  Engineering
Prime University
Dhaka, Bangladesh

Shabuj Hossen
Department of Computer Science and
  Engineering
Prime University
Dhaka, Bangladesh

Feroza Naznin
Department of Computer Science and
  Engineering
Prime University
Dhaka, Bangladesh

Zahidul Islam
Department of Information Communication
  Technology
Islamic University
Kushtia, Bangladesh

Mohammad Ali Moni
School of Health and Rehabilitation
  Sciences
Faculty of Health and Behavioural Sciences
University of Queensland
St. Lucia, Australia

Jakaria Islam Emon
Department of Computer Science
American International University
Dhaka, Bangladesh

M. M. Manjurul Islam
School of Computing, Engineering and
  Intelligent Systems
University of Ulster
Ulster, United Kingdom

Syeda Amina Abedin
Department of Computer Science
American International University
Dhaka, Bangladesh

Hossain Shakhawat
Department of Computer Science
American International University
Dhaka, Bangladesh

Rupam Kumar Das
College of Science and Engineering
University of Glasgow
Glasgow, UK

Harsha Kumara Kalutarage
School of Computing
Robert Gordon University
Aberdeen, UK

Muhammad Shadi Hajar
School of Computing
Robert Gordon University
Aberdeen, UK

M. Omar Al-Kadri
School of Computing and Digital
  Technology
Birmingham City University
Birmingham, UK

M. Humayun Kabir
Department of Electrical and Electronic
  Engineering
Islamic University
Kushtia, Bangladesh

Introduction to Artificial Intelligence (AI) for Disease Diagnosis and Prognosis in Smart Healthcare

S. M. Riazul Islam

Department of Computer Science, University of Huddersfield, Huddersfield, UK

Faisal Tariq

James Watt School of Engineering, University of Glasgow, UK

Ghita Kouadri Mostefaoui

Department of Computer Science, University College London, UK

CONTENTS

THE INTELLECTUAL capability of animals, including humans, is known as natural intelligence. In contrast, intelligence demonstrated by machines is termed artificial intelligence (AI). An AI-driven system can perceive its environment and take the most favorable actions to achieve its goals. As such, AI-driven systems can independently find ways to solve problems, be able to conclude, and make decisions. In this chapter, we first provide an overview of AI applications, advantages, and technologies, followed by a discussion on AI in healthcare. Then, we present a short description of each chapter.

1.1 AI APPLICATIONS AND ADVANTAGES

Due to advancements in AI, we can find a significant collaboration between humans and machines in our daily lives. Like electricity or computers, AI is now a versatile technology with endless applications. AI has impacted various fields such as

DOI: 10.1201/9781003251903-1

agriculture, autonomous vehicles, gaming and home appliances. In addition, it has enhanced language translation, image recognition, customer service, finance, sales and marketing and many other domains. When it comes to human health, many healthcare service providers and related organisations are relying on AI.

1.2 AI TECHNOLOGIES

AI encompasses a wide range of tools and technologies. Whereas some of them are proven, others are emerging. Major subfields of AI are machine learning (ML), natural language processing (NLP) and computer vision (CV). It is worth noting that ML algorithms process large amounts of data and learn from them. There are different types of ML models and approaches such as supervised learning, unsupervised learning and reinforcement learning. Besides, deep learning (DL) models (another prominent family of ML methods), built on artificial neural networks, are widely used in AI applications. In addition, various other digital technologies and soft computing approaches, including the Internet of Things (IoT), fuzzy logic and genetic algorithms, play a significant role in developing AI-driven systems [1, 2]. When it comes to the components of AI applications, it usually contains three key elements: Data, algorithms and human feedback. Therefore, ensuring each of these components is adequately structured and validated for developing and implementing AI applications is highly important.

1.3 AI IN SMART HEALTHCARE

The term "AI in healthcare" refers to the systematic use of AI technologies to analyze medical and healthcare data and extract actionable insights for improved healthcare services. The adoption of AI in healthcare simplifies the lives of patients, physicians, and care providers in less time, with reduced errors at lower cost. AI in healthcare has facilitated pharmaceutical companies to accelerate their drug discovery process. It has made clinical trial automation more straightforward and shortened the required time. AI-enabled medical robots can assist patients. In the era of big data, data-driven medicine has the potential to improve disease diagnosis and open the door to targeted therapy [4]. It has made brain diseases such as Alzheimer's disease prediction and detection more accurate and precise [3,4]. Researchers have uncovered many medical diagnostic systems using AI algorithms [5]. AI-driven systems are even effective in detecting, monitoring, and managing contagious diseases like coronavirus (COVID-19) [6].

Because of its immense and broad business opportunities, big high-tech firms are also studying and implementing AI technologies to solve practical problems in healthcare. For example, Amazon Comprehend Medical is an NLP service that uses ML for disease diagnosis from health data. IBM Watson Health is a digital tool that provides cloud, data and AI solutions for many medical problems. Also, Google Health and Samsung Health can track various aspects of daily life, contributing to people's well-being.

However, the application of AI in the healthcare field requires special attention, given the direct implications for human life and well-being. Rapid progress in AI leads to the possibility of exploiting healthcare data for designing practical tools for the automated diagnosis of chronic diseases such as dementia and diabetes. Data is the fuel for any AI application. On the other hand, in the medical domain, data can be challenging to obtain, mainly for privacy concerns. That makes the combination of both AI and healthcare a significant research challenge. The role of this book is to explore the advances, opportunities, and challenges of AI in healthcare solutions. We highlight the current research trends in applying AI models to various disease diagnoses and prognoses to make healthcare more effective and efficient.

1.4 ORGANISATION OF THIS BOOK

In Chapter 2, we provide a comprehensive overview of ML approaches for human disease assessment along with their benefits and shortcomings. In addition, the chapter explains the challenges associated with healthcare applications in real-world settings.

Chapter 3 provides an overview of the potential applications of precision medicine. It explains the interplay between data science, AI, and ML in the context of precision medicine realization. Moreover, the chapter suggests a useful road map for the developers to apply precision medicine to futuristic healthcare solutions.

Chapter 4 presents the recent developments in drug response prediction methods using ML algorithms. It discusses both the monotherapy and drug combination prediction methods. The chapter also provides an overview of pharmacogenomic data selection methods and relevant ML algorithms.

Chapter 5 presents a thorough tutorial on AI-based skin disease diagnosis. It first introduces the skin diseases and then subsequently describes the image acquisition methods, available datasets and required pre-processing techniques. The chapter then explains the feature extraction and classification of skin diseases in detail. At each tutorial step, this chapter provides some sample *Python* codes.

Chapter 6 provides a comprehensive survey of DL approaches to COVID-19 identification and lung segmentation. It reviews the X-ray and computed tomography (CT) images for feature extraction and classification based on the $COVID_X$ dataset. Finally, the chapter identifies the hurdles in detecting DL approaches for COVID-19 and offers some avenues to address the challenges.

Chapter 7 discusses an AI-assisted whole slide imaging (WSI) system to grade invasive breast cancer patients, enabling therapeutic decisions automatically. In addition, the chapter demonstrates how the proposed framework overcomes the weaknesses of current practices in the selection of patients with breast cancer for therapeutic purposes.

Chapter 8 reviews the significance of CALD1 (Caldesmon 1) gene expression in brain cancer using multiomics analysis. The chapter investigates the expression patterns, functions and prognostic values of CALD1 in brain cancer by analyzing gene expression data via a set of online bioinformatics tools. It provides some important insights into whether CALD1 expression has any prognostic significance in the context of brain cancer.

Chapter 9 illustrates the basic principles for Parkinson's disease (PD) detection using ML algorithms and provides a survey of AI-based PD identification. The chapter reviews various feature extension and reduction methods to find reliable solutions for detecting PD in its early stages.

Chapter 10 reviews several commonly used breast cancer datasets. It also provides an overview of the efficacy of several DL architectures for breast cancer screening. The chapter helps us to realize the current developments in DL approaches for breast tumor diagnosis.

Chapter 11 explains the working principles of digital pathology based on the WSI system. It illustrates how the application of image processing and AI extracts meaningful insights that are largely beyond the scope of traditional pathology, which eventually leads to precise disease diagnosis and improved patient care.

Chapter 12 provides an overview of how ML and DL techniques can be applied to extract various features related to diabetes mellitus and exploit them for early prediction and management of diabetes in a cost-effective way. This early warning mechanism can aid in the prevention of the disorder by taking appropriate measures and, at minimum, to reduce the severity of the disease and prolong its onset.

Chapter 13 describes how IoT and DL can be applied to assist physicians and radiologists in remotely diagnosing skin lesions. In addition, the chapter explains the importance of trustworthy applications via a demonstration of cloud-based computer-assisted diagnosis.

Chapter 14 presents an automated facemask detection scheme in a real-time video stream. The technique uses a deep convolutional neural network (DCNN)-based transfer learning approach. The experimental results suggest that the proposed framework can be advantageous in public health management during the crisis in COVID-19-like pandemics.

Chapter 15 illustrates the importance of wireless technologies to fulfill the requirements of AI-enabled medical systems in the era of 5G and beyond. Furthermore, it explains how we can protect wireless health systems and networks from external adversaries and internal intruders with respect to the wireless body area network (WBAN).

Chapter 16 presents the AI-based security and privacy solutions for physiological data protection following an end-to-end signal flow. It highlights the shortcomings of the associated ML approaches and provides some indications of overcoming the identified weaknesses.

Finally, in Chapter 17, we provide a short overview of the future challenges in AI for disease diagnosis and prognosis.

Bibliography

1. S. M. R. Islam, D. Kwak, M. H. Kabir, M. Hossain, and K.-S. Kwak, "The internet of things for health care: a comprehensive survey," *IEEE Access*, vol. 3, pp. 678–708, 2015.

2. F. Ali, S. M. R. Islam, D. Kwak, P. Khan, N. Ullah, S.-j. Yoo, and K. S. Kwak, "Type-2 fuzzy ontology-aided recommendation systems for IoT-based healthcare," *Computer Communications*, vol. 119, pp. 138–155, 2018.

3. S. K. Saha, S. M. R. Islam, K.-S. Kwak, M. Rahman, S.-G. Cho, F. Ali, "Prom1 and prom2 expression differentially modulates clinical prognosis of cancer: a multiomics analysis," *Cancer Gene Therapy*, vol. 27, no. 3, pp. 147–167, 2020.

4. P. Khan, M. F. Kader, S. M. R. Islam, A. B. Rahman, M. S. Kamal, M. U. Toha, and K.-S. Kwak, "Machine learning and deep learning approaches for brain disease diagnosis: principles and recent advances," *IEEE Access*, vol. 9, pp. 37622–37655, 2021.

5. S. Kaur, J. Singla, L. Nkenyereye, S. Jha, D. Prashar, G. P. Joshi, S. El-Sappagh, M. S. Islam, and S. M. R. Islam, "Medical diagnostic systems using artificial intelligence (AI) algorithms: principles and perspectives," *IEEE Access*, vol. 8, pp. 228049–228069, 2020.

6. N. El-Rashidy, S. El-Sappagh, S. M. R. Islam, H. M. El-Bakry, and S. Abdelrazek, "End-to-end deep learning framework for coronavirus (COVID-19) detection and monitoring," *Electronics*, vol. 9, no. 9, p. 1439, 2020.

## CHAPTER 2

Machine Learning for Disease Assessment

Sujay Saha

Department of Electrical and Electronic Engineering, University of Dhaka

Shekh Mahmudul Islam

Department of Electrical and Electronic Engineering, University of Dhaka

CONTENTS

MACHINE LEARNING (ML) has attracted significant interest toward developing smart healthcare systems by providing intelligent clinical decision making especially on disease assessment and its associated risk prediction. At present, many diseases need to be identified at early stages to prevent the associated risk which can also play an important role in smart digital healthcare interventions. When ML is applied to clinical data it tries to interpret the data to suggest the best action that needs to be performed to improve the individual's health. However, there remain significant challenges to incorporate the ML-based approach in a real-world setting such as performance between trained model and computational time, and so on. This chapter focuses on discussing the basics of the methodology of the ML approach for disease assessment and it also highlights different assessment aspects. Additionally, it will also discuss different disease detection and assessment-related application areas of the ML-based approach and its associated pros and cons. It will also elucidate different challenges of this approach and potential solutions to bring this ML approach in real-world settings for the benefit of humanity.

DOI: 10.1201/9781003251903-2

2.1   INTRODUCTION

In the modern world, the number of incurable and deadly diseases is increasing rapidly even though there is a significant advancement in the medical field [5]. For the betterment of human health incurable and deadly diseases need to be identified at their early stage. Therefore, there is a need for analyzing medical data and images in a minimum amount of time with greater accuracy. Analyzing medical data with greater efficiency can also help to predict and detect the disease promptly. Implementing and designing new effective algorithms for processing medical data is always challenging as decision-making from medical data is difficult due to the complex nature of the medical data. Additionally, processing large medical data requires a huge amount of time as a huge amount of data needs to be processed for decision making. In the healthcare field, starting from information from medical data and then for disease diagnosis machine learning (ML) is gaining attention and has significant progress.

ML is being applied in different techniques and algorithms in various healthcare activities [3]. For clinical decision-making, machine learning is being mostly employed. When ML is applied to medical data, then the learning model can interpret the data to suggest the best actions that need to be performed to improve an individual's health. In general, the ML model needs to learn the problem statement and the quality of the data is also important for decision making [5]. In general, the ML model is not very strong and concrete when it is first used but with repetition, it gains experience from the previous iterations. Additionally, the learning model can also become robust and automated with the previous training experience. There are two ways that clinical decision-making is performed by comparing the previous knowledge that is included in the dataset. The first one is based on intuition on the underlying pattern of the medical dataset and is mostly used in medical emergencies. Another approach is a slow and reasoned approach where a large medical dataset is used to train the learning model. The second approach is more accurate but it takes a little bit more time as a large dataset needs to be handled to train the learning model. With the advancement of the processing power of computing devices, a reasoned approach is being used in the healthcare sector to improve quality.

Artificial intelligence (AI) is a branch of computer science where the intelligence of the machine is being integrated to perform any task. ML is a subset of AI and it is training algorithms for decision making. ML is being used in different areas of healthcare where the goal to integrate ML is to interpret the medical data in a meaningful way for the prediction and assessment of a disease [6]. ML models are being used in different areas such as heart disease prediction, type 2 diabetes risk prediction, and so on [6]. ML model is trying to bring a paradigm shift to healthcare due to the rapid progress of the analytical technique [6]. Most recently, the world is undergoing unprecedented challenges due to the COVID pandemic [4]. Different researchers are also trying to integrate AI and ML models for predicting the risk factor of COVID-19 infected patients [2]. For disease assessment, it is not always about detecting or predicting disease but it covers the broader spectrum including medical treatment research, public healthcare policy-making, patient care, hospital volume, and so on. Thus, AI/ML can play an important role in digital healthcare intervention because it can cover a broader range of applications.

AI is playing an important role in the transformation of healthcare but some ethical aspects need to be considered for pragmatic implementation of this technology in the real world. For example, decision-making is performed by the algorithm so the decision may be biased due to the algorithm bias. So, it may cause the issue of accountability and transparency of the system and at the same time, it will also question the integrity of the clinicians if they depend on ML. Therefore, it is very important to consider the robustness and efficacy of the AI/ML model for any specific disease assessment. Moreover, algorithms in AI can perform analysis of and prediction from big datasets to provide fast and informed decisions. At present in the healthcare sectors mostly AI/ML model is employed as an aid for diagnosis. To integrate those methods in real-world settings and for direct decision-making, several challenges need to be overcome.

This chapter focuses on discussing the basics of integrating AI/ML modes for disease assessment. For instance, it will highlight the basics of ML methodology for disease assessment, disease risk prediction, and evaluation method of the ML model. In addition, it will also elucidate some of the potential applications for disease assessment using ML and AI. It also concentrates on highlighting the challenges of this method and its associated potential solutions for bringing this into real-world settings for the benefit of humanity.

2.1.1 Methodology for Disease Assessment

Identifying a disease accurately is a huge challenge in the field of medical treatment. It is felt more severely in recent days with the introduction of COVID in our lives. The detection of COVID and its effect on our bodies may take a long time. The state of the patient can degrade significantly within that period, which may cause permanent damage to the body or even death. Overcoming this challenge can remarkably decrease the risk of diseases and create the scope for initial treatment that may save one from permanent damage. The procedure of identifying a disease and its intensity is formally known as *disease assessment* [7]. The difficulty of the entire process of disease assessment can vary based on the disease itself or the approach taken. With the advancement of technology, most approaches are ML-based regardless of the disease type. X-ray images are now classified using a computer vision-based AI model. The model needs to be built systematically to do such assessments using the ML/AI model. The approach can differ depending on the application and disease, but most follow a usual course.

First is data collection. Data collection is referred to as the activity of gathering data from different types of sources according to our needs and entering the data into one single framework. It is one of the most important works in the entire process, and can be a bit tedious depending on the approach and need. If we want to determine the effect of COVID on a patient from chest X-rays, we need to collect X-ray images of thousands of people. Images from patients who are in different stages of COVID need to be collected. Data of those that don't have a record of COVID is also necessary to

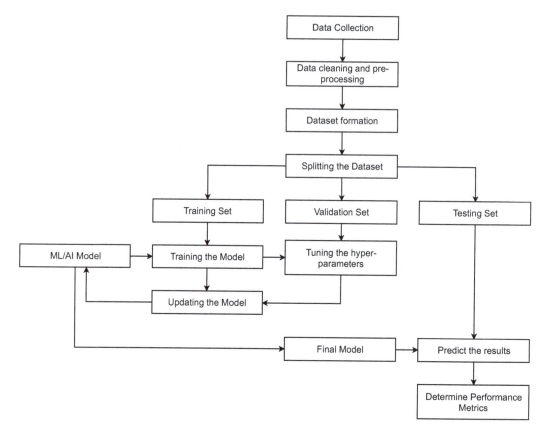

Figure 2.1 Methodology for disease assessment.

collect as it will help make the dataset more robust. One can make a state-of-the-art AI model, but the model won't perform well if the data is not collected decently.

Data can come in different shapes and forms. It can include age, gender, nation/race, address, location, heart rate, and more. Depending on the need, those data can be quantitative data like integers, floats, strings, characters, dates, Booleans, images, sounds, etc., and qualitative data like feelings and emotions. All these data can be collected in many different ways—via sensors, forms, live questionnaires, cameras, automated scripts, etc.

Data can also be classified as primary data and secondary data depending on the necessity of the particular use case. It is a good practice to keep the time and date stamp for each data. But, in many cases, we may not need that date and time. Here the date and time are considered as secondary data. On the other hand, time can include primary data where the disease assessment or medication is highly time-dependent. Finally, the data collection process should be non-biased and neutral throughout the entire process.

During the data collection process, chances are the collected data may not be perfect or somewhat accurate. There might be some missing data; some data may have the wrong value type; some data may not be correct at all; some data may not be in between the range it should be. That's why the data needs to be cleaned and pre-processed before feeding the data into the model.

Let us say we want to recognize human activity or gait from images or videos. A lot of things can go wrong here. The camera might fall or shake slightly making the images blurry and out of focus. And the worst-case scenario, the camera might stop working altogether. We don't want these types of anomalies in our data. So, we need to exclude such data. The data cleaning process helps eliminate such anomalies in the data, making the dataset more robust and real-life accurate.

The data cleaning process can be conducted in a few different ways—manually, automated, or a combination of both. Manual cleaning of the data is pretty laborious as it requires going through every data. For our example, we need to go through each frame to check if they are accurate. If not, we have to discard those frames. Researchers tend to set fixed binding clauses or threshold values for automated data cleaning that confines the data into those rules. While automated data cleaning may be significantly faster, it can also be inaccurate at times. That is where a mixture of both comes in handy. Using a combination of both makes a dataset more robust and accurate.

Features are those measurable variables that define the property or characteristics of any situation, occurrence, event, or fact. They are considered the building block of any dataset. For any model to function most efficiently and accurately, it is necessary to define the appropriate features of a given dataset.

While defining the features we should keep a few things in consideration. Redundant features should not be a part of the feature set at any cost. This makes the model perform worse not only in the sense of accuracy but also latency. For example, we are using temperature data for monitoring a patient. A particular temperature sensor might record data in both Celsius and Fahrenheit. Using both of them as features will create redundancy as they have a linear relation between them. So, we should include only one of them as a feature for that particular case.

Another factor is the normalization of the features. Before feeding the features to any ML or deep neural network, it is necessary to normalize the data to a particular range of values, usually 0 to 1. To avoid any more issues, we should make sure that all the units of the feature set are consistent. Let us consider our previous example, but with a caveat. Instead of using one temperature sensor, we will be using multiple temperature sensors for different parts of the body. We should make sure that all the data of various parts of the body are in either Celsius or Fahrenheit. Moreover, other features which have temperature in their unit should also use the same unit. In brief, consistency is a key factor.

Also, we should take such features into consideration that have their values distributed somewhat evenly. Let us say we want to find the symptoms of a disease. If we have a model that takes age as a feature, we should make sure that the data consists of people of all ages. If the dataset consists of data for 2 people that are 50 years old and data for 100 people that are 55 years old, then age should not be considered as a feature for that model. So, we need an even distribution of data. It helps the model to learn how much value that feature value carries for that particular label.

After all this, we need to make a proper dataset. A dataset should be well organized. People should be able to expand upon it later with their contributions. Moreover, we should make sure that people can use the dataset using their methods and models. A dataset is generally split into three parts—training set, validation set, and testing set. One author [9] has defined those three sets. The training set is a set of examples used for learning, that is to fit the parameters of the classifier. The validation set is a set of examples used to tune the parameters of a classifier, for example, to choose the number of hidden units in a neural network. Finally, the testing set is a set of examples used only to assess the performance of a fully specified classifier.

After building the dataset, we need to train our model on the training set and tune the hyperparameters on the validation set. Then the final model needs to be tested on the testing set. The testing set is only for determining the performance matrices of the model. Finally, the model will be deployed in real life.

2.1.2 Dataset and Model Description

For disease prediction, many different datasets are required for studying certain diseases. Most of the dataset used in the different study comes from the hospital. The hospital dataset is sometimes stored in the data center. The inpatient data is mostly composed of structured and unstructured text data. The structured data mainly consists of laboratory data and patients' basic information such as the patient's age, gender, habits, and so on. On the other hand, unstructured data includes the patient's information about their previous illness, doctor's record about diagnosis, and so on. For the study, different data may be used for analysis such as for patients' geographic area-related disease analysis structured data is required whereas for the previous history-related study unstructured data analysis is required. For instance, for disease risk prediction, supervised ML is employed on the attribute of the dataset. Let's say $X = x_1, x_2, x_3, \ldots x_n$ where X includes the personal information such as age, gender, the prevalence of symptoms, and so on. The output value is C that indicates the patient is amongst the high-risk population $C = C_0, C_1$, where C_0 indicates the patient is at high risk of cancer or any other disease and C_1 indicates patient is at low risk of a certain disease. Below we describe the dataset, experiment settings, dataset characteristics, and learning algorithm.

1. *Structured data (S-data)*: Structured data uses structured data to predict whether the patient is at high risk of a different disease or not. It also elucidates the probability of a patient being infected with a certain disease. It gives an idea of being infected with a certain disease.

2. *Text data (T-data)*: Text data uses unstructured text data to predict whether the patient is at high risk of a certain disease or not. It mostly focuses on unstructured information.

3. *Structured and text data (S&T-data)*: It uses both S-data and T-data to predict the risk of a certain disease. By integrating data fusion of both the structured and text data the risk of certain diseases can be detected and it provides better

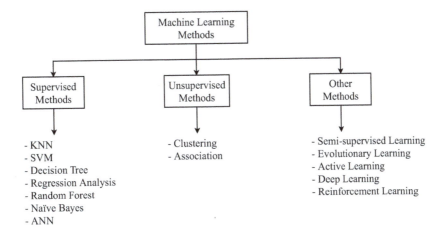

Figure 2.2 Machine learning approaches for disease assessment.

accuracy than structured data and text data alone as it takes the combined dataset.

2.2   MACHINE LEARNING ALGORITHMS

2.2.1   Supervised Learning

As the name refers, the learning procedure of the algorithm will be under supervision. But, under the supervision of what? Under the supervision of structured output. That means, the inputs to the model during the training process will have specific outputs.

In the simplest of sense, supervised learning can be considered similar to a function. There is an input X and output Y, where you have to map the input and the output using an algorithm that is the function $f()$.

$$Y = f(X)$$

The training procedure does the mapping of the function according to the input and output. That function is a programmed algorithm that is used to train and optimize on labeled training data. Later, this trained model or algorithm is used to predict or classify data based on the input of the model of unlabeled test data.

In supervised learning, there are two types of predictions—classification and regression. Classification is allocating outputs into different discrete classes. Accuracy is the main parameter to evaluate the model. Regression is predicting a continuous value. Various types of error values are calculated to evaluate a model.

- *K-Nearest Neighbor (KNN)*: K-Nearest Neighbor is a super-simple supervised way to classify data. As it does not have any specific algorithm and specialized training procedure like other algorithms it is considered a lazy learning algorithm. It is also called a non-parametric algorithm as no parameter is defined or assumed about the underlying data. Instead of depending on different parameters, it uses distance to classify any testing data. We can use multiple techniques to calculate the distance value. The most common approach is Euclidean distance.

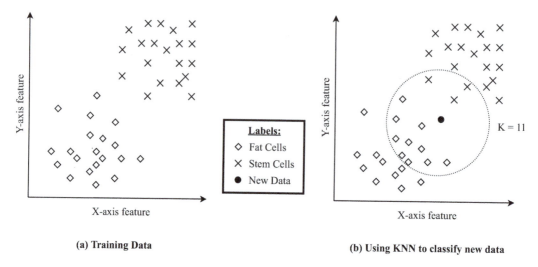

(a) Training Data (b) Using KNN to classify new data

Figure 2.3 K-Nearest neighbor.

Let us consider we have a lot of data on different cell types—stem cells and fat cells. The data is plotted in Figure 2.3. We want to use KNN to classify new cell data for K = 11. Then the new data is compared with the rest of the data to find the distance between the new data and the training data. Then K number of least distances from training instances is summarized. Among those data, the label with the highest number of occurrences is then classified for the new data. In this case, in the range of K = 11, there are data for 8 fat cells and data for 3 in the boundary. So, the new data gets labeled as a fat cell. If there is a tie anytime, then a random label is selected.

In KNN, the larger the dataset, the more accurate the output will be. Although the accuracy of KNN is very high, it has a few drawbacks as well. The computational and memory cost is considerably higher for KNN as it needs to load all the training data, which is vectors in a multi-dimensional free space, to the memory and compute the distance of new data from all those points.

- *Support Vector Machine (SVM)*: SVM is a simple ML model that classifies both linear and nonlinear data. Both linear and nonlinear classification is possible using SVM. SVM maps data to n-dimensional feature space. Then the data among different classes are separated using a hyperplane with some degree of tolerance. A hyperplane is a function or a plane that separates data of different labels and has one less dimension in contrast to the dimension of the data space itself. The hyperplane is defined in such a way that the distance between the classes is maximum. It helps to insert new data point into the data pool. A hyperplane can also be termed as "Decision Boundary." The data points that have the minimum distance to the hyperplane are known as "Support

Vector" as they are used as supports while determining the hyperplane. There may be instances where the data is not linearly separable. In such cases, the data space is transformed into a higher-dimension space using some nonlinear transformation.

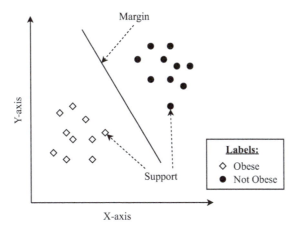

Figure 2.4 Support vector machine.

For example, we want to find out if a person is obese or not. We have two features – one on the X-axis, another on the Y-axis that can be height, weight, etc. We apply the SVM algorithm to determine the best hyperplane that separates obese data from not obese. If new test data is plotted, we can determine its class based on its location relative to the hyperplane.

SVM is highly effective in high-dimensional spaces. Even if the number of dimensions is greater than the number of samples, SVM is still effective. SVM is very versatile as different kernel functions can be specified for the decision function that allows using custom kernels alongside common kernels. But it is necessary to avoid over-fitting while choosing kernel functions if the number of features is much greater than the number of samples. If we want to calculate probability estimates, we need to use expensive five-fold cross-validation as SVM does not provide those directly.

- *Decision Trees (DT)*: Decision trees are a prediction-based supervised ML technique designed like a tree where the decisions and possible outcomes are taken into consideration in each node of the tree to find the final consequence of any given input. Both classification and regression problems can be solved using decision trees.

  Depending on different conditions, a dataset is usually split in many different ways to make a decision tree. The tree can be categoric or numeric, or a combination of both. The classification can be categoric or numeric. Moreover, final classifications can be repetitive.

  A decision tree works from starting at the top. Then it works its way down and finally gets to a point where there are no more decisions to take. There are three different types of nodes in a decision tree. The top of the tree is called the root node or root. Arrows are positioned pointing from them. The nodes in the middle are called internal nodes. Arrows pointing to and from them. The last nodes that provide the classified results are called leaf nodes or leaves. Arrows will be pointing to them.

When there are many stacked nodes, each node separates the data in the best possible way depending on different parameters. This can be determined in many different ways among which the "Gini impurity" approach is considered the best.

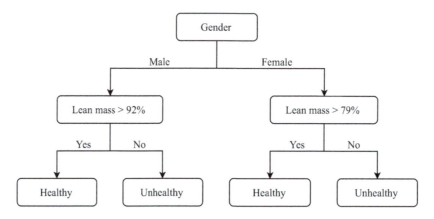

Figure 2.5 Decision trees.

Let us assume we can find out if a person is healthy based on gender and bone density. First, we need to ask if the person is male or female. Then we want to ask if the lean mass is greater than 92% for males and greater than 79% for females. The decision tree is figured in Figure 2.5. From this, we can determine whether a person is healthy or not.

- *Regression Analysis*: Regression analysis is an estimation process that is formed using a set of statistical methods. In general, there is a dependent variable that is to be estimated and there are one or more independent variables that will determine the output of the dependent variable. It is very effective for modeling the future relationship between those variables and predicting the future outcome of any event.

  The first step for training a dataset using regression analysis is to plot the data points on a chart. This gives an idea of the dataset and helps figure out if there is any actual relationship between the dependent and independent data. The dependent variable is plotted on the Y-axis and the independent variable is plotted on the X-axis. After plotting the data, the correlation between the variables is figured out. Then a line is drawn to define the relationship between the dependent and independent variables. This line is called the regression line. Based on the type of regression analysis it can be a linear function or a polynomial function. Moreover, if the number of dependent variables increases, then the dimension of the data pool also increases making the function more complex. There are three types of regression analysis–linear regression, multiple linear regression, and nonlinear regression.

In linear regression, there is only one independent function. There is a linear relationship between the slope and the intercept due to dependent and independent variables. The independent value is not random. The value of residual error is zero, constant across all observations, not correlated among observations, and follows the normal distribution. If we formulate simple linear regression, the equation will be

$$Y = a + bX + \epsilon$$

where:

- Y : Dependent variable
- X : Independent (explanatory) variable
- a : Intercept
- b : Slope
- ϵ : Residual (error)

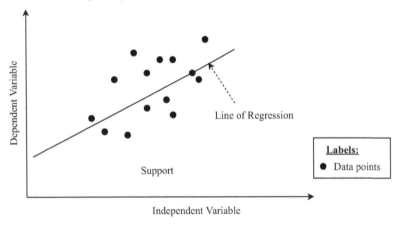

Figure 2.6 Simple linear regression analysis.

Multiple linear regression analysis is a lot like the simple linear model. The only crucial difference is that there are multiple independent variables used in the model. If we formulate multiple linear regression, the equation will be

$$Y = a + bX_1 + cX_2 + \ldots\ldots + dX_n + \epsilon$$

where:

- Y : Dependent variable
- $X_1, X_2 \ldots \ldots X_3$: Independent (explanatory) variable
- a : Intercept
- b, c, d : Slopes
- ϵ : Residual (error)

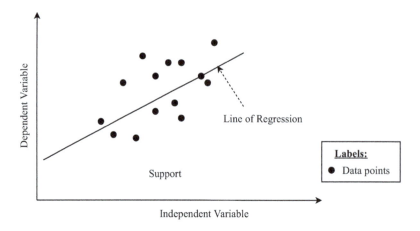

Figure 2.7 Multiple linear regression

In nonlinear regression, the dependent variable is correlated with the independent variables using nonlinear functions. Nonlinear functions like logarithm, exponential, polynomial, etc., are used to define the relationship between the variables.

Suppose, we want to find the number of people who will be affected by COVID in the upcoming days based on the places that are in lockdown, the number of people properly using hand sanitizer, and the number of people properly wearing a mask. In this case, we took three dependent variables and the independent variable is the number of people that could get infected in the upcoming week or month. Now, based on the previous data we can create a multiple linear regression model and find the value of the people that could get infected. This will help us to maintain the medical supplies according to the need so that the hospitals don't get overwhelmed or saturated.

- *Random Forest (RF)*: Random forest is a popular ML technique that is considered a stepped-up version of decision trees. It is used to solve both regression and classification problems. The ensemble learning technique is used as the base of the random forest algorithm. Ensemble learning is a technique that is used to solve a complex problem by combining many classifiers.

Many decision trees make up a random forest algorithm. Bagging, also known as bootstrap aggregating, is an ensemble meta-algorithm that is used to train the "forest" generated by the random forest algorithm. Bagging reduces the variance within a noisy dataset that increases the accuracy of ML algorithms.

The main building block of the random forest algorithm is decision trees. The random forest provides the output based on different predictions made by each decision tree. The mean or average output from all the decision trees is taken into consideration to predict the outcome of the entire model. The more the number of trees, the more the accuracy of the random forest algorithm. Figure 2.8 shows an illustration of Random Forest.

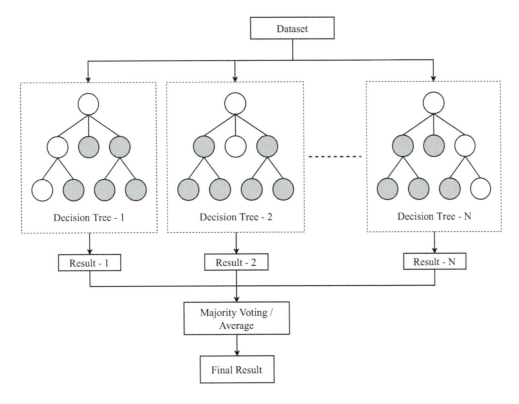

Figure 2.8 Random forest.

Random forest is good at handling missing values as it can accomplish this automatically. It uses a rule approach instead of a distance-based approach which does not require scaling of the features. Moreover, it is very efficient at handling nonlinear parameters. It can also handle outliers pretty well and is less affected by noise which makes it a lot more stable. But, making a random forest is complex and the training time is very long.

COVID-19 is a relatively new disease in the medical field. So, detection of COVID-19 is not easy, and unreliable using one ML method as the dataset is not yet that vast. In this case, we can use a few types of ML algorithms for each decision tree. Then we can combine them using a random forest algorithm. This will provide the final output which will be more accurate compared to other individual methods.

- *Naïve Bayes*: The Naïve Bayes algorithm is based on the probability theory, Bayes theorem. To find the probability of an event occurring given the probability of another event that has already occurred is the main purpose of Bayes theorem. We can state Bayes theorem as:

$$Posterior = prior x likelihood / evidence$$

We can state that mathematically as the following equation:

$$P(a|x) = P(x|a)P(c)/P(a)$$

$$P(a|X) = P(x_1|a)P(x_2|a)...P(x_n|a)P(a)$$

where:

$P(c|x)$ is the posterior probability of class (c, target) given predictor (x, attributes)

$P(x|c)$ is the likelihood which is the probability of predictor given class

$P(c)$ is the prior probability of class

$P(x)$ is the prior probability of predictor

Naïve Bayes is not a single algorithm. This is a group of algorithms that are basically based on Bayes theorem. It is assumed that all the predictors are independent. That means the presence of one certain feature in a class does not have any impact on another one.

Naïve Bayes is a lot faster than other algorithms and saves a lot of time. Solving a multi-class prediction problem is a highly suitable use case for Naïve Bayes. If the features are actually independent of each other, they can actually outperform other models and less training data is required. But there may be cases where the features are not independent of each other and one feature relies on another feature. In such cases, Naïve Bayes may underperform compared to other models. It works better on categorical input variables rather than numerical input variables. If the probability of an event can become zero anytime, which means the instance of the event is not present in the training dataset. In such cases, the accuracy of the outcome may suffer. Moreover, probability-based outputs should not be taken seriously.

In the medical field, Naïve Bayes can be used to find the probability of a disease occurring in a person. Depending on the different attributes and data of a person, we can assume the probability of whether a person will contract the disease or not.

- *Artificial Neural Network (ANN)*: The concept of ANNs took birth from the thought of how biological neural networks work. Here, the term biological neural network refers to the human brain. In a human brain, the smallest unit or cells are called neurons. They are interconnected to one another assembling the entire structure of the brain. Similarly, an ANN has neurons that are interconnected in several layers. These neurons work together as one unit according to our need to classify data into various classes.

  Let us compare biological neural networks and ANN. Figures 2.9 and 2.10 show both biological neural networks and ANN. The dendrites in a biological neural network are comparable to the inputs of the ANN. The cell nucleus works as the nodes, synapses are equivalent to the weights, and the axons work as output.

  A single node of an ANN works by taking the individual inputs from the previous layer and multiplying it with some weight values and summing them up. Sometimes a bias value is also added to the summation. Then the value is

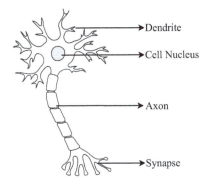

Figure 2.9 Biological neural network.

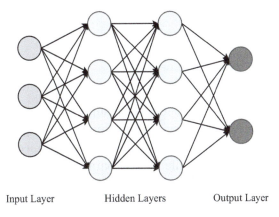

Input Layer   Hidden Layers   Output Layer

Figure 2.10 Artificial neural network.

passed through an activation function. It has a certain threshold value and can consist of a function that modifies the value if necessary.

ANNs are currently extensively used to diagnose patients. Electrical signal analysis, medical image analysis, and radiology are the major sectors that benefit from ANN. For example, we have an X-ray of a heart patient. Now we want to diagnose the patient using an automated system and detect the disease. For such cases, other ML approaches fail as they cannot process image data. But, a convolutional neural network, a type of ANN, can operate on image data prominently and efficiently. It can be trained to detect diseases and it is more accurate and structured in classifying diseases.

2.2.2   Unsupervised Learning

The main idea of unsupervised learning is to deal with unlabeled data. In unsupervised learning, the labels of the data are unknown as the data is not labeled manually. It is more of a data-driven approach. In comparison to supervised learning, it provides the ability to perform more complex processing tasks. But it can be more unpredictable.

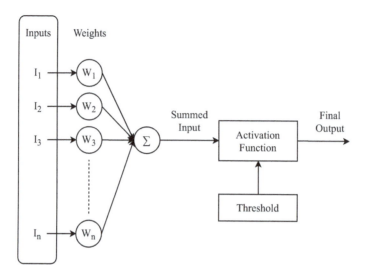

Figure 2.11 Work process of a single neuron.

Based on the parameters set by the user, unsupervised learning can be categorized into two parts—Parametric unsupervised learning and Non-parametric unsupervised learning. In parametric unsupervised learning, it is assumed that the data source has a probability distribution. While in non-parametric unsupervised learning it is assumed that data is clustered into groups.

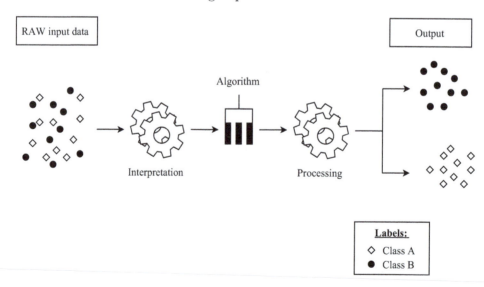

Figure 2.12 Unsupervised learning.

Based on the training procedure there are two types of unsupervised learning algorithms.

- *Clustering*: Grouping objects into a cluster with similar attributes is known as clustering. The instances with similarities stay in a single cluster and the instances with different attributes or features stay in different clusters. Here data is processed by the algorithm and finds natural groups if it exists in the data. Clustering unsupervised learning can possess different techniques and types. Here we will discuss a few.

- *Partition clustering*: Here, data is partitioned or clustered in such a way that one data can belong to one cluster only.

  Agglomerative clustering: In this method, data is initially clustered into a fixed number of clusters and does not require the number of clusters where each data is a single cluster. As the data is trained, the number of clusters decreases as they merge into one another. Finally, it forms a single cluster containing all the instances.

  Probabilistic clustering: Here data is clustered based on the probability distribution of the objects.

  Overlapping: In this process, data is clustered using fuzzy sets where every point might belong to two or more clusters. Each piece of data will have a separate degree of membership.

  Based on these techniques there are different algorithms—Hierarchical clustering (builds a hierarchy of clusters), K-means clustering (clusters the data based on the highest value of K), Dendrogram clustering (height of the dendrogram is the level similarity between clusters), K-Nearest neighbor clustering (similar to supervised learning KNN but with unlabeled data).

- *Association*: The Association rule is unsupervised learning based on the if-else condition. It checks if one data instance is dependent on another data instance. The goal is to find some relation or association among the data points. The "if" element in an association is called antecedent and the statement part is called the consequent.

  There are three types of association learning rules:

  Apriori Algorithm: Breadth-first search and the Hash tree is used for this algorithm. Datasets that contain transaction are a popular use case for this algorithm.

  Eclat Algorithm: Eclat stands for equivalent class transformation. This relies on the depth-first search technique to find an association among the instances inside the dataset.

  F-P Growth Algorithm: F-P stands for the frequent pattern. It is an improved version of the Apriori algorithm. It works by representing the dataset as a tree structure. This structure is known as a frequent pattern whose purpose is to extract the most frequent patterns.

Unsupervised learning is very good at detecting anomalies. It can detect fraud transactions pretty well. Moreover, it is good at working with fewer features and decomposing the dataset into multiple components. On the contrary, the output of the model is not labeled and not known. This leads to less accurate data for the general user.

The user will need to spend extra time labeling the classes after classification. Moreover, the spectral features can change with time that may result in some inaccurate classification.

In the medical field, unsupervised learning is very effective in tissue differentiation. It can be used to decern different types of tissues at a certain place in the body. This is very helpful for treating people who have muscle injuries.

2.2.3 Other Methods

Many other methods are different from traditional methods and take very different approaches. Here we will discuss a few of the most popular ones.

- *Semi-Supervised Learning*: In supervised learning, a ML engineer or data scientist has to hand-label the data individually. This entire process gets very tedious, time-consuming, and costly as the number of data increases. On the other hand, real-life application for unsupervised learning is limited. To tackle these disadvantages of both worlds, semi-supervised learning comes in handy.

  The main target of unsupervised learning is to cluster a set of data into homogeneous subgroups. In a certain group, all the data are similar to each other and different from the rest of the data. It is required to have any previous information to achieve the final goal that is to find similarities and dissimilarities among the data points. But, there may be cases where a some of the data is labeled and the rest are not. This is when semi-supervised learning is effective.

  Semi-supervised learning uses a technique called pseudo-labeling. It can train data with less labeled data compared to supervised learning. It usually uses a combination of different neural network models and training methods.

  Semi-supervised learning starts by training the labeled data similar to supervised learning. Then it predicts the output for unlabeled data. These outputs are called pseudo-labels and may not be very accurate. Then a link is created between the labels from labeled training data and the pseudo-labels created by the model. Then the inputs of both the labeled training data and unlabeled data are linked. Finally, the model is trained again to decrease the error and improve the accuracy of the model.

  The assumptions on the data for semi-supervised learning can be of different types—continuity assumption (outputs are more likely to have the same output label if the points are close to each other), cluster assumption (discrete clusters are used to divide the data points and points inside a single cluster are prone to have the same output label) and manifold assumption (the distance and densities of the data are used inside a manifold which has a much lower dimension compared to the input space).

  In the medical sector, protein sequence classification is a significant emerging field as it can help understand and solve diseases. It can help change the way we view any disease and understand even better than ever before. Generally, DNA strands are very large. This is where semi-supervised makes things easy to understand the protein sequence of DNA.

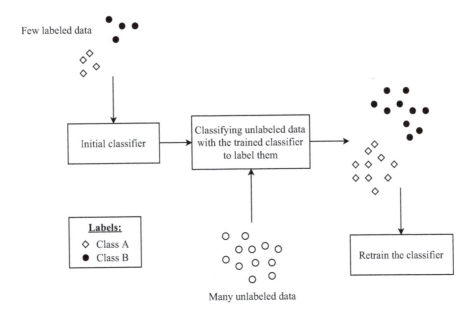

Few labeled data

Initial classifier

Classifying unlabeled data with the trained classifier to label them

Labels:
◇ Class A
● Class B

Many unlabeled data

Retrain the classifier

Figure 2.13 Semi-supervised learning.

- *Reinforcement Learning*: "Good things get rewarded, bad things get punished" —based on this concept reinforcement learning works. A reinforcement learning algorithm learns via a trial-and-error procedure. Trial and error is used during the training process. The desired behavior is rewarded while the undesired behavior gets punished. This is done by assigning a positive value to the desired action to encourage the agent and a negative value to the undesired action. The goal of the algorithm is to accumulate as many reward points as possible. This encourages the model to go after the maximum possible overall reward and find the most optimal solution. This helps the model learn to go after positive things and avoid negative things. As the model gets more trained, it learns to seek the positive and avoid the negative.

Reinforcement learning is a combination of several algorithms. The algorithms vary from one another based on the application and the environment in which the model will be trained. A few popular algorithms for reinforcement learning are—State-action-reward-state-action (SARSA), Q-learning, deep Q-Networks, etc.

In reinforcement learning, the input is the initial state from which the model will start learning. For the output of a particular problem, there may be a lot of possible solutions. The training process starts from the input and tries to find a solution. The model will return a state. Based on the output of the model it will be rewarded or punished by the user. In this way, the model will learn in a loop until the best solution is decided based on the threshold value of the maximum reward.

Although reinforcement learning has very high potential, a major challenge is that it is difficult to deploy. Moreover, it is confined by the environment and

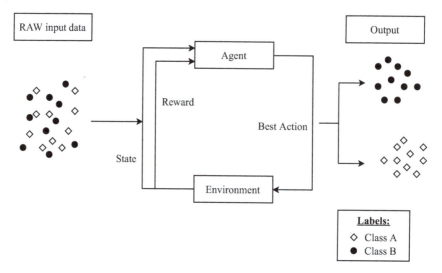

Figure 2.14 Reinforcement learning.

application. This means it cannot learn new things that might produce a better result eventually even if it wants to. Also, due to time confinement, it may not learn about new possible outcomes that can be discovered from a particular dataset. In contrast to supervised learning, it can be less efficient and slower if the data provided to the supervised model is large enough. It is also difficult to build the entire environment from scratch for complex situations. Scaling and tweaking the model is not easy due to fact that the only way to communicate with the model is via the reward values. Moreover, reaching the local optima is also a challenge for reinforcement learning. The model may even be sometimes rewarded positively despite not reaching the ultimate goal.

Reinforcement can be of two types—Positive and Negative. Positive reinforcement increases the strength and frequency after each event. It helps maximize the performance, sustain change for a long period. On the other hand, negative reinforcement works by strengthening the behavior as it stops or avoids a negative condition. It helps to increase the behavior, provide defiance to a minimum standard of performance.

In the medical field, reinforcement learning can help patients receive treatments from policies. From previous experience, the model will provide the best policy making the application more of a control-based system.

- *Evolutionary Learning*

    There is grandeur in this view of life, with its several powers, having been originally breathed into a few forms or into one; and that, whilst this planet has gone cycling on according to the fixed law of gravity, from so simple a beginning endless forms most beautiful and most wonderful have been, and are being, evolved.

    —Charles Darwin, The Origin of Species

This quote from Charles Darwin is the basis of evolutionary learning. It is not a particular single algorithm, but a combination of multiple algorithms. Problems that cannot be easily solved in polynomial time are solved using evolutionary algorithms. Evolutionary algorithms are usually combined with other algorithms to find an optimal solution in a shorter time. Without the presence of an evolutionary algorithm in such cases, generic algorithms might struggle to find an optimal solution within the time limit.

In simple words, "survival of the fittest" is the main motivation for evolutionary algorithms. The steps in evolutionary algorithms are made mimicking the scenario of evolution found in nature. It is divided into four basic steps—initialization, selection, genetic operators, and termination.

After initialization, the process of natural selection is a method of killing all living beings. Those living beings are considered unfit for their environments. Beings that can survive the tough environment are naturally "selected." These beings can sustain a rough environment. Then comes crossover and mutation. In nature, living beings usually reproduce and propagate to their descendants. Similarly, the algorithms reproduce and propagate. The crossover of the genes produces completely new living beings. Similarly, evolutionary algorithms can crossover and create a new outcome. Fitness is a measurement using which we can understand how well a model is performing. As genes are carried from one generation to another, they might deform and change their qualities. This is known as mutation. Depending on the environment it can enhance the abilities of the beings. Mutation introduces diversity in the beings as well as the algorithm. Due to mutation, a new solution to a problem may arise. The main target is to optimize the fitness function so that the score for fitness is maximum during termination.

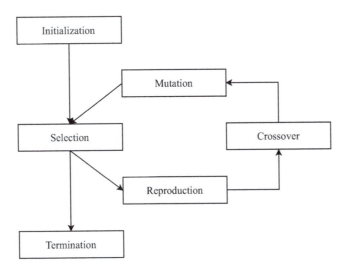

Figure 2.15 Evolutionary learning.

In the realm of medical treatment, we can see the use of evolutionary algorithms to detect some neurologic diseases. Diseases like Alzheimer's, depression, and Parkinson's are detected using evolutionary algorithms based on the feature extraction from speech signals.

- *Active Learning*: Active Learning is a special type of supervised machine learning algorithm that can interactively query a user to label data with the desired outputs. The size of the training dataset is kept at a minimum constructing a high-performance classifier. The creation of active learning was done keeping a few points in mind—it will work for a dataset with a huge and small number of data; annotation of unlabeled data costs human effort, time, and money; the computation power is limited.

  Active learning is based on the concept of achieving higher accuracy with a small number of labeled training data. The model itself will select the data from which the model will learn. This is done by proactively selecting the next set of examples that will be labeled from the unlabeled data. That's why during the training stage, the active learner can interactively ask queries. Those queries are usually unlabeled that are asked to humans for labeling the instances.

  Implementation of active learning can be done using three main approaches.

  - *Stream-based selective sampling approach*: During the training process, the model is presented with data and the model calculates if labeling the data will be beneficial for the model. That is, whether the model wants the data to be labeled or not. The main issue in this approach is high human labor. Moreover, it is very hard to hire a data scientist to label data within the budget.

  - *Pool-based sampling approach*: This is the most applicable method for active learning. The active learner goes through the entire dataset and evaluates the dataset. Then the best group of queries are selected. In some cases, the active learner is even trained on the fully labeled training dataset. Based on that the learner detects which of the unlabeled data will be beneficial for that scenario if that data is inserted into the data pool. Then the model has to be trained once again. The only drawback of this approach is high memory consumption.

  - *Membership query synthesis approach*: This approach is not applicable for all scenarios. Here data is generated synthetically. The creation of new data is allowed to the active learner. This is effective in situations where it is easy to create new data instances.

In healthcare, in the field of medical imaging data and other cases, there may be a limited amount of labeled data. In such cases, a human annotate can label the data to assist the algorithm. It may be a bit tedious and slow process to label those data and readjust the model accordingly. But it saves time compared to traditional data collection methods.

- *Deep Learning*: Deep learning is a subset of ML. This sector studies algorithm that is inspired by the working process and structure of the human brain. It is a lot similar to ANN. But the most significant difference is that the number of layers is three or more than three. The depth of networks is very high and the architecture can be very complex. Usually, a lower number of layers may provide some accuracy and the system might function as intended. Having a higher number of layers help to improve the model, optimize the model, and increase the accuracy of the model.

  Deep learning shines when the input (and even output) of the system is an analog value. Analog data can be images, text documents, audio, etc. Deep learning works by mimicking the way our brain functions using a combination of data inputs, bias, and weights. These ingredients work together to classify data according to our needs. Another key ingredient in deep learning is forward and backpropagation. This helps to find the error value of the model and update the bias and weight values to a more optimal value. With each iteration, this helps to update the values gradually making the model more accurate.

  Deep learning helps to accelerate automation. It can help machines do a lot of complex analytical or physical works without the intervention of humans. It is really difficult for normal ML algorithms to process image and audio data. Deep learning helps to process such data and provide insights. Deep learning can do some pretty impressive things with some impressive results as it gives high accuracy.

  Though deep learning is effective in many fields, it has some drawbacks as well. It requires a huge amount of labeled data to function as intended. The size of the data increases even more if the data has multiple dimensions, like images or videos. This huge number of data requires substantially powerful computing horsepower. To compute such a huge number of complex calculations using such a huge amount of data can be computationally costly. High-performance GPU (graphic processing units) are required to train such models and algorithms. Those devices are power hungry and costly. Even if we have such devices, it can take hours or even days to train a model properly.

  The medical industry has been greatly benefited from the introduction of deep learning. Due to the digitalization of hospital records, images, and radiology, image recognition has played a vital role to help radiologists and medical imaging specialists. It has made disease detection and assessment easier than ever before as it takes almost no time to process such complex data.

2.3 DISEASE-RISK PREDICTION

Diseases can happen to any person at any time. It can transmit via viruses, bacteria or can even occur genetically. But we can at least take precautions if we know how much risk we have of having a disease. To know about the risk we have, it is necessary to perform disease risk prediction.

This can be done using different ML approaches. For example, if we daily monitor our blood pressure, blood sugar, heart rate, etc., we can determine by comparing standard data, previous data, and current data to find out the risk of having any heart disease. We can also determine the current condition of the disease and monitor whether the situation is improving or degrading. It can also mobilize treatment to some extent.

Not only that, we can find the probability of some having cancer or diabetes. If the probability is high, the doctor can recommend medicines and lifestyles accordingly. It can greatly decrease the number of people affected by complex diseases across the world.

2.4 EVALUATION METHODS

Building an ML model is not everything. Making sure that the model works properly is also a big deal. But how we can determine that the model we designed is working properly? To find that, we need some variables. These variables are called evaluation metrics that help us understand and evaluate the model. There can be various types of evaluation metrics based on whether the model is designed for classification or regression.

- *Confusion Matrix*: The confusion matrix is an N × N matrix. Here N is the number of classes that will be predicted. From this matrix, we find the accuracy of the model, positive predictions, negative predictions, precision, recall, and specificity. Usually, we care about any one of those metrics.

- *F1-Score*: From the confusion matrix we can find the recall and precision. And from those two values, we can find the value of the F1-Score. In some cases, it is necessary to have a balance between precision and recall value. For those cases, we calculate the F_1-Score to understand the model.

$$F_1 = \left(\frac{recall^{-1} + precision^{-1}}{2} \right)^{-1}$$

- *Gain and Lift Charts*: Checking the rank ordering of the probabilities is the main concern for gain and lift charts. It is effective in cases where campaigning is involved.

- *Kolomogorov Smirnov Chart*: The Kolomogorov Smirnov or K-S chart is used to measure the performance of classification models. The higher the value, the better the model can decern between positives and negatives.

- *Area Under the ROC Curve (AUC-ROC)*: This metric is more often used in the industry. ROC curve provides us the graph of sensitivity vs. (1-specificity). Having a higher value means the model is performing well. In some cases, the lift chart will fail to provide the current data if the parameters change. But, this makes more sense.

- *Log Loss*: AUC ROC works with the probability of classification. But sometimes the model can perform even better which is not taken into account. To solve this issue, log loss calculates the log of correctly predicted probabilities for each instance. This value gradually decreases with time as the model is trained more and more.

- *Gini Coefficient*: Gini coefficient is used for classification models. Gini is the ratio between the area between the ROC curve and the diagonal line and the area of the above triangle.

- *Concordant-Discordant Ratio*: The concordant-discordant ratio is used to define the relationship between pairs of classes. This is one of the most important metrics for classification problems.

- *Root Mean Squared Error (RMSE)*: RMSE is the most important metric to evaluate any regression model. The idea is to find the error for each instance, square them, and find the mean root of that. The equation for RMSE is:

$$RMSE = \sqrt{\frac{\sum_{i=1}^{N}(Predicted_i - Actual_i)^2}{N}}$$

- *Root Mean Squared Logarithmic Error*: This is also one of the most important metrics to evaluate any regression model. It is very similar to RMSE. But the major difference is that we take the logarithmic value after calculating the error. The equation for RMSLE is:

$$RMSLE = \sqrt{\frac{\sum_{i=1}^{N}(log(Predicted_i + 1) - log(Actual_i + 1))^2}{N}}$$

It is used when we do not want to penalize the model with huge differences that we can find from RMSE.

2.5   APPLICATIONS IN REAL-LIFE SCENARIOS

The application of ML in the medical field is enormously visible in recent days. As the capability of storing and processing data is ever-increasing, the use of ML is also increasing. One may think, ML is used only to detect diseases. But it can be used to keep track and even cure the disease. To utilize ML in the medical field, it is necessary to have a dataset. We can find a lot of such practical use cases. We will discuss a few of them here.

Alzheimer's disease (AD) is a common type of dementia that contributes to 70% of dementia cases. People tend to suffer from mild memory loss. People can even lose the ability to communicate or respond to the environment. Mild cognitive impairment (MCI) is one of the first signs of AD. Functional magnetic resonance imaging (fMRI) is an effective way of detecting AD. Researchers from Lithuania [8] worked on the Alzheimer's Disease Neuroimaging Initiative (ADNI) fMRI dataset to predict MCI, early MCI (EMCI), late MCI (LMCI), and AD. The dataset itself consists of data from 138 subjects. Due to the complexity of brain imaging and the late development of AD, it is difficult to predict AD at an early stage. But, this group of researchers used an AI model based on the ResNET18 network to diagnose AD. They achieved an accuracy of 99.99% on EMCI vs. AD, 99.95% on LMCI vs. AD, and 99.95% on MCI and EMCI.

Parkinson's disease (PD) is a similar disease that affects movement. It is a progressive nervous system disorder that leads to shaking, stiffness, and difficulty with walking and keeping the balance of the body. It may not be noticeable in the early stages. As time passes, it can cause tremors, slow movements, rigid muscles, impaired posture and balance, loss of automatic movements, speech changes, and writing changes. Though it can happen to both males and females, about 50% more men are affected compared to women. People usually develop PD at about the age of 60 years. Researchers [1] used facial expressions to diagnose PD. They collected 1812 videos from 604 individuals. Among those individuals 61 had PD, 543 did not. And the subjects had a mean age of 63.9 years. They utilized the micro-expressions of the face and applied SVM on that data to train their model. They achieved an accuracy of 95.6%.

In 2020, Wang [10] published a dataset on COVID-19 to fight against the disease that shook the world. The dataset is used to generate new insights on the disease utilizing. Natural Language Processing and other AI techniques. The dataset is ever-increasing. Data from all over the world is being added to the dataset. In addition, many other research projects are going on to detect and diagnose COVID-19 using ML. Researchers are using chest X-ray data to learn the damage by COVID-19, as well as different datasets to predict the spread of COVID depending on the geological position and past spread trend.

Different types of cancer can also be detected and diagnosed using ML. Different types of sensor and radiography imaging data are combined with AI models to identify cancer in a split second. These AI models are powerful enough to detect cancers and predict their future spread. This can help boost the treatment significantly. One group of researchers have shown that image analysis can help in lung cancer pathology with the proper utilization of AI models. Tasks such as tumor region identification, prognosis prediction, tumor microenvironment characterization, and metastasis detection have great potential due to the recent progress in AI and computer vision.

A recently growing sector in medical treatment is telemedicine. The process of diagnosing a patient remotely is called telemedicine. This helps a patient to get medical assistance without being physically present in front of the doctor. There may be cases, where a doctor is not available for consultancy. In such cases, AI has revolutionized telemedicine. AI chatbots and assistants can do the job of doctors to help with the process of initial treatment of the patient. Researchers are training AI models using natural language processing (NLP) with the previous data records from different doctors. Electronic health records generate huge data for this purpose. This data is then processed to create those AI assistants and chatbots.

2.6 CONCLUSION

ML has become an undeniable part of medical research in the past few years. With the advances in radiology, computer vision, sensor-based data, etc. data logging and processing have become easier than ever before. ML is also contributing to producing new drugs. From finding new drugs to drug repurposing and performing clinical trials, all can be done using ML. In the future, the use of ML will increase even more. With the increase in computational power of modern-day computers, processes like gnome sequencing and protein synthesis are increasing. This helps to understand how the disease is evolving with time. If we can understand the disease, we can find the cure accordingly.

Bibliography

1. Mohammad Rafayet Ali, Taylor Myers, Ellen Wagner, Harshil Ratnu, E. Ray Dorsey, and Ehsan Hoque. Facial Expressions Can Detect Parkinson's Disease: Preliminary Evidence from Videos Collected Online. *NPJ Digital Medicine*, 4(1), 2021.

2. Gopi Battineni, Getu Gamo Sagaro, Nalini Chinatalapudi, and Francesco Amenta. Applications of Machine Learning Predictive Models in the Chronic Disease Diagnosis. *Journal of Personalized Medicine*, 10(2):21, March 2020.

3. Min Chen, Yixue Hao, Kai Hwang, Lu Wang, and Lin Wang. Disease Prediction by Machine Learning over Big Data from Healthcare Communities. *IEEE Access*, 5:8869–8879, 2017.

4. Shekh M. M. Islam, Francesco Fioranelli, and Victor M. Lubecke. Can Radar Remote Life Sensing Technology Help Combat COVID-19? *Frontiers in Communications and Networks*, 2:648181, May 2021.

5. Senerath Mudalige Don Alexis Chinthaka Jayatilake and Gamage Upeksha Ganegoda. Involvement of Machine Learning Tools in Healthcare Decision Making. *Journal of Healthcare Engineering*, 2021:1–20, January 2021.

6. Naresh Kumar, Nripendra Narayan Das, Deepali Gupta, Kamali Gupta, and Jatin Bindra. Efficient Automated Disease Diagnosis Using Machine Learning Models. *Journal of Healthcare Engineering*, 2021:1–13, May 2021.

7. Forrest W Nutter, Paul D Esker, and Rosalee A Coelho Netto. Disease Assessment Concepts and the Advancements Made in Improving the Accuracy and Precision of Plant Disease Data. *European Journal of Plant Pathology*, 115(1):95–103, 2006.

8. Modupe Odusami, Rytis Maskeliūnas, Robertas Damaševičius, and Tomas Krilavičius. Analysis of Features of Alzheimer's Disease: Detection of Early Stage from Functional Brain Changes in Magnetic Resonance Images Using a Fine-Tuned ResNet-18 Network. *Diagnostics*, 11(6), 2021.

9. Brian Ripley. *Pattern Recognition and Neural Networks*. Cambridge University Press, 1 edition, 2008.

10. Lucy Lu Wang, Kyle Lo, Yoganand Chandrasekhar, Russell Reas, Jiangjiang Yang, Doug Burdick, Darrin Eide, Kathryn Funk, Yannis Katsis, Rodney Kinney, Yunyao Li, Ziyang Liu, William Merrill, Paul Mooney, Dewey Murdick, Devvret Rishi, Jerry Sheehan, Zhihong Shen, Brandon Stilson, Alex Wade, Kuansan Wang, Nancy Xin Ru Wang, Chris Wilhelm, Boya Xie, Douglas Raymond, Daniel S. Weld, Oren Etzioni, and Sebastian Kohlmeier. CORD-19: The COVID-19 Open Research Dataset, 2020. https://www.kaggle.com/datasets/allen-institute-for-ai/CORD-19-research-challenge

11. Shidan Wang, Donghan M. Yang, Ruichen Rong, Xiaowei Zhan, Junya Fujimoto, Hongyu Liu, John Minna, Ignacio Ivan Wistuba, Yang Xie, and Guanghua Xiao. Artificial Intelligence in Lung Cancer Pathology Image Analysis. *Cancers*, 11(11):1673, 2019.

## CHAPTER 3

Precision Medicine and Future Healthcare

Muhammad Afzal and Maqbool Hussain

Department of Software, Sejong University, South Korea

CONTENTS

PRECISION MEDICINE applications are growing and creating tremendous opportunities to shape the future of healthcare. Though it is currently more visible in oncology, other domains such as precision medication and rare/genetic diseases are no exception. It also demonstrated some promise in treating COVID-19. The data-driven approaches, artificial intelligence (AI), and machine learning (ML) are rightfully improving the traditional symptom-driven practice of medicine and changing the mode of medicine from "reactive" to "predictive." Advancements in genetics, the availability of genetic data, fast and accurate genome and transcriptome sequencing, and computational tools allow creating personalized treatment plans. We are motivated to inspect the potential applications of precision medicine in oncology and beyond. We intend to describe the features of famous and powerful bioinformatics tools and techniques. What role data science, AI, and machine learning can play in boosting the practicality of precision medicine will be an important thesis of this chapter. Finally, we aim to propose a comprehensive roadmap for a developer and implementer community to apply for precision medicine in futuristic healthcare solutions.

3.1 INTRODUCTION

Precision medicine is a fledging paradigm that holds the potential to revolutionize medicine and healthcare [1]. Various people describe precision medicine differently since related terms personalized, stratified, and individualized medicine are used across multiple disciplinary domains, but recently precision medicine has become

DOI: 10.1201/9781003251903-3

a ubiquitous term [1, 2]. Precision medicine also referred to as precision health, is an approach to consider disease based on patient-individual data, including diagnoses, phenotype, biologic findings, environmental factors, demographic information, and lifestyle factors [3].

Precision medicine is an attention-grabbing area of research that has the potential to bring diverse communities together to deal with heterogeneity in clinical, genomic, and environmental information that influences patient care. It is an exciting area of research for the research community to exploit technological advancements for common goals. However, it is challenging to grasp the technicalities involved in drawing relationships among different prospects in this cross-disciplinary research field.

The paradigm shift from traditional general and guideline-specific therapy toward personalized and patient-specific precision medicine is possible due to advancements in investigation tools, biomarker technologies, and AI use in healthcare [4]. Traditionally, a general approach is considered to leverage the average conditions and clinical outcomes for the patients of interest, while precision medicine entails the individual lifestyle, gene variability, and environment [5]. As a result, traditional approaches may work for one cohort of patients and not for the other, whereas precision medicine is highly likely to be effective for each cohort of patients based on their individualistic characteristics and conditions. In addition, precision medicine has changed the signs or symptoms approach that deals with patients in the same way if they have the same symptoms with a target-oriented approach that allows each group of patients to be treated correctly with the proper treatment.

3.2   THE EVOLVING PARADIGM OF PRECISION MEDICINE

Precision medicine is rapidly evolving, with the advent of innovative genome sequencing technologies [6], especially the advances in next-generation sequencing (NGS) technology have spurred precision medicine applications to gauge personalized therapies and improve outcomes of patient care [7]. NGS allows sequencing of long sequences consisting of a high number of nucleotides in a short period at a reasonable cost [8, 9]. Nevertheless, the success of practicing precision medicine at the point of care requires smooth access to the genomic variant data and their correct interpretations [10]. Among the three main types of NGS sequencing, targeted sequencing (TS) is more practical in the context of precision medicine due to the lower costs and the higher depth, particularly for the discovery of new druggable targets [9, 11]. Furthermore, blasting services such as NCBI BLAST (Basic Local Alignment Search Tool) [12] facilitates target sequence matching with reference genome sequences passed in a query. Depending on the size of input and type of sequence, alignment results are obtained relatively quickly, making the processes suitable in actual setups.

Other bioinformatics tools are available to make knowledge-based interpretation of variant data a reality at the point of care. For instance, TBtools is a stand-alone tool with over a hundred functions and a user-friendly interface designed for big data analyses, ranging from bulk sequence processing to interactive data visualization [13]. GMAP (Genomic Mapping and Alignment Program) [14] is another tool for NGS sequence mapping working on genomic, transcriptomic, and epigenomic data.

Similarly, for omics data alignments, such as BWA (Burrows-Wheeler aligner) [15], and GATK (Genome Analysis Toolkit) [16], and other tools are in service.

Gene Set Enrichment Analysis (GSEA) is one of the analytical methods used to extract biological insights from the RNA expression dataset [17]. This analytical method was initially used in leukemia and lung cancer, demonstrating related gene sets regulations in correlation with the intended target gene, p53. GSEA method is implemented in a comprehensive toolset and thousands of related gene sets in different cancer domains [18].

3.3 ROLE OF ARTIFICIAL INTELLIGENCE AND MACHINE LEARNING IN PRECISION MEDICINE

AI applications in medicine are categorized into virtual and physical divisions. Whereas the virtual branch involves machine learning (ML) and deep learning (DL), the physical branch refers to medical devices, sophisticated robots, and other physical objects [19]. Here we focus on the role of the virtual branch in precision medicine because its importance and utility in precision medicine are enormous, where some even phrased as "there is no precision medicine without AI." After all, these approaches use the cognitive capabilities of physicians on a new scale [20]. Furthermore, AI techniques are changing the way physicians make diagnostic and treatment decisions. Therefore, physicians in everyday practice are motivated to look at advances and innovations spreading over faster than ever using disruptive AI tools and techniques and exponential increase of healthcare data [21]. Some areas are underlined here, where applications of ML and DL have been widely used.

1. **Biomedical data processing**: Electronic health records (EHRs) and other health systems generate clinical data in structured and unstructured formats. It is straightforward to incorporate in the clinical workflow; however, the unstructured data is more challenging to process, obtaining meaningful information that might help solve clinical questions about patient health conditions, clinical reasoning, and inferencing. Traditionally, clinically relevant information from clinical documents is extracted through manual methods with the support of clinical domain experts, which creates hurdles in terms of scalability and costs. ML algorithms have been used to ease the human job on manual assessment of unstructured clinical and scientific data. In the last few years, ML algorithms, especially DL, have made several successes in the text mining domain, including text annotations, classification, clustering, summarization, named entity recognition, and many others. The modern DL algorithms armed with word embeddings took over traditional ML or shallow neural network models. The latest state-of-the-art model called BERT—Bidirectional Encoder Representations from Transformers is considered a big success in text processing for a wide range of tasks, such as question answering, text quality assessment, and language inference [22]. The BERT model's exhibition of higher accuracy on different text processing tasks and other variations like BioBERT encourages

various precision medicine applications like automatic concept extraction for cancer, genomics, and precision oncology [23].

2. **Phenotype discovery**: The biological process may involve a related set of genes—known as pathways that play critical roles in regulating various behavior. For example, Hippo pathways include genes that regulate cell proliferation and growth in living organisms [24]. The Hippo pathway was initially postulated to have a key role in the egregious overgrowth of cells in cancer. However, recently, the postulate has been validated by discovering mutations and altered expressions of a subset of Hippo pathway genes in human cancers [25]. These analyses are based on enhanced statistical methods, such as GSEA, to find the association of different clinical characteristics into the expression level of a subset of gene sets in the Hippo pathway. GSEA is one of the key computational methods that allow whether an a priori defined set of genes shows statistically significant, concordant differences between two biological states (e.g., phenotypes) [17]. For example, comparing the expression status of genes in a particular pathway associated with stage-I compared to advanced stage-IV cancer, GSEA determines and ranks these genes to be expressed (either higher or lower).

3. **Image analysis**: Medical images are a source of information necessary for disease diagnosis and treatment and due to the patient-specific nature, they can be an important component for precision medicine [26]. Various technologies have been developed for image information processing, such as photorealistic visualization of images with cinematic rendering, artificial agents for in-depth image understanding, and patient-specific computational models with enhanced predictive capabilities that enables precision medicine to step forward. Also, AI and DL can add value and speed to a pathologist's diagnosis in the domain of digital pathology to learn complex morphological patterns in various diagnostic or prognostic tasks [27].

4. **Next-generation sequencing**: The improvements in NGS to generate hundreds of millions to multigigabit levels of DNA sequence has significantly expanded the possibilities for patient care and research for diagnostic genome sequencing for personalized healthcare and precision medicine [28]. With the integration of this powerful genome testing technology into standards of patient care across multiple specialties, the role of professionals like clinicians will broaden. The NGS technology supports the very claim of precision medicine—the right treatment to the right patient at the right time, by allowing the rapid and accurate sequencing of many genes at once [29]. Most of the steps of the NGS bioinformatics pipeline, like sequencing alignment, variant calling, variant filtering, variant annotation, and prioritization, for performing appropriate analyses, involves AI and ML in processing and integrating diverse datasets for enhanced diagnosis and treatment decisions, especially in case of rare diseases [30]. Some of the examples using DL for variant calling are DeepVariant [31] that uses a deep convolutional neural network (CNN) for variant calling from

short-read sequencing by reconstructing DNA alignments as an image. Clair-voyante [32] has used deep CNN for predicting variant type, and DeepNano [33] that uses a deep recurrent neural network (RNN) for base calling in MinION nanopore reads. Similarly, models such as Skyhawk [34] and DANN [35] use deep neural networks (DNNs) for variant optimization and annotations, while DeepSea [36] uses deep CNNs for the same task.

3.4 FUTURE IMPLEMENTATION ROADMAP OF PRECISION MEDICINE

Precision medicine is broadly welcomed around the world. However, the area is still in its infancy, and many aspects are untouched, with many challenges lying ahead. Tech companies such as IBM, Google, Apple, and Amazon invest heavily in health-care analytics to facilitate precision medicine. However, despite the assistance and advancements fueled by AI for omic data processing and analysis, various challenges still exist.

- *Scarcity of data validation*: Obtaining statistically significant and valid patient data.

- *Transparency and reproducibility*: Business companies, commercial platforms, and research publications offer limited information for the public to reuse the data and models.

- *Training and education*: There is a lack of a comprehensive mechanism for patients and physicians to get updated, educated, and trained on emerging tools and technologies used in precision medicine.

It is still a big challenge to construct an infrastructure that supports the prevalent sharing and effective use of health and genomic data to advance a healthcare system that is the least reliant on external sponsored resources. A framework proposed in our previous work to address the issue of data integration and staged analysis involves four significant modules: Primary analysis, secondary analysis, knowledge management for knowledge services, and data analytics for data services [1]. As shown in Figure 3.1, diverse data is acquired from different input sources, including clinical, molecular, and -omic, sensory, environmental, and published literature. The collected data is pre-processed to remove the undesirable data items. The secondary analysis module uses the pre-processed data to analyze further, where some parts of the data are integrated for a combined analysis. At this stage, correlations among various data items among the genes are found. The analyzed data is saved to reuse for various services. The knowledge management (KM) and data analytic (DA) modules employ the analyzed data generated at the secondary analysis. The DM module constructs KBs by creating, maintaining, and validating knowledge rules from the analyzed data, whereas the DA module targets designing descriptive, predictive, and prescriptive services models.

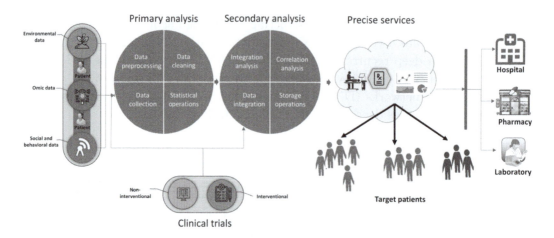

Figure 3.1 Conceptual framework architecture for the implementation of precision medicine.

The existing EHR infrastructure can be integrated with precision medicine infrastructure by transferring the patient specimen information from EHR to the laboratory information management system (LIMS). The LIMS component sends back the report formed over the specimen data after processing, sequencing, and analyzing to the pathology system of the EHR infrastructure. Clinical informatics, bioinformatics, and participatory health informatics can build a holistic ecosystem to exercise precision medicine in standard healthcare by considering the following ideas:

1. To curate data from various mobile and sensor devices, social media platforms, IoT gadgets, and environmental factors can be combined with clinical and genomic data to enable one-point decision precisely.

2. To develop infrastructure, tools, techniques, and applications to create synergy between bioinformatics and clinical informatics and allow the sharing of data to integrate individual patient data into the clinical research environment.

3. To enable multi-scale population data, including the phenome, the genome, the exposome, and interconnections.

There can be multiple opinions and considerations for implementing a precision medicine platform.

1. A single-clinic solution owned by clinical researchers and physicians with the backing of an enterprise-level developing team,

2. A centralized solution supported and ran on cloud-based infrastructure, and

3. An intermediary solution allows an institution-wide platform to support innovations across multiple clinics via a single secure and cost-effective technological foundation. This strategy balances single-clinic and centralized solutions by taking advantage of the former's fast development and flexibility and the latter's scalability and sustainability.

3.5   CONCLUSION

Achieving a common goal of better-quality patient care is the need of the day; therefore, it is high time for collaboration of the informatics community with the medical community to build a workable solution for implementing a scalable and sustainable precision medicine platform. First, we described the significant areas of research and development for the realization of precision medicine from the perspective of informatics. We have provided a broader overview of enabling tools and techniques for biomedical informatics and explained the role of AI and ML in precision medicine. Finally, we described the futuristic roadmap by proposing a conceptual framework architecture to implement precision medicine. Future works are required to see precision medicine at its maximum potential. Of the essentials, the low computational complexity solutions for molecular and -omics data analysis appears to be a big challenge. Processing and mining heterogeneous datasets such as omics, imaging, and clinical data required for associated predictions also requires much research. Furthermore, secure mechanisms have to be introduced to protect genomic data from being maliciously used.

Bibliography

1. M. Afzal, S. M. Riazul Islam, M. Hussain, and S. Lee, "Precision medicine informatics: Principles, prospects, and challenges," *IEEE Access*, vol. 8, pp. 13593–13612, 2020, doi: 10.1109/ACCESS.2020.2965955.

2. E. Faulkner, Holtorf AP, Walton S, Liu CY, Lin H, Biltaj E. et al., "Being precise about precision medicine: What should value frameworks incorporate to address precision medicine? A report of the personalized precision medicine special interest group," *Value in Health*, vol. 23, no. 5, pp. 529–539, May 2020, doi: 10.1016/J.JVAL.2019.11.010.

3. S. J. Maceachern and N. D. Forkert, "Machine learning for precision medicine," *Genome*, vol. 64, no. 4, pp. 416–425, 2021, doi: 10.1139/GEN-2020-0131/ASSET/IMAGES/LARGE/GEN-2020-0131F1.JPEG.

4. T. H. Su, C. H. Wu, and J. H. Kao, "Artificial intelligence in precision medicine in hepatology," *Journal of Gastroenterology and Hepatology*, vol. 36, no. 3, pp. 569–580, Mar. 2021, doi: 10.1111/JGH.15415.

5. I. R. König, O. Fuchs, G. Hansen, E. von Mutius, and M. v. Kopp, "What is precision medicine?" *European Respiratory Journal*, vol. 50, no. 4, p. 1700391, Oct. 2017, doi: 10.1183/13993003.00391-2017.

6. F. Passiglia and G. v. Scagliotti, "The evolving paradigm of precision medicine in lung cancer," *Current Opinion in Pulmonary Medicine*, vol. 27, no. 4, pp. 249–254, Jul. 2021, doi: 10.1097/MCP.0000000000000778.

7. D. A. Dominguez and X. W. Wang, "Impact of next-generation sequencing on outcomes in hepatocellular carcinoma: How precise are we really?" *Journal of Hepatocellular Carcinoma*, vol. 7, p. 33, Mar. 2020, doi: 10.2147/JHC.S217948.

8. F. Mosele, Remon, J. Mateo, C.B. Westphalen, F. Barlesi et al., "Recommendations for the use of next-generation sequencing (NGS) for patients with metastatic cancers: A report from the ESMO Precision Medicine Working Group," *Annals of Oncology*, vol. 31, no. 11, pp. 1491–1505, Nov. 2020, doi: 10.1016/J.ANNONC.2020.07.014.

9. V. de Falco, L. Poliero, P. P. Vitello, D. Ciardiello, P. Vitale. et al., "Feasibility of next-generation sequencing in clinical practice: results of a pilot study in the Department of Precision Medicine at the University of Campania "Luigi Vanvitelli,'" *ESMO Open*, vol. 5, no. 2, p. e000675, Jan. 2020, doi: 10.1136/ESMOOPEN-2020-000675.

10. G. Alterovitz, B. Heale, J. Jones, D. Kreda, F. Lin, L. Liu et al., "FHIR Genomics: enabling standardization for precision medicine use cases," *NPJ Genomic Medicine 2020 5:1*, vol. 5, no. 1, pp. 1–4, Mar. 2020, doi: 10.1038/s41525-020-0115-6.

11. F. Bewicke-Copley, E. Arjun Kumar, G. Palladino, K. Korfi, and J. Wang, "Applications and analysis of targeted genomic sequencing in cancer studies," *Computational and Structural Biotechnology Journal*, vol. 17, pp. 1348–1359, Jan. 2019, doi: 10.1016/J.CSBJ.2019.10.004.

12. M. Johnson, I. Zaretskaya, Y. Raytselis, Y. Merezhuk, S. McGinnis, and T. L. Madden, "NCBI BLAST: A better web interface," *Nucleic Acids Research*, vol. 36, no. suppl_2, pp. W5–W9, Jul. 2008, doi: 10.1093/NAR/GKN201.

13. C. Chen, C. Hao, Y. Zhang, H. R. Thomas, M.t H. Frank, Y. He, et al., "TBtools: An integrative toolkit developed for interactive analyses of big biological data," *Molecular Plant*, vol. 13, no. 8, pp. 1194–1202, Aug. 2020, doi: 10.1016/J.MOLP.2020.06.009.

14. T. D. Wu and C. K. Watanabe, "GMAP: A genomic mapping and alignment program for mRNA and EST sequences," *Bioinformatics*, vol. 21, no. 9, pp. 1859–1875, May 2005, doi: 10.1093/BIOINFORMATICS/BTI310.

15. H. Li and R. Durbin, "Fast and accurate short read alignment with Burrows–Wheeler transform," *Bioinformatics*, vol. 25, no. 14, pp. 1754–1760, Jul. 2009, doi: 10.1093/BIOINFORMATICS/BTP324.

16. A. McKenna, M. Hanna, E. Banks, A. Sivachenko, K. Cibulskis, A. Kernytsky, K. Garimella, et al., "The Genome Analysis Toolkit: A MapReduce framework for analyzing next-generation DNA sequencing data," *Genome Research*, vol. 20, no. 9, pp. 1297–1303, Sep. 2010, doi: 10.1101/GR.107524.110.

17. A. Subramanian, P. Tamayo, V. K. Mootha, S. Mukherjee, L. E. Benjamin, A. G. Michael, et al., "Gene set enrichment analysis: A knowledge-based approach for interpreting genome-wide expression profiles," *Proceedings of the National Academy of Sciences*, vol. 102, no. 43, pp. 15545–15550, Oct. 2005, doi: 10.1073/PNAS.0506580102.

18. "GSEA." http://www.gsea-msigdb.org/gsea/index.jsp (accessed Jan. 05, 2022).

19. P. Hamet and J. Tremblay, "Artificial intelligence in medicine," *Metabolism*, vol. 69, pp. S36–S40, Apr. 2017, doi: 10.1016/J.METABOL.2017.01.011.

20. B. Mesko, "The role of artificial intelligence in precision medicine," *http://dx.doi.org/10.1080/23808993.2017.1380516*, Taylor & Francis Online, vol. 2, no. 5, pp. 239–241, 2017, doi: 10.1080/23808993.2017.1380516.

21. C. Krittanawong, "The rise of artificial intelligence and the uncertain future for physicians," *European Journal of Internal Medicine*, vol. 48, pp. e13–e14, Feb. 2018, doi: 10.1016/J.EJIM.2017.06.017.

22. M. v. Koroteev, "BERT: A review of applications in natural language processing and understanding," Mar. 2021, Accessed: Jan. 05, 2022. [Online]. Available: https://arxiv.org/abs/2103.11943v1

23. N. Greenspan, Y. Si, and K. Roberts, "Extracting concepts for precision oncology from the biomedical literature," *AMIA Summits on Translational Science Proceedings*, vol. 2021, p. 276, 2021, Accessed: Jan. 05, 2022. [Online]. Available: /pmc/articles/PMC8378653/

24. N. Tapon, K. F. Harvey, D. W. Bell, D. C. Wahrer, T. A. Schiripo, D. Haber, I. K. Hariharan, et al., "Salvador promotes both cell cycle exit and apoptosis in drosophila and is mutated in human cancer cell lines," *Cell*, vol. 110, no. 4, pp. 467–478, Aug. 2002, doi: 10.1016/S0092-8674(02)00824-3.

25. K. F. Harvey, X. Zhang, and D. M. Thomas, "The Hippo pathway and human cancer," *Nature Reviews Cancer 2013 13:4*, vol. 13, no. 4, pp. 246–257, Mar. 2013, doi: 10.1038/nrc3458.

26. D. Comaniciu, K. Engel, B. Georgescu, and T. Mansi, "Shaping the future through innovations: From medical imaging to precision medicine," *Medical Image Analysis*, vol. 33, pp. 19–26, Oct. 2016, doi: 10.1016/J.MEDIA.2016.06.016.

27. P. D. Caie, N. Dimitriou, and O. Arandjelović, "Precision medicine in digital pathology via image analysis and machine learning," *Artificial Intelligence and Deep Learning in Pathology*, pp. 149–173, Jan. 2021, doi: 10.1016/B978-0-323-67538-3.00008-7.

28. J. Gibson, "Precision medicine and next-generation sequencing: Challenges and opportunities Culling Memorial Lecture," *Journal of Histotechnology*, vol. 40, no. 4, p. 99, Oct. 2017, doi: 10.1080/01478885.2017.1383138.

29. M. Morash, H. Mitchell, H. Beltran, O. Elemento, and J. Pathak, "The role of next-generation sequencing in precision medicine: A review of outcomes in oncology," *Journal of Personalized Medicine*, vol. 8, no. 3, Sep. 2018, doi: 10.3390/JPM8030030.

30. Z. Liu, L. Zhu, R. Roberts, and W. Tong, "Toward clinical implementation of next-generation sequencing-based genetic testing in rare diseases: Where are we?" *Trends in Genetics*, vol. 35, no. 11, pp. 852–867, Nov. 2019, doi: 10.1016/J.TIG.2019.08.006.

31. R. Poplin et al., "A universal SNP and small-indel variant caller using deep neural networks," *Nature Biotechnology*, vol. 36, no. 10, pp. 983–987, Sep. 2018, doi: 10.1038/nbt.4235.

32. R. Luo, F. J. Sedlazeck, T. W. Lam, and M. C. Schatz, "A multi-task convolutional deep neural network for variant calling in single molecule sequencing," *Nature Communications*, vol. 10, no. 1, pp. 1–11, Mar. 2019, doi: 10.1038/s41467-019-09025-z.

33. V. Boža, B. Brejová, and T. Vinař, "DeepNano: Deep recurrent neural networks for base calling in MinION nanopore reads," *PLoS ONE*, vol. 12, no. 6, p. e0178751, Jun. 2017, doi: 10.1371/journal.pone.0178751.

34. R. Luo, T. W. Lam, and M. C. Schatz, "Skyhawk: An artificial neural network-based discriminator for reviewing clinically significant genomic variants," *International Journal of Computational Biology and Drug Design*, vol. 13, no. 5–6, pp. 431–437, 2020, doi: 10.1504/IJCBDD.2020.113818.

35. D. Quang, Y. Chen, and X. Xie, "DANN: A deep learning approach for annotating the pathogenicity of genetic variants," *Bioinformatics*, vol. 31, no. 5, pp. 761–763, Mar. 2015, doi: 10.1093/BIOINFORMATICS/BTU703.

36. J. Zhou and O. G. Troyanskaya, "Predicting effects of noncoding variants with deep learning–based sequence model," *Nature Methods*, vol. 12, no. 10, pp. 931–934, Aug. 2015, doi: 10.1038/nmeth.3547.

## CHAPTER 4

AI-Driven Drug Response Prediction for Personalized Cancer Medicine

Julhash U. Kazi

Division of Translational Cancer Research
Lund University Cancer Centre

CONTENTS

CANCER TREATMENT remains challenging due to heterogeneous drug responses and primary or acquired resistance to treatments. Technological advancements have allowed cost-effective genetic and transcriptomic profiling of individual cancer patients that can be used for better disease classification and personalized treatment. However, the use of complex data in clinical practice remains challenging. Artificial intelligence (AI) has been used to develop drug sensitivity prediction models, raising the opportunity of using it for personalized cancer medicine. Incorporating large-scale pharmacogenomic data to the various machine learning (ML) algorithms, several models have been developed for preclinical assessment or research purposes. Although these advancements clearly show promise, several challenges remain to be

DOI: 10.1201/9781003251903-4

addressed for clinical benefits. This chapter will describe recent advancements in drug response prediction methods using ML algorithms. Monotherapy and drug combination prediction methods will be discussed. A brief overview of pharmacogenomic data selection methods and appropriate ML algorithms will also be included. Furthermore, the limitations of current approaches will be addressed. The ability to predict drug responses for individual cancer patients will improve the quality of treatment by reducing unnecessary drug-associated side effects and financial burdens.

4.1 INTRODUCTION

In the past decades, artificial intelligence (AI) has been widely applied to cancer research in diagnosis, treatment planning, drug development, drug sensitivity prediction, and prognosis determination. Advancements in digital data collection have made a massive amount of data available for the development of computational programs. Improvements in computer hardware and algorithms have made AI more accessible to researchers for the development of better in silico tools using experimental data.

Due to the heterogeneous response to drugs, cancer treatment remains challenging. Although we would expect a drug to trigger a similar response in the same type of cancer, the response can be very different for different patients that share a similar cancer subtype. Therefore, there is an unmet need for development tools that can guide whether a patient will benefit from a drug prior to applying it. Such an approach would be beneficial in determining personalized cancer medicine.

Machine learning (ML), a subset of AI, has been applied widely and some of the earlier studies have been documented their applications in cancer research [1-9]. These studies offer valuable opinions on the advancement of the field, relevant computational challenges, and illustrate the importance of the field for cancer treatment. The area of cancer research is very broad, and ML can be used for several different aspects. In this chapter, we will briefly discuss ML algorithms that are currently applied to drug response prediction.

4.2 ML IS A SUBSET OF AI

Any technique that enables computers to mimic human intelligence can be categorized as AI, and if a method allows a computer to perform a task from experience, it can be categorized as ML (Figure 4.1a). Therefore, ML is a subset of AI that has been widely applied in cancer research, particularly drug sensitivity prediction. ML algorithms can be further classified into four subgroups: (1) Supervised learning—where algorithms are trained with a dataset linked to known outcomes to build hypothetical models that allow prediction of the outcome from unknown samples. (2) Unsupervised learning—which identifies hidden patterns from unlabeled data that allow clustering and pattern detection in unlabeled biological samples. (3) Semi-supervised learning—where data with known and unknown outcomes are mixed and algorithms learn from partially labeled data to develop a prediction model. (4) Reinforcement learning—where the algorithm learns how to predict a situation using completely unlabeled data (Figures 4.1b and 4.1c). In drug sensitivity prediction, supervised learning is widely used, and

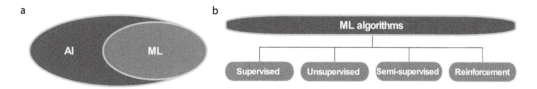

Data type	Learning algorithm	Examples
Labeled	Supervised learning	Decision trees, Logistic Regression, k-Nearest Neighbors, Support vector machine, Random forest, Naive Bayes, etc.
Unlabeled	Unsupervised learning	Hierarchical clustering, Hidden Markov models, k-means clustering, Principal component analysis, etc.
Partially labled	Semi-supervised learning	ADE-CoForest, Co-Adaboost, Co-Forest, Co-NB-SVM, Rasco, etc.
Unlabeled	Reniforcement learning	Markov decision process, Dynamic programming, Monte Carlo methods, Q-learning, QR-DQN, Temporal Difference algorithms, etc.

Figure 4.1 AI and ML. (a) ML is a subset of AI algorithms. (b) Possible subgroups of ML algorithms. (c) Examples of ML algorithms.

Figure 4.2 Algorithms that have been introduced over time. Earlier studies preferentially used ML algorithms while later studies mainly focused on deep learning (DL).

therefore this chapter will focus on different supervised learning algorithms that have been used to develop drug sensitivity prediction models.

4.3 ALGORITHMS USED FOR DRUG SENSITIVITY PREDICTION

Over the past decade, drug sensitivity prediction models have evolved with the improvement of computational power and the development of advanced algorithms. Although initial studies primarily focused on sparse linear regression models, later studies introduced more sophisticated models. Figure 4.2 presents a timeline graph of different models that have been introduced in drug sensitivity prediction studies, and it is clear that current studies are more biased toward deep neural network architecture.

Acronyms Used in the Text

BM-MKL – Bayesian multitask multiple kernel learning (also known as BMT-MKL)

CNN – convolutional NN

DNN – deep NN

EN – elastic net

GCN – graph convolutional network

LASSO – least absolute shrinkage and selection operator

MF – matrix factorization

NN – neural network

RF – random forest

RR – ridge regression

SVM – support vector machine

VAE – variational autoencoder

VNN – visible NN

4.4   PHARMACOGENOMIC DATASETS AND DATA PROCESSING

To build a prediction model, high-quality pharmacogenomic data are necessary. Several studies mentioned in Table 4.1 have generated or curated high-quality pharmacogenomic data, initially from cell lines and later from patient-derived xenografts and primary patient materials. These datasets primarily offer drug sensitivity data as well as baseline transcriptomic and genomic data.

Data pre-processing is one of the initial but important steps in any computational analysis. It includes several steps such as data selection, imputation of missing values, noise filtering, feature selection, and normalization. The selection of appropriate datasets remains challenging due to possible inconsistencies between different datasets [28]. Two elegant studies compared the largest publicly available pharmacogenomic datasets and suggested that individual datasets independently demonstrate considerable predictive power [15, 29]. Furthermore, pharmacogenomic datasets come with different types of biological data including genetic mutations, gene expression, copy number variations, protein expression, epigenetic modifications, and metabolomic data. Gene expression data can have more than 20,000 features, and then combining all features from a biological sample will result in several thousand variables. However, the largest pharmacogenomic dataset may contain around 1000 samples. Therefore, systematic selection of important features is crucial to developing an ideal predictive model [30, 31]. Several techniques for feature selection have been

TABLE 4.1   Pharmacogenomic Datasets

Dataset	Data Type							Ref.	Link to Dataset
CCLE	M	T	E	P	C	Me	D	[10, 11]	https://sites.broadinstitute.org/ccle/datasets
NCI-60							D	[12]	https://dtp.cancer.gov/discovery_development/nci-60/
GDSC	M	T	E		C		D	[13, 14]	https://www.cancerrxgene.org/
gCSI							D	[15]	https://www.gene.com/gCSI_GRvalue
CTRP							D	[16, 17]	https://portals.broadinstitute.org/ctrp/
PharmacoDB							D	[18]	https://pharmacodb.pmgenomics.ca/
CellMinerCDB	M	T	E	P	C		D	[19, 20]	https://discover.nci.nih.gov/rsconnect/cellminercdb/
DrugComb							D	[21]	https://drugcomb.fimm.fi/
DrugCombDB							D	[22]	http://drugcombdb.denglab.org/main
PDX finder	M	T			C		D	[23]	https://www.cancermodels.org/
PRoXE	M	T						[24]	https://www.proxe.org/
PDMR	M	T			C		D	[25]	https://pdmr.cancer.gov/
Vizome	M	T					D	[26, 27]	http://www.vizome.org

Abbreviations: C: copy number variation, D: drug sensitivity, E: epigenetic, Me: metabolomic, M: genetic mutation, P: proteomic, T: transcriptomic.

applied, including principal component analysis, backward feature elimination, high-correlation filters, and minimum redundancy maximum relevance (mRMR) [32-40]. Furthermore, due to the wide variation in biological features, one or multiple data scaling methods (normalization) such as data standardization, min-max normalization, cross-correlation, rank-invariant set normalization, or scaling to unit length may need to be applied [41].

4.5   DRUG SENSITIVITY PREDICTION STUDIES

Drug sensitivity prediction problems can be categorized as classification problems or regression problems. In a classification model, an algorithm learns whether a particular drug would display considerable sensitivity or resistance to a cancerous object, for example, a cancer cell line or patient-derived xenograft (PDX). In such a scenario, cell viability or death is measured in the presence of a drug covering a wide range of concentrations to fit a drug response curve that determines drug sensitivity. A fitted curve can later be used to determine quantitative parameters such as half-inhibitory concentrations (IC_{50}—half-maximal inhibitory concentration, EC_{50}—half-maximal effective concentration, or GR_{50}—half-maximal growth rate inhibition), the area under the curve, the area above the curve, or other drug sensitivity scores. Such scores can be used to develop a regression model where the prediction model is trained against drug sensitivity scores. Datasets summarized in Table 4.1 provide drug sensitivity scores that can be used to develop both classification models and regression models.

TABLE 4.2   Notable Drug Sensitivity Prediction Studies

Algorithms	Dataset Used	Features	Ref.
Elastic net	GDSC	Mutation, CNV, gene expression	[14]
	CCLE	Mutation, CNV, gene expression	[10]
	CTRP	Mutation, CNV	[16]
	CCLE, GDSC	Gene expression (1000 selected genes)	[32]
Ridge regression	GDSC, clinical data	Gene expression	[42]
	van de Wetering et al. [43], Lee et al. [44]	Gene expression, pathway	[45]
Random forest	GDSC, CCLE, NCI	Gene expression (1000 selected genes)	[33]
	PDXGEM	Gene expression	[46]
Elastic net and ridge regression	CCLE, GDSC	Mutation, CNV, gene expression	[47]
Elastic net and random forest	GDSC 2	Mutation, CNV, gene expression, methylation	[13]
Bayesian multitask multiple kernel learning	NCI-DREAM	Mutation, CNV, gene expression, proteomic	[48]
Random forest and support vector machine	NCI	Mutation, CNV, gene expression, RPLA, miRNA	[49]
Cell line-drug network model	GDSC, CCLE	Gene expression	[50]
Network-based gene-drug associations	AML patient and cell line data	Gene expression, mutation, CNV, methylation	[51]
Neural networks and random forests	GDSC	Selected genomic features	[52]
Recommender systems	CCLE, GDSC	Gene expression	[53]
DNN	LINCS	Gene expression	[54]
	GDSC	Gene expression	[55]
	GDSC	Mutation, CNV, gene expression.	[56]
	GDSC, KEGG, STITCH	Gene expression, pathway	[57]
	CCLE, Shah et al. [58]	Gene expression	[58]
Convolutional neural network	GDSC	Genomic fingerprints	[59]
	GDSC	Mutations and CNV	[60]
Autoencoder	GDSC, CCLE	Gene expression	[39]
Variational autoencoder	PharmacoDB, CMap	Gene expression	[37]
	TCGA, CCLE	Mutation, gene expression	[38]
	AML patient and cell line data	Gene expression	[61]
Visible neural network	GDSC, CCLE, CTRP	Gene expression, mutation, CNV, methylation	[40]
Graph convolutional network	GDSC	Mutations and CNV	[62]

As depicted in Figure 4.2, over time drug sensitivity prediction studies shifted from classical ML algorithms to deep learning (DL) algorithms. Table 4.2 provides a summary of notable drug prediction studies grouped according to the algorithm implemented. A detailed description will be provided below separately. Overall, those studies primarily used pharmacogenomic data from landmark cell line studies to develop their models. A few studies used PDX and primary patient data.

4.6 USE OF RIDGE REGRESSION, ELASTIC NET AND LASSO MODELS

Linear regression algorithms are widely used ML algorithms. The ordinary linear least squares regression is the most common algorithm, in which the cost function minimizes the sum of squared residuals (Figure 4.3a). Ordinary linear least squares regression may be useful for simplified data but pharmacogenomic data are complicated and, therefore, this regression should not be used for drug sensitivity predictions. The major drawback is overfitting, where the model displays low bias but suffers from higher variance. The addition of extra factors in the cost function has been shown to be useful to minimize overfitting. For example, in the ridge regression, a so-called L2 regularization factor has been added, which is the squared value of the slope multiplied by λ (Figure 4.3b). Similarly, in the least absolute shrinkage and selection operator (lasso) model L1 regularization (λ * absolute value of slope) has been added to the ordinary linear least-squares cost function (Figure 4.3c). Finally, the elastic net algorithm applies both L1 and L2 regularizations (Figure 4.3d). By regulating the value of λ, these models can be regulated [9].

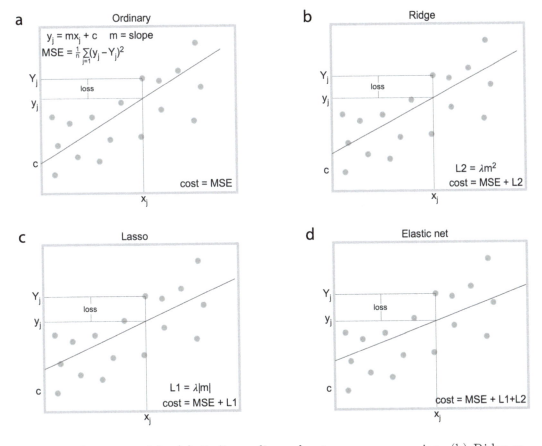

Figure 4.3 Linear models. (a) Ordinary linear least squares regression. (b) Ridge regression. (c) Least absolute shrinkage and selection operator (Lasso). (d) Elastic net algorithm.

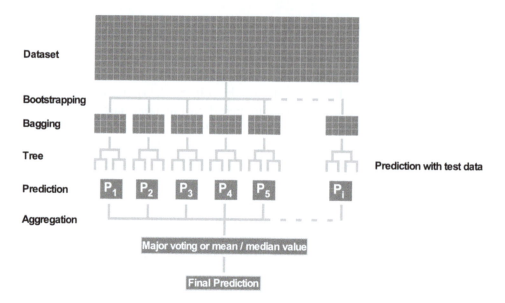

Figure 4.4   Random forest model. In a random forest model, a relatively large dataset is divided into several small datasets. This process is called bootstrapping and bagging. Then small datasets are used to develop the decision trees. The test dataset is applied on all decision trees and the predicted values are aggregated to make the final prediction.

Several initial studies developed sparse linear regression models to predict drug sensitivity. These studies usually applied ridge regression [42, 45, 47] and elastic net algorithms [10, 14, 16, 32]. Although the performance of those models was poorly generalized, those studies generated a massive amount of pharmacogenomic data that were useful for follow-up studies.

4.7   RANDOM FOREST AND KERNEL FUNCTIONS IN DRUG SENSITIVITY PREDICTION

Decision trees are popular supervised learning algorithms and are widely used for clinical decision analysis [63]. In a decision tree, model data are taken in the root node of the tree and, according to a test rule, keep growing until a decision can be drawn [64]. The beauty of a decision tree is that each step can be understood, interpreted, and visualized. A random forest uses a group of decision trees to solve relatively complex problems. In this method, data are randomly divided into small sets and then applied to decision trees (Figure 4.4). Prediction is made by the most chosen value (classification) or by the mean/median value (regression) from the decision of all trees. This process eventually increases the prediction accuracy [65].

Stetson et al. used pharmacogenomic data from the Genomics of Drug Sensitivity in Cancer (GDSC), Cancer Cell Line Encyclopedia (CCLE), and the National Cancer Institute (NCI) datasets and used 1000 selected genes to develop and compare ML models [33]. They found that the random forest model outperforms several other models including elastic net and kernel functions. While nonlinear models may be relatively better than linear models, particularly for classification problems, they can display poor performance with regressions.

The kernel functions can handle not only classification and regression problems for supervised learning but also can be applied to unsupervised learning. A wide range of functions, such as linear, nonlinear, sigmoid, radial, and polynomial, can be used to transform data to a high-dimensional space. Support vector machines (SVMs) are among the most popular kernel-based algorithms to be applied in drug sensitivity prediction [49]. Bayesian multitask multiple kernel learning (BM-MKL or BMT-MKL) is another such type of algorithm that has shown promising outcomes in an NCI-DREAM challenge summarized by Costello et al. [48].

4.8 MATRIX FACTORIZATION ALGORITHM FOR DRUG SENSITIVITY PREDICTION

Matrix factorization (MF) is one of the widely used unsupervised learning algorithms for dimensionality reduction but can also be used for supervised learning problems. MF is a popular method in e-commerce recommender systems [66]. The beauty of MF is that it can process high-dimensional data with missing information. MF decomposes the input matrix into lower-dimensional matrices that have the same number of latent factors [9]. The drawback of the system is that each specific model needs to be customized to serve a specific purpose. Using a modified recommender system, Suphavilai et al. developed Cancer Drug Response prediction using a Recommender System (CaDRReS) [53]. CaDRReS calculates cell line features (CCLE/GDSC data) using gene expression information, correlates the drug sensitivity information, and then factorizes into cell line and drug matrices. Besides drug sensitivity prediction, the method can be applied to explore the mechanisms of drug action, the interaction between drug and signaling pathways, and subtypes of cell lines or patients [53].

4.9 USE OF NEURAL NETWORK ALGORITHMS TO PREDICT DRUG SENSITIVITY

DL algorithms are a subtype of ML algorithms in which the backbone is developed by biologically inspired neural network algorithms. The algorithm can autonomously learn classification or regression problems from suitably labeled data. The algorithm applies nonlinear activation to process input data through hidden layers by forming a complex neural network. This process allows it to learn complex functions and at the same time amplifies important features to minimize noise. Figure 4.5 summarizes a simple DL model where training data are processed through a hidden layer and then a dropout layer. It is not mandatory to introduce a dropout layer, but a random dropout layer can help the model to learn important features and to reduce overfitting. The algorithm minimizes a cost function by defining values of weights and biases for each layer. Usually, gradient descent is used to find the minima, where gradients are determined by the chain rule of derivatives (also known as backpropagation) [67]. Such types of algorithms are widely used for image classification and have also recently been used for drug sensitivity prediction [68]. However, the classification of images is an easier task due to the availability of data, whereas currently, the availability of pharmacogenomic data is relatively limited when compared to the number of dimensions of such type of data. Therefore, the development of a DL model using pharmacogenomic data remains challenging [9]. Initial studies applied feedforward neural networks to develop a drug sensitivity prediction model [52, 54]. Instead of using all available or transcriptomic features, those studies reduced the number of features by selecting important genomic [52] or transcriptomic [54] variables. Follow-up studies have applied a similar approach to building deep neural network (DNN) models and have shown that those models perform better than classical ML models [55, 57].

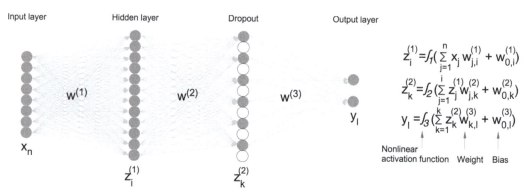

Figure 4.5 Deep neural network. In a DNN model, every node of the input layer is fully connected to the nodes of the hidden layer. A dropout layer can be included to remove some nodes randomly.

Convolutional neural networks (CNNs) are yet another domain of neural networks that are useful to find important features. The method applies several convolution and pooling steps from which it learns these features [67]. CNNs are widely used for image classification but recently have been used for drug sensitivity prediction. The Cancer Drug Response Profile scan (CDRscan) applied convolutions separately to molecular fingerprints of drugs and genomic fingerprints of cell lines and then merged the output results to develop a prediction model [59]. Using pharmacogenomic data from GDSC, particularly transcriptomic, somatic mutations and copy number variations, a similar model was built by Sharifi-Noghabi et al. and then validated by PDXs and primary patient materials [56]. Due to the nature of pooling important features during learning [67], models developed by CNN displayed considerably higher robustness and generalizability.

To develop a general model of therapy response prediction, chemical and structural information of multiple drugs is required to be incorporated during training. The two-dimensional molecular fingerprint is widely used for this purpose, in which partial structures of the molecules are determined and converted into a binary representation. The extended-connectivity fingerprint (ECFP) algorithm is a popular algorithm for determining two-dimensional molecular fingerprints [69]. The 3-dimensional structural information can also be collected in a similar fashion by a 3-dimensional fingerprint descriptor. Such types of converted information can be used as the input to a neural network. Weininger developed the Simplified Molecular-Input Line-Entry System (SMILES), a linear notation method [70] that can be used in CNN. Molecular graphs are a nonstructural way of storing atomic features from small-molecule drug structures with multiple atoms and bonds [71]. Because conventional CNNs are usually used for structural data, a modified CNN known as graph convolutional networks (GCN) has been developed to use molecular graphs [72]. Nguyen et al. used GCN to extract important molecular features from small-molecule drugs, combined that data with the CNN-extracted genomic features, and fed it into the fully connected feedforward neural network [62]. Use of graph representation of drugs seems to provide improved predictive performance over SMILES [60].

Although DL models displayed a promise in the prediction of drug sensitivity, model interpretation remains challenging due to the difficulties in understanding the model's internal rationality for a prediction. By incorporating genomic and transcriptomic data with the prior knowledge of signaling pathways and cellular architecture, Kuenzi et al. developed a visible neural network model (VNN) that can be relatively easily explained [40]. The model processes the chemical information of drugs separately and later combines it with the embedded genotype data. Although the model developed using VNN displayed similar or slightly better predictive performance, it holds the distinction of being easily explainable.

4.10 DIMENSIONALITY REDUCTION PRIOR TO DRUG SENSITIVITY PREDICTION: AUTOENCODER AND VARIATIONAL AUTOENCODER

Because pharmacogenomic data contains many features in comparison to sample numbers, an autoencoder has been used to reduce the dimension of features. An autoencoder is used to map high-dimensional data onto a lower-dimensional latent space in which the algorithm learns latent variables from the observed data. It contains two different types of layers: An encoding layer that projects the high-dimensional data to the lower-dimensional latent space and a decoding layer that decodes data from the latent space (Figure 4.6a). The loss function is calculated as the least square difference between input and decoded data. An autoencoder can take unlabeled data to develop the model, which can be used to reduce features that later can be used for drug sensitivity prediction [39, 73]. An improved form of an autoencoder (known as a variational autoencoder or VAE) in which the deterministic bottleneck layer has been replaced with stochastic sampling vectors (mean and standard deviation) (Figure 4.6b). Due to the stochastic nature of the sampling vector, the loss function can be updated by adding a Kullback-Leibler (KL) divergence term. This reparameterization trick allows the algorithm for backpropagation optimization [74].

Chiu et al. applied an autoencoder on unlabeled mutation and gene expression data from TCGA (The Cancer Genome Atlas) to develop a model that can be used to determine the latent variables from labeled data from CCLE (Cancer Cell Line Encyclopedia) [38]. The latent variables were used to develop a feedforward neural network to perform the drug sensitivity prediction. Due to the abundance of unlabeled data compared to labeled pharmacogenomic data, the implementation of an autoencoder

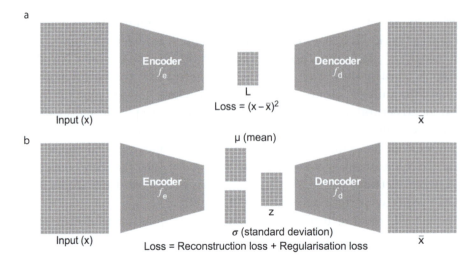

Figure 4.6 Schematic representation of autoencoders. (a) In an autoencoder, latent variables are determined by reducing dimensions, and then data are decoded from the latent variables. (b) In a VAE, similar encoding/decoding processes are used unless the latent variables are coded by the mean and standard deviation.

prior to the development of a prediction model increases prediction performance. Besides the autoencoder, VAE has also been applied to develop a monotherapy prediction model. Dincer et al. used gene expression data from acute myeloid leukemia (AML) patients to determine an eight-dimensional latent representation using a VAE and then a lasso model for the final drug sensitivity prediction [61]. A similar model was also described in which drug-induced gene expression perturbation was used in a VAE [37]. These are classical examples where a simple linear model can be developed for prediction from high-dimensional complex data. The use of drug-induced gene expression perturbation can be a viable approach for the reduction of features and can provide useful insights in determining pathways that regulate drug response [58].

4.11 DRUG COMBINATION PREDICTIONS

Although most studies have focused on the development of monotherapy prediction, the use of monotherapy for cancer treatment is very limited in real life. Due to the complex nature of the disease, most cancers are treated with a combination of drugs. Therefore, the development of an algorithm that can suggest a combination of therapies, or whether a therapeutic combination will display an enhanced efficacy (also known as synergy), will have a great impact in clinical settings. Table 4.3 summarizes notable studies that have attempted to develop methods for drug combination prediction.

The development of an ideal model to determine drug combinations required a large amount of genetic/transcriptomic information from biological samples, drug combination response, and structural information of the drug. As summarized in Table 4.3, several used a limited amount of published or publicly available data

TABLE 4.3 Studies Predicting Drug Combinations

Algorithms	Dataset Used	References
Semi-supervised learning	DCDB [75], KEGG, NCI-DREAM	[76]
Random forest	DREAM [77]	[78]
	Held et al. [79]	[80]
Regularized least squares	DrugBank, SIDER, OFFSIDES	[81]
Naïve Bayes, support vector machines, and random forest	DCDB [75], Cmap [82]	[83]
Elastic net, random forest, support vector machine	O'Neil 2016 [84]	[85]
Network-based	LINCS L1000, Held et al. [79]	[86]
	LINCS L1000	[87]
Ordinary differential equations	Perturbation data [88]	[89]
Multiple	DREAM	[90]
DL	NCI-ALMANAC [91]	[92]
	O'Neil 2016 [84]	[93]
Transformer boosted DL	O'Neil 2016 [84]	[94]
Graph convolutional network	O'Neil 2016 [84]	[95]
Autoencoder	O'Neil 2016 [84]	[96]
Restricted Boltzmann machine	DREAM [77]	[97]

to develop a drug combination prediction model. Like the monotherapy prediction models, several initial studies used ML algorithms to build a prediction model using pharmacogenomic data from cell lines [76–89]. Although these methods, including a study initiated by the DREAM challenge, were not optimal for clinical development, they provide a substantial framework for further development [77, 98].

Several studies have used feedforward neural networks to develop drug combination prediction models using previously published or publicly available data. For example, Xia et al. used data from transcriptomic and proteomic data from the NCI-ALMANAC database [91] to develop a feedforward neural network [92]. They used two separate submodels to process biological data and drug descriptors. The use of submodels allowed them to reduce the dimensionality before introducing data to the final prediction model. Furthermore, two studies used data from O'Neil et al. [84] to develop a deep learning model [93] or a modified DL model [94]. Although these models display considerable improvement in prediction performance compared to classical ML approaches, they also need to be upgraded with clinically relevant pharmacogenomic data.

The incorporation of drug descriptors with pharmacogenomic data remains challenging due to the substantial increase in the number of features. Recent studies have focused on methods that allow a reduction of variables prior to building the prediction model. Jiang et al. implemented a GCN to build a cell-line-specific drug combination prediction model [95]. Several other studies used autoencoders [96] or a restricted Boltzmann machine (RBM) [97] to extract meaningful features from high-dimensional pharmacogenomic data. Nevertheless, the extraction of meaningful features can improve the predictive performance of a model.

4.12   CLINICALLY RELEVANT PREDICTION MODELS

Because pharmacogenomic data for cell lines can be relatively easily obtained, most studies have used this type of data to develop prediction models. Although cell line models are widely used to study several aspects of cancer, they may not always represent cancer patients well. Therefore, the model developed based on cell line data may perform poorly with the primary clinical materials. To overcome such limitations, recent studies have used clinically relevant PDXs [46] or three-dimensional organoid culture models [45] to develop ML models. These studies have developed clinically relevant prediction models that increase the probability of successful clinical applications.

4.13   LIMITATIONS IN CURRENT APPROACHES

Although we have experienced rapid development in computational power, we still have a limited amount of clinically relevant pharmacogenomic data to develop a drug sensitivity prediction model. Most current studies have used pharmacogenomic data from cell lines. While cell line data are useful for hypothesis generation, they do not recapitulate clinical phenotypes. In humans, tumors form in a multicellular environment and interact with surrounding cells. A model that does not consider possible interactions of other cells with tumor cells may not achieve clinical level accuracy. To overcome such limitations, recently several studies have used PDXs and tumor organoids [45, 46]. Furthermore, several attempts have been taken to develop PDX repositories that provide clinically relevant pharmacogenomic data [23–25, 99].

Any genetic or epigenetic changes may reflect the baseline gene expression level in cancer cells. Therefore, the use of baseline gene expression data are useful to understand cancer biology. However, during drug treatment, although cells can hold similar genetic or epigenetic features, the gene expression pattern can become deregulated and that in turn can alter drug sensitivity. Currently, the majority of pharmacogenomic datasets provide base gene expression data that has been widely used to develop prediction models [10, 14, 16, 52]. Only a few studies have provided insight into drug perturbations [37, 88]. Therefore, in the future, we may need a large amount of clinically relevant drug perturbation data. For example, data about how a patient responded to a specific drug in the clinic and how the gene expression pattern has been changed after drug treatment would provide valuable information for future predictive models.

The computational challenges in the implementation of AI in precision medicine have been highlighted previously, including medical data processing, environmental data collection, clinical text, and data processing [9, 100, 101]. Besides those limitations, the incorporation of pharmacogenomic data in the clinical decision system can be challenging. Furthermore, as discussed partly above, the development of a prediction model in which each step can be explained remains challenging.

4.14 CONCLUSION

Drug sensitivity prediction for any single drug or a combination of drugs is challenging but has the utmost clinical interest in order to improve the quality of treatment by reducing the unnecessary use of drugs, particularly for cancer treatment. Although recent studies in the development of predictive models have displayed promise, a large amount of work still needs to be done for clinical applications. Long-term clinical trials to collect clinically relevant pharmacogenomic data for specific drugs could be useful but would be very expensive. Perhaps pharmacogenomic data from cell lines, PDXs, tumor organoids, and primary patients can be incorporated to develop clinically relevant models. Therefore, despite very promising inventions in drug sensitivity prediction models, we need to put more effort in generating pharmacogenomic data from primary patients to develop clinically applicable models.

Bibliography

1. K. Kourou, T.P. Exarchos, K.P. Exarchos, M.V. Karamouzis, D.I. Fotiadis, Machine learning applications in cancer prognosis and prediction, Comput Struct Biotechnol J, 13 (2015) 8–17.

2. A. Sharma, R. Rani, A systematic review of applications of machine learning in cancer prediction and diagnosis, archives of computational methods in engineering, 10.1007/s11831-021-09556-z (2021).

3. R. Hamamoto, K. Suvarna, M. Yamada, K. Kobayashi, N. Shinkai, M. Miyake, M. Takahashi, S. Jinnai, R. Shimoyama, A. Sakai, K. Takasawa, A. Bolatkan, K. Shozu, A. Dozen, H. Machino, S. Takahashi, K. Asada, M. Komatsu, J. Sese, S. Kaneko, Application of artificial intelligence technology in oncology: Towards the establishment of precision medicine, cancers (Basel), 12 (2020).

4. P.M. Putora, M. Baudis, B.M. Beadle, I. El Naqa, F.A. Giordano, N.H. Nicolay, Oncology informatics: Status quo and outlook, Oncology, 98 (2020) 329–331.

5. H. Shimizu, K.I. Nakayama, Artificial intelligence in oncology, Cancer Sci, 111 (2020) 1452–1460.

6. S. Huang, J. Yang, S. Fong, Q. Zhao, Artificial intelligence in cancer diagnosis and prognosis: Opportunities and challenges, Cancer Lett, 471 (2020) 61–71.

7. C. Nardini, Machine learning in oncology: A review, Ecancermedicalscience, 14 (2020) 1065.

8. F.V. Filipp, Opportunities for artificial intelligence in advancing precision medicine, Curr Genet Med Rep, 7 (2019) 208–213.

9. R. Rafique, S.M.R. Islam, J.U. Kazi, Machine learning in the prediction of cancer therapy, Comput Struct Biotechnol J, 19 (2021) 4003–4017.

10. J. Barretina, G. Caponigro, N. Stransky, K. Venkatesan, A.A. Margolin, S. Kim, C.J. Wilson, J. Lehar, G.V. Kryukov, D. Sonkin, A. Reddy, M. Liu, L. Murray, M.F. Berger, J.E. Monahan, P. Morais, J. Meltzer, A. Korejwa, J. Jane-Valbuena, F.A. Mapa, J. Thibault, E. Bric-Furlong, P. Raman, A. Shipway, I.H. Engels, J. Cheng, G.K. Yu, J. Yu, P. Aspesi, Jr., M. de Silva, K. Jagtap, M.D. Jones, L. Wang, C. Hatton, E. Palescandolo, S. Gupta, S. Mahan, C. Sougnez, R.C. Onofrio, T. Liefeld, L. MacConaill, W. Winckler, M. Reich, N. Li, J.P. Mesirov, S.B. Gabriel, G. Getz, K. Ardlie, V. Chan, V.E. Myer, B.L. Weber, J. Porter, M. Warmuth, P. Finan, J.L. Harris, M. Meyerson, T.R. Golub, M.P. Morrissey, W.R. Sellers, R. Schlegel, L.A. Garraway, The Cancer Cell Line Encyclopedia enables predictive modelling of anticancer drug sensitivity, Nature, 483 (2012) 603–607.

11. M. Ghandi, F.W. Huang, J. Jane-Valbuena, G.V. Kryukov, C.C. Lo, E.R. Mc-Donald, 3rd, J. Barretina, E.T. Gelfand, C.M. Bielski, H. Li, K. Hu, A.Y. Andreev-Drakhlin, J. Kim, J.M. Hess, B.J. Haas, F. Aguet, B.A. Weir, M.V. Rothberg, B.R. Paolella, M.S. Lawrence, R. Akbani, Y. Lu, H.L. Tiv, P.C. Gokhale, A. de Weck, A.A. Mansour, C. Oh, J. Shih, K. Hadi, Y. Rosen, J. Bistline, K. Venkatesan, A. Reddy, D. Sonkin, M. Liu, J. Lehar, J.M. Korn, D.A. Porter, M.D. Jones, J. Golji, G. Caponigro, J.E. Taylor, C.M. Dunning, A.L. Creech, A.C. Warren, J.M. McFarland, M. Zamanighomi, A. Kauffmann, N. Stransky, M. Imielinski, Y.E. Maruvka, A.D. Cherniack, A. Tsherniak, F. Vazquez, J.D. Jaffe, A.A. Lane, D.M. Weinstock, C.M. Johannessen, M.P. Morrissey, F. Stegmeier, R. Schlegel, W.C. Hahn, G. Getz, G.B. Mills, J.S. Boehm, T.R. Golub, L.A. Garraway, W.R. Sellers, Next-generation characterization of the Cancer Cell Line Encyclopedia, Nature, 569 (2019) 503–508.

12. R.H. Shoemaker, The NCI60 human tumour cell line anticancer drug screen, Nat Rev Cancer, 6 (2006) 813–823.

13. F. Iorio, T.A. Knijnenburg, D.J. Vis, G.R. Bignell, M.P. Menden, M. Schubert, N. Aben, E. Goncalves, S. Barthorpe, H. Lightfoot, T. Cokelaer, P. Greninger, E. van Dyk, H. Chang, H. de Silva, H. Heyn, X. Deng, R.K. Egan, Q. Liu, T. Mironenko, X. Mitropoulos, L. Richardson, J. Wang, T. Zhang, S. Moran, S. Sayols, M. Soleimani, D. Tamborero, N. Lopez-Bigas, P. Ross-Macdonald, M. Esteller, N.S. Gray, D.A. Haber, M.R. Stratton, C.H. Benes, L.F.A. Wessels, J. Saez-Rodriguez, U. McDermott, M.J. Garnett, A landscape of pharmacogenomic interactions in cancer, Cell, 166 (2016) 740–754.

14. M.J. Garnett, E.J. Edelman, S.J. Heidorn, C.D. Greenman, A. Dastur, K.W. Lau, P. Greninger, I.R. Thompson, X. Luo, J. Soares, Q. Liu, F. Iorio, D. Surdez, L. Chen, R.J. Milano, G.R. Bignell, A.T. Tam, H. Davies, J.A. Stevenson, S. Barthorpe, S.R. Lutz, F. Kogera, K. Lawrence, A. McLaren-Douglas, X. Mitropoulos, T. Mironenko, H. Thi, L. Richardson, W. Zhou, F. Jewitt, T. Zhang, P. O'Brien, J.L. Boisvert, S. Price, W. Hur, W. Yang, X. Deng, A. Butler, H.G. Choi, J.W. Chang, J. Baselga, I. Stamenkovic, J.A. Engelman, S.V.

Sharma, O. Delattre, J. Saez-Rodriguez, N.S. Gray, J. Settleman, P.A. Futreal, D.A. Haber, M.R. Stratton, S. Ramaswamy, U. McDermott, C.H. Benes, Systematic identification of genomic markers of drug sensitivity in cancer cells, Nature, 483 (2012) 570–575.

15. P.M. Haverty, E. Lin, J. Tan, Y. Yu, B. Lam, S. Lianoglou, R.M. Neve, S. Martin, J. Settleman, R.L. Yauch, R. Bourgon, Reproducible pharmacogenomic profiling of cancer cell line panels, Nature, 533 (2016) 333–337.

16. A. Basu, N.E. Bodycombe, J.H. Cheah, E.V. Price, K. Liu, G.I. Schaefer, R.Y. Ebright, M.L. Stewart, D. Ito, S. Wang, A.L. Bracha, T. Liefeld, M. Wawer, J.C. Gilbert, A.J. Wilson, N. Stransky, G.V. Kryukov, V. Dancik, J. Barretina, L.A. Garraway, C.S. Hon, B. Munoz, J.A. Bittker, B.R. Stockwell, D. Khabele, A.M. Stern, P.A. Clemons, A.F. Shamji, S.L. Schreiber, An interactive resource to identify cancer genetic and lineage dependencies targeted by small molecules, Cell, 154 (2013) 1151–1161.

17. B. Seashore-Ludlow, M.G. Rees, J.H. Cheah, M. Cokol, E.V. Price, M.E. Coletti, V. Jones, N.E. Bodycombe, C.K. Soule, J. Gould, B. Alexander, A. Li, P. Montgomery, M.J. Wawer, N. Kuru, J.D. Kotz, C.S. Hon, B. Munoz, T. Liefeld, V. Dancik, J.A. Bittker, M. Palmer, J.E. Bradner, A.F. Shamji, P.A. Clemons, S.L. Schreiber, Harnessing connectivity in a large-scale small-molecule sensitivity dataset, Cancer Discov, 5 (2015) 1210–1223.

18. P. Smirnov, V. Kofia, A. Maru, M. Freeman, C. Ho, N. El-Hachem, G.A. Adam, W. Ba-Alawi, Z. Safikhani, B. Haibe-Kains, PharmacoDB: An integrative database for mining in vitro anticancer drug screening studies, Nucleic Acids Res, 46 (2018) D994–D1002.

19. V.N. Rajapakse, A. Luna, M. Yamade, L. Loman, S. Varma, M. Sunshine, F. Iorio, F.G. Sousa, F. Elloumi, M.I. Aladjem, A. Thomas, C. Sander, K.W. Kohn, C.H. Benes, M. Garnett, W.C. Reinhold, Y. Pommier, CellMinerCDB for integrative cross-database genomics and pharmacogenomics analyses of cancer cell lines, iScience, 10 (2018) 247–264.

20. A. Luna, F. Elloumi, S. Varma, Y. Wang, V.N. Rajapakse, M.I. Aladjem, J. Robert, C. Sander, Y. Pommier, W.C. Reinhold, CellMiner Cross-Database (CellMinerCDB) version 1.2: Exploration of patient-derived cancer cell line pharmacogenomics, Nucleic Acids Res, 49 (2021) D1083–D1093.

21. B. Zagidullin, J. Aldahdooh, S. Zheng, W. Wang, Y. Wang, J. Saad, A. Malyutina, M. Jafari, Z. Tanoli, A. Pessia, J. Tang, DrugComb: An integrative cancer drug combination data portal, Nucleic Acids Res, 47 (2019) W43–W51.

22. H. Liu, W. Zhang, B. Zou, J. Wang, Y. Deng, L. Deng, DrugCombDB: A comprehensive database of drug combinations toward the discovery of combinatorial therapy, Nucleic Acids Res, 48 (2020) D871–D881.

23. N. Conte, J.C. Mason, C. Halmagyi, S. Neuhauser, A. Mosaku, G. Yordanova, A. Chatzipli, D.A. Begley, D.M. Krupke, H. Parkinson, T.F. Meehan, C.C. Bult, PDX Finder: A portal for patient-derived tumor xenograft model discovery, Nucleic Acids Res, 47 (2019) D1073–D1079.

24. E.C. Townsend, M.A. Murakami, A. Christodoulou, A.L. Christie, J. Koster, T.A. DeSouza, E.A. Morgan, S.P. Kallgren, H. Liu, S.C. Wu, O. Plana, J. Montero, K.E. Stevenson, P. Rao, R. Vadhi, M. Andreeff, P. Armand, K.K. Ballen, P. Barzaghi-Rinaudo, S. Cahill, R.A. Clark, V.G. Cooke, M.S. Davids, D.J. DeAngelo, D.M. Dorfman, H. Eaton, B.L. Ebert, J. Etchin, B. Firestone, D.C. Fisher, A.S. Freedman, I.A. Galinsky, H. Gao, J.S. Garcia, F. Garnache-Ottou, T.A. Graubert, A. Gutierrez, E. Halilovic, M.H. Harris, Z.T. Herbert, S.M. Horwitz, G. Inghirami, A.M. Intlekofer, M. Ito, S. Izraeli, E.D. Jacobsen, C.A. Jacobson, S. Jeay, I. Jeremias, M.A. Kelliher, R. Koch, M. Konopleva, N. Kopp, S.M. Kornblau, A.L. Kung, T.S. Kupper, N.R. LeBoeuf, A.S. LaCasce, E. Lees, L.S. Li, A.T. Look, M. Murakami, M. Muschen, D. Neuberg, S.Y. Ng, O.O. Odejide, S.H. Orkin, R.R. Paquette, A.E. Place, J.E. Roderick, J.A. Ryan, S.E. Sallan, B. Shoji, L.B. Silverman, R.J. Soiffer, D.P. Steensma, K. Stegmaier, R.M. Stone, J. Tamburini, A.R. Thorner, P. van Hummelen, M. Wadleigh, M. Wiesmann, A.P. Weng, J.U. Wuerthner, D.A. Williams, B.M. Wollison, A.A. Lane, A. Letai, M.M. Bertagnolli, J. Ritz, M. Brown, H. Long, J.C. Aster, M.A. Shipp, J.D. Griffin, D.M. Weinstock, The public repository of xenografts enables discovery and randomized phase II-like trials in mice, Cancer Cell, 29 (2016) 574–586.

25. PDMR, NCI's Patient-derived models repository, https://pdmr.cancer.gov/, (2021). Accessed: 16 November, 2022.

26. J.W. Tyner, C.E. Tognon, D. Bottomly, B. Wilmot, S.E. Kurtz, S.L. Savage, N. Long, A.R. Schultz, E. Traer, M. Abel, A. Agarwal, A. Blucher, U. Borate, J. Bryant, R. Burke, A. Carlos, R. Carpenter, J. Carroll, B.H. Chang, C. Coblentz, A. d'Almeida, R. Cook, A. Danilov, K.T. Dao, M. Degnin, D. Devine, J. Dibb, D.K.t. Edwards, C.A. Eide, I. English, J. Glover, R. Henson, H. Ho, A. Jemal, K. Johnson, R. Johnson, B. Junio, A. Kaempf, J. Leonard, C. Lin, S.Q. Liu, P. Lo, M.M. Loriaux, S. Luty, T. Macey, J. MacManiman, J. Martinez, M. Mori, D. Nelson, C. Nichols, J. Peters, J. Ramsdill, A. Rofelty, R. Schuff, R. Searles, E. Segerdell, R.L. Smith, S.E. Spurgeon, T. Sweeney, A. Thapa, C. Visser, J. Wagner, K. Watanabe-Smith, K. Werth, J. Wolf, L. White, A. Yates, H. Zhang, C.R. Cogle, R.H. Collins, D.C. Connolly, M.W. Deininger, L. Drusbosky, C.S. Hourigan, C.T. Jordan, P. Kropf, T.L. Lin, M.E. Martinez, B.C. Medeiros, R.R. Pallapati, D.A. Pollyea, R.T. Swords, J.M. Watts, S.J. Weir, D.L. Wiest, R.M. Winters, S.K. McWeeney, B.J. Druker, Functional genomic landscape of acute myeloid leukaemia, Nature, 562 (2018) 526–531.

27. H. Zhang, S. Savage, A.R. Schultz, D. Bottomly, L. White, E. Segerdell, B. Wilmot, S.K. McWeeney, C.A. Eide, T. Nechiporuk, A. Carlos, R. Henson, C. Lin, R. Searles, H. Ho, Y.L. Lam, R. Sweat, C. Follit, V. Jain, E. Lind,

G. Borthakur, G. Garcia-Manero, F. Ravandi, H.M. Kantarjian, J. Cortes, R. Collins, D.R. Buelow, S.D. Baker, B.J. Druker, J.W. Tyner, Clinical resistance to crenolanib in acute myeloid leukemia due to diverse molecular mechanisms, Nat Commun, 10 (2019) 244.

28. B. Haibe-Kains, N. El-Hachem, N.J. Birkbak, A.C. Jin, A.H. Beck, H.J. Aerts, J. Quackenbush, Inconsistency in large pharmacogenomic studies, Nature, 504 (2013) 389–393.

29. C. Cancer Cell Line Encyclopedia, C. Genomics of Drug Sensitivity in Cancer, Pharmacogenomic agreement between two cancer cell line datasets, Nature, 528 (2015) 84–87.

30. K. Koras, D. Juraeva, J. Kreis, J. Mazur, E. Staub, E. Szczurek, Feature selection strategies for drug sensitivity prediction, Sci Rep, 10 (2020) 9377.

31. M. Ali, S.A. Khan, K. Wennerberg, T. Aittokallio, Global proteomics profiling improves drug sensitivity prediction: Results from a multi-omics, pan-cancer modeling approach, Bioinformatics, 34 (2018) 1353–1362.

32. S. Papillon-Cavanagh, N. De Jay, N. Hachem, C. Olsen, G. Bontempi, H.J. Aerts, J. Quackenbush, B. Haibe-Kains, Comparison and validation of genomic predictors for anticancer drug sensitivity, J Am Med Inform Assoc, 20 (2013) 597–602.

33. L.C. Stetson, T. Pearl, Y. Chen, J.S. Barnholtz-Sloan, Computational identification of multi-omic correlates of anticancer therapeutic response, BMC Genomics, 15 Suppl 7 (2014) S2.

34. C. Ding, H. Peng, Minimum redundancy feature selection from microarray gene expression data, J Bioinform Comput Biol, 3 (2005) 185–205.

35. T.H. Lin, H.T. Li, K.C. Tsai, Implementing the Fisher's discriminant ratio in a k-means clustering algorithm for feature selection and dataset trimming, J Chem Inf Comput Sci, 44 (2004) 76–87.

36. M. Nakajo, M. Jinguji, A. Tani, D. Hirahara, H. Nagano, K. Takumi, T. Yoshiura, Application of a machine learning approach to characterization of liver function using (99m)Tc-GSA SPECT/CT, Abdom Radiol (NY), (2021).

37. L. Rampasek, D. Hidru, P. Smirnov, B. Haibe-Kains, A. Goldenberg, Dr.VAE: Improving drug response prediction via modeling of drug perturbation effects, Bioinformatics, 35 (2019) 3743–3751.

38. Y.C. Chiu, H.H. Chen, T. Zhang, S. Zhang, A. Gorthi, L.J. Wang, Y. Huang, Y. Chen, Predicting drug response of tumors from integrated genomic profiles by deep neural networks, BMC Med Genomics, 12 (2019) 18.

39. M. Li, Y. Wang, R. Zheng, X. Shi, Y. Li, F. Wu, J. Wang, DeepDSC: A deep learning method to predict drug sensitivity of cancer cell lines, IEEE/ACM Trans Comput Biol Bioinform, (2019).

40. B.M. Kuenzi, J. Park, S.H. Fong, K.S. Sanchez, J. Lee, J.F. Kreisberg, J. Ma, T. Ideker, Predicting drug response and synergy using a deep learning model of human cancer cells, cancer cell, 38 (2020) 672–684 e676.

41. X. Liu, N. Li, S. Liu, J. Wang, N. Zhang, X. Zheng, K.S. Leung, L. Cheng, Normalization methods for the analysis of unbalanced transcriptome data: A review, Front Bioeng Biotechnol, 7 (2019) 358.

42. P. Geeleher, N.J. Cox, R.S. Huang, Clinical drug response can be predicted using baseline gene expression levels and in vitro drug sensitivity in cell lines, Genome Biol, 15 (2014) R47.

43. M. van de Wetering, H.E. Francies, J.M. Francis, G. Bounova, F. Iorio, A. Pronk, W. van Houdt, J. van Gorp, A. Taylor-Weiner, L. Kester, A. McLaren-Douglas, J. Blokker, S. Jaksani, S. Bartfeld, R. Volckman, P. van Sluis, V.S. Li, S. Seepo, C. Sekhar Pedamallu, K. Cibulskis, S.L. Carter, A. McKenna, M.S. Lawrence, L. Lichtenstein, C. Stewart, J. Koster, R. Versteeg, A. van Oudenaarden, J. Saez-Rodriguez, R.G. Vries, G. Getz, L. Wessels, M.R. Stratton, U. McDermott, M. Meyerson, M.J. Garnett, H. Clevers, Prospective derivation of a living organoid biobank of colorectal cancer patients, Cell, 161 (2015) 933–945.

44. S.H. Lee, W. Hu, J.T. Matulay, M.V. Silva, T.B. Owczarek, K. Kim, C.W. Chua, L.J. Barlow, C. Kandoth, A.B. Williams, S.K. Bergren, E.J. Pietzak, C.B. Anderson, M.C. Benson, J.A. Coleman, B.S. Taylor, C. Abate-Shen, J.M. McKiernan, H. Al-Ahmadie, D.B. Solit, M.M. Shen, Tumor evolution and drug response in patient-derived organoid models of bladder cancer, cell, 173 (2018) 515–528 e517.

45. J. Kong, H. Lee, D. Kim, S.K. Han, D. Ha, K. Shin, S. Kim, Network-based machine learning in colorectal and bladder organoid models predicts anti-cancer drug efficacy in patients, Nat Commun, 11 (2020) 5485.

46. Y. Kim, D. Kim, B. Cao, R. Carvajal, M. Kim, PDXGEM: Patient-derived tumor xenograft-based gene expression model for predicting clinical response to anticancer therapy in cancer patients, BMC Bioinformatics, 21 (2020) 288.

47. I.S. Jang, E.C. Neto, J. Guinney, S.H. Friend, A.A. Margolin, Systematic assessment of analytical methods for drug sensitivity prediction from cancer cell line data, Pac Symp Biocomput, (2014) 63–74.

48. J.C. Costello, L.M. Heiser, E. Georgii, M. Gonen, M.P. Menden, N.J. Wang, M. Bansal, M. Ammad-ud-din, P. Hintsanen, S.A. Khan, J.P. Mpindi, O. Kallioniemi, A. Honkela, T. Aittokallio, K. Wennerberg, N.D. Community, J.J. Collins,

D. Gallahan, D. Singer, J. Saez-Rodriguez, S. Kaski, J.W. Gray, G. Stolovitzky, A community effort to assess and improve drug sensitivity prediction algorithms, Nat Biotechnol, 32 (2014) 1202–1212.

49. I. Cortes-Ciriano, G.J. van Westen, G. Bouvier, M. Nilges, J.P. Overington, A. Bender, T.E. Malliavin, Improved large-scale prediction of growth inhibition patterns using the NCI60 cancer cell line panel, Bioinformatics, 32 (2016) 85–95.

50. N. Zhang, H. Wang, Y. Fang, J. Wang, X. Zheng, X.S. Liu, Predicting anticancer drug responses using a dual-layer integrated cell line-drug network model, PLoS Comput Biol, 11 (2015) e1004498.

51. S.I. Lee, S. Celik, B.A. Logsdon, S.M. Lundberg, T.J. Martins, V.G. Oehler, E.H. Estey, C.P. Miller, S. Chien, J. Dai, A. Saxena, C.A. Blau, P.S. Becker, A machine learning approach to integrate big data for precision medicine in acute myeloid leukemia, Nat Commun, 9 (2018) 42.

52. M.P. Menden, F. Iorio, M. Garnett, U. McDermott, C.H. Benes, P.J. Ballester, J. Saez-Rodriguez, Machine learning prediction of cancer cell sensitivity to drugs based on genomic and chemical properties, PLoS One, 8 (2013) e61318.

53. C. Suphavilai, D. Bertrand, N. Nagarajan, Predicting cancer drug response using a recommender system, Bioinformatics, 34 (2018) 3907–3914.

54. A. Aliper, S. Plis, A. Artemov, A. Ulloa, P. Mamoshina, A. Zhavoronkov, Deep learning applications for predicting pharmacological properties of drugs and drug repurposing using transcriptomic data, Mol Pharm, 13 (2016) 2524–2530.

55. T. Sakellaropoulos, K. Vougas, S. Narang, F. Koinis, A. Kotsinas, A. Polyzos, T.J. Moss, S. Piha-Paul, H. Zhou, E. Kardala, E. Damianidou, L.G. Alexopoulos, I. Aifantis, P.A. Townsend, M.I. Panayiotidis, P. Sfikakis, J. Bartek, R.C. Fitzgerald, D. Thanos, K.R. Mills Shaw, R. Petty, A. Tsirigos, V.G. Gorgoulis, A deep learning framework for predicting response to therapy in cancer, Cell Rep, 29 (2019) 3367–3373 e3364.

56. H. Sharifi-Noghabi, O. Zolotareva, C.C. Collins, M. Ester, MOLI: Multi-omics late integration with deep neural networks for drug response prediction, Bioinformatics, 35 (2019) i501–i509.

57. L. Deng, Y. Cai, W. Zhang, W. Yang, B. Gao, H. Liu, Pathway-Guided deep neural network toward interpretable and predictive modeling of drug sensitivity, J Chem Inf Model, 60 (2020) 4497–4505.

58. K. Shah, M. Ahmed, J.U. Kazi, The Aurora kinase/beta-catenin axis contributes to dexamethasone resistance in leukemia, NPJ Precis Oncol, 5 (2021) 13.

59. Y. Chang, H. Park, H.J. Yang, S. Lee, K.Y. Lee, T.S. Kim, J. Jung, J.M. Shin, Cancer Drug Response Profile scan (CDRscan): A deep learning model that predicts drug effectiveness from cancer genomic signature, Sci Rep, 8 (2018) 8857.

60. P. Liu, H. Li, S. Li, K.S. Leung, Improving prediction of phenotypic drug response on cancer cell lines using deep convolutional network, BMC Bioinformatics, 20 (2019) 408.

61. A.V. Dincer, S. Celik, N. Hiranuma, S.I. Lee, DeepProfile: Deep learning of cancer molecular profiles for precision medicine, bioRxiv, 10.1101/278739 (2018).

62. T.T. Nguyen, G.T.T. Nguyen, T. Nguyen, D.H. Le, Graph convolutional networks for drug response prediction, IEEE/ACM Trans Comput Biol Bioinform, PP (2021).

63. J.M. Bae, The clinical decision analysis using decision tree, Epidemiol Health, 36 (2014) e2014025.

64. V. Podgorelec, P. Kokol, B. Stiglic, I. Rozman, Decision trees: An overview and their use in medicine, J Med Syst, 26 (2002) 445–463.

65. L. Breiman, Random forests, Machine Learning, 45 (2001) 5–32.

66. Y. Koren, R. Bell, C. Volinsky, Matrix factorization technique for recommender system, Computer, 42 (2009) 30–37.

67. Y. LeCun, Y. Bengio, G. Hinton, Deep learning, Nature, 521 (2015) 436–444.

68. N.T. Issa, V. Stathias, S. Schurer, S. Dakshanamurthy, Machine and deep learning approaches for cancer drug repurposing, Semin Cancer Biol, (2020).

69. D. Rogers, M. Hahn, Extended-connectivity fingerprints, J Chem Inf Model, 50 (2010) 742–754.

70. D. Weininger, SMILES, a chemical language and information system. 1. Introduction to methodology and encoding rules, J. Chem. Inf. Comput. Sci., 28 (1988) 31–36.

71. M. Sun, S. Zhao, C. Gilvary, O. Elemento, J. Zhou, F. Wang, Graph convolutional networks for computational drug development and discovery, Brief Bioinform, 21 (2020) 919–935.

72. T.N. Kipf, M. Welling, Semi-supervised classification with graph convolutional networks, arXiv, 1609.02907v4 (2017).

73. G.P. Way, C.S. Greene, Extracting a biologically relevant latent space from cancer transcriptomes with variational autoencoders, Pac Symp Biocomput, 23 (2018) 80–91.

74. D.P. Kingma, M. Welling, Auto-encoding variational bayes, arXiv, 1312.6114 (2014).

75. Y. Liu, Q. Wei, G. Yu, W. Gai, Y. Li, X. Chen, DCDB 2.0: A major update of the drug combination database, Database (Oxford), 2014 (2014) bau124.

76. Y. Sun, Z. Sheng, C. Ma, K. Tang, R. Zhu, Z. Wu, R. Shen, J. Feng, D. Wu, D. Huang, D. Huang, J. Fei, Q. Liu, Z. Cao, Combining genomic and network characteristics for extended capability in predicting synergistic drugs for cancer, Nat Commun, 6 (2015) 8481.

77. M. Bansal, J. Yang, C. Karan, M.P. Menden, J.C. Costello, H. Tang, G. Xiao, Y. Li, J. Allen, R. Zhong, B. Chen, M. Kim, T. Wang, L.M. Heiser, R. Realubit, M. Mattioli, M.J. Alvarez, Y. Shen, N.-D. Community, D. Gallahan, D. Singer, J. Saez-Rodriguez, Y. Xie, G. Stolovitzky, A. Califano, N.-D. Community, A community computational challenge to predict the activity of pairs of compounds, Nat Biotechnol, 32 (2014) 1213–1222.

78. X. Li, Y. Xu, H. Cui, T. Huang, D. Wang, B. Lian, W. Li, G. Qin, L. Chen, L. Xie, Prediction of synergistic anti-cancer drug combinations based on drug target network and drug induced gene expression profiles, Artif Intell Med, 83 (2017) 35–43.

79. M.A. Held, C.G. Langdon, J.T. Platt, T. Graham-Steed, Z. Liu, A. Chakraborty, A. Bacchiocchi, A. Koo, J.W. Haskins, M.W. Bosenberg, D.F. Stern, Genotype-selective combination therapies for melanoma identified by high-throughput drug screening, Cancer Discov, 3 (2013) 52–67.

80. K.M. Gayvert, O. Aly, J. Platt, M.W. Bosenberg, D.F. Stern, O. Elemento, A computational approach for identifying synergistic drug combinations, PLoS Comput Biol, 13 (2017) e1005308.

81. C. Yan, G. Duan, Y. Pan, F.X. Wu, J. Wang, DDIGIP: Predicting drug-drug interactions based on Gaussian interaction profile kernels, BMC Bioinformatics, 20 (2019) 538.

82. J. Lamb, The connectivity map: A new tool for biomedical research, Nat Rev Cancer, 7 (2007) 54–60.

83. A. Cuvitoglu, J.X. Zhou, S. Huang, Z. Isik, Predicting drug synergy for precision medicine using network biology and machine learning, J Bioinform Comput Biol, 17 (2019) 1950012.

84. J. O'Neil, Y. Benita, I. Feldman, M. Chenard, B. Roberts, Y. Liu, J. Li, A. Kral, S. Lejnine, A. Loboda, W. Arthur, R. Cristescu, B.B. Haines, C. Winter, T. Zhang, A. Bloecher, S.D. Shumway, An unbiased oncology compound screen to identify novel combination strategies, Mol Cancer Ther, 15 (2016) 1155–1162.

85. A. Malyutina, M.M. Majumder, W. Wang, A. Pessia, C.A. Heckman, J. Tang, Drug combination sensitivity scoring facilitates the discovery of synergistic and efficacious drug combinations in cancer, PLoS Comput Biol, 15 (2019) e1006752.

86. K.E. Regan, P.R.O. Payne, F. Li, Integrative network and transcriptomics-based approach predicts genotype-specific drug combinations for melanoma, AMIA Jt Summits Transl Sci Proc, 2017 (2017) 247–256.

87. K.E. Regan-Fendt, J. Xu, M. DiVincenzo, M.C. Duggan, R. Shakya, R. Na, W.E. Carson, 3rd, P.R.O. Payne, F. Li, Synergy from gene expression and network mining (SynGeNet) method predicts synergistic drug combinations for diverse melanoma genomic subtypes, NPJ Syst Biol Appl, 5 (2019) 6.

88. A. Korkut, W. Wang, E. Demir, B.A. Aksoy, X. Jing, E.J. Molinelli, O. Babur, D.L. Bemis, S. Onur Sumer, D.B. Solit, C.A. Pratilas, C. Sander, Perturbation biology nominates upstream-downstream drug combinations in RAF inhibitor resistant melanoma cells, Elife, 4 (2015).

89. B. Yuan, C. Shen, A. Luna, A. Korkut, D.S. Marks, J. Ingraham, C. Sander, CellBox: Interpretable machine learning for perturbation biology with application to the design of cancer combination therapy, Cell Syst, 12 (2021) 128–140 e124.

90. M.P. Menden, D. Wang, M.J. Mason, B. Szalai, K.C. Bulusu, Y. Guan, T. Yu, J. Kang, M. Jeon, R. Wolfinger, T. Nguyen, M. Zaslavskiy, D.C. AstraZeneca-Sanger Drug Combination, I.S. Jang, Z. Ghazoui, M.E. Ahsen, R. Vogel, E.C. Neto, T. Norman, E.K.Y. Tang, M.J. Garnett, G.Y.D. Veroli, S. Fawell, G. Stolovitzky, J. Guinney, J.R. Dry, J. Saez-Rodriguez, Community assessment to advance computational prediction of cancer drug combinations in a pharmacogenomic screen, Nat Commun, 10 (2019) 2674.

91. S.L. Holbeck, R. Camalier, J.A. Crowell, J.P. Govindharajulu, M. Hollingshead, L.W. Anderson, E. Polley, L. Rubinstein, A. Srivastava, D. Wilsker, J.M. Collins, J.H. Doroshow, The National Cancer Institute ALMANAC: A comprehensive screening resource for the detection of anticancer drug pairs with enhanced therapeutic activity, Cancer Res, 77 (2017) 3564–3576.

92. F. Xia, M. Shukla, T. Brettin, C. Garcia-Cardona, J. Cohn, J.E. Allen, S. Maslov, S.L. Holbeck, J.H. Doroshow, Y.A. Evrard, E.A. Stahlberg, R.L. Stevens, Predicting tumor cell line response to drug pairs with deep learning, BMC Bioinformatics, 19 (2018) 486.

93. K. Preuer, R.P.I. Lewis, S. Hochreiter, A. Bender, K.C. Bulusu, G. Klambauer, DeepSynergy: Predicting anti-cancer drug synergy with deep learning, Bioinformatics, 34 (2018) 1538–1546.

94. Q. Liu, L. Xie, TranSynergy: Mechanism-driven interpretable deep neural network for the synergistic prediction and pathway deconvolution of drug combinations, PLoS Comput Biol, 17 (2021) e1008653.

95. P. Jiang, S. Huang, Z. Fu, Z. Sun, T.M. Lakowski, P. Hu, Deep graph embedding for prioritizing synergistic anticancer drug combinations, Comput Struct Biotechnol J, 18 (2020) 427–438.

96. T. Zhang, L. Zhang, P.R.O. Payne, F. Li, Synergistic drug combination prediction by integrating multiomics data in deep learning models, Methods Mol Biol, 2194 (2021) 223–238.

97. G. Chen, A. Tsoi, H. Xu, W.J. Zheng, Predict effective drug combination by deep belief network and ontology fingerprints, J Biomed Inform, 85 (2018) 149–154.

98. C.P. Goswami, L. Cheng, P.S. Alexander, A. Singal, L. Li, A new drug combinatory effect prediction algorithm on the cancer cell based on gene expression and dose-response curve, CPT Pharmacometrics Syst Pharmacol, 4 (2015) e9.

99. M. Hidalgo, F. Amant, A.V. Biankin, E. Budinska, A.T. Byrne, C. Caldas, R.B. Clarke, S. de Jong, J. Jonkers, G.M. Maelandsmo, S. Roman-Roman, J. Seoane, L. Trusolino, A. Villanueva, Patient-derived xenograft models: An emerging platform for translational cancer research, Cancer Discov, 4 (2014) 998–1013.

100. M. Afzal, S.M.R. Islam, M. Hussain, S. Lee, Precision medicine informatics: Principles, prospects, and challenges, IEEE Access, 8 (2020) 13593–13612.

101. S. Kaur, J. Singla, L. Nkenyereye, S. Jha, D. Prashar, G.P. Joshi, Medical diagnostic systems using artificial intelligence (AI) algorithms: Principles and perspectives, IEEE Access, 8 (2020) 228049–228069.

## CHAPTER 5

Skin Disease Recognition and Classification Using Machine Learning and Deep Learning in *Python*

Masum Shah Junayed and Arezoo Sadeghzadeh

Department of Computer Engineering, Bahcesehir University, Istanbul, Turkey

Baharul Islam

Department of Computer Engineering, Bahcesehir University, Istanbul, Turkey
College of Data Science & Engineering, American University of Malta, Bormla, Malta

CONTENTS

DOI: 10.1201/9781003251903-5

R ECENTLY, SKIN DISEASES have extensively affected people's lives worldwide. Being one of the most serious and common diseases among people, timely and accurate diagnosis of skin diseases particularly the most threatening, such as skin cancer, is vital to receiving the effective treatment. In the last decade, automated computerized skin disease systems based on machine learning and deep learning have gained extreme attention among researchers due to their outstanding performance. This chapter is allocated to the overall process and trends toward skin disease recognition and classification based on machine learning and deep learning. The main purpose of this chapter is to provide knowledge useful for both beginners and more advanced-level researchers in this field. To this end, after presenting a brief introduction to skin diseases, main steps of automated skin disease recognition systems such as image acquisition and available datasets, pre-processing, segmentation, augmentation, feature extraction, and classification are all explained in detail. The general concept of deep learning-based architectures is presented. Additionally, some samples of *Python* codes in each step are provided so that you can comfortably and easily apply the Keras, Tensorflow, Scikit-learn, and OpenCV libraries with *Python* programming language to design your own automated system by following the steps and then train and evaluate them.

5.1 INTRODUCTION

Skin is the most extensive organ (around 3.6 kg and 2 square meters for an adult) covering the human body as a waterproof shield to protect it from harmful UV lights, dangerous chemical substances, extreme temperatures, adventitious viruses, bacteria, and potentially dangerous diseases. It also stabilizes the body temperature and produces vitamin D from sun light. Skin contains nerves, muscles, and blood and lymphatic vessels and is formed by three layers of epidermis, dermis, and subcutaneous tissues (also called the hypodermis). Among these three layers, the epidermis is the outer layer of the skin with no blood supply and two main functionalities of being as a barrier to the environmental infections and regulating the water quantity that the body loses as transepidermal water loss (i.e. preventing dehydration). Dermis is the inner layer of the skin which is located between the epidermis and hypodermis and composed of connective tissue, blood vessels, and hair follicles. The sense of touch and heat are provided through this layer. The lowermost layer of the skin under the dermis is the hypodermis containing the fat of the body whose blood vessels and nerves are larger than those of the dermis.

Due to the vital role of the skin in protecting the healthy life of people with numerous important functions, accurate identification of skin conditions is essential. Skin disease/disorder is an atypical skin condition likely to occur in people from almost all age ranges. As one of the most common diseases among people, almost 20% of Americans (one person out of five) are suffering from some type of skin disease [14]. Some skin diseases and disorders have external and situational causes such as

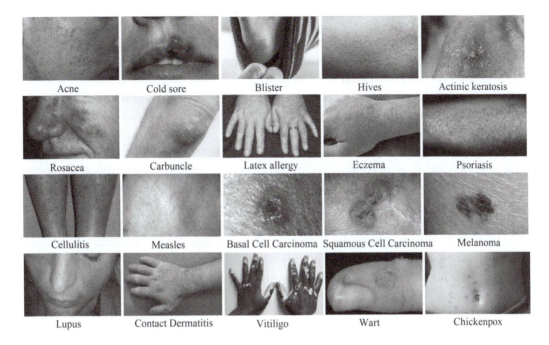

Figure 5.1 Some example images from different types of skin diseases [6].

profession and variations in environmental conditions and climate, while some others are caused by internal factors such as genetics, a different diet, hormones, and immune system. Even based on the variations in the lifestyle factors and environments, the rate of skin disease has increased [26]. Poor sanitation and overpopulation in underdeveloped nations cause skin disorders to spread quickly. There are various types of skin diseases (with various symptoms and severity) some of which are illustrated in Figure 5.1[1]. Each of these diseases may also contain some sub-classes, e.g. basal cell carcinoma, squamous cell carcinoma, and melanoma are different types of skin cancers among which melanoma is the most fatal [37]. Some of these diseases are temporary and some others are permanent and even incurable and life-threatening. However, based on recent research, it has been discovered that these dangerous diseases may have the possibility to be cured if they are recognized in the early stages and even if they are not completely curable, timely diagnosis and treatment can increase the five-year survival rate by about 14% [9]. Dermatologists usually diagnose skin diseases by a simple visual inspection with naked-eye exams or using dermoscopy images which is the most popular one for clinical diagnosis. However, acquisition of these images is time-consuming due to employing a noninvasive method. Another limitation of these diagnoses is that they are also inaccurate or irreproducible as they are extremely subjective to the dermatologists' judgment and vary among different experts based on their experience. There are numerous skin analysis methods and systems, e.g., ABCD

---

[1]The images have been provided via Wikimedia Commons under Creative Commons License which means that they are free to be copied and used in any medium [http://creativecommons.org/licenses/by/2.0].

rules, pattern analysis, Menzies method and 7-Point Checklist, VISIA10 from Canfield, and ANTERA 3D11 from Miravex, which can be employed for diagnoses only by well-experienced experts. On the other hand, there is a lack of dermatologists, especially in underdeveloped countries, due to which disease diagnosis is delayed and the patients cannot receive timely treatments. Even in developed countries, the treatment procedure is delayed as an appointment with a dermatologist takes an average of 32 days for a patient [13].

Consequently, providing an automated computerized diagnostic system as a more objective method is vital to timely recognition of the skin disease without the use of expensive equipment or being affected by the subjective opinions of experts. Furthermore, identifying different types of specific skin diseases and their severity grading based on an automated system is significant in choosing the effective and appropriate treatment and its related protocols for that disease. These systems can also be used as remote screening tools in underdeveloped countries. However, providing such systems with high accuracy is challenging due to several main reasons:

- Existence of too many different skin diseases with high diversity of lesion types for each disease.

- Similar visual characteristics for different skin diseases which make the visual diagnosis difficult not only for the computerized systems but also for the dermatologists.

- Dependency of the skin disease appearance to the human skin tones and types, age, and gender.

- Lack of appropriate datasets with sufficient number of images considering all possible variations that may occur in a real-world application such as differing camera angles, uneven zooming, and lighting effects.

To overcome these limitations and provide a highly accurate system, numerous methods have been carried out based on machine learning techniques and deep learning models in the last decade for both tasks of diagnosis and classification, e.g. diagnosis of skin lesions [10], skin cancer classification [22, 23], eczema classification [24], etc.

The general process of the machine learning- and deep learning-based skin disease recognition and classification systems is illustrated in Figure 5.2. Firstly, almost all approaches in both categories pass three steps of image acquisition and dataset preparation, image pre-processing, and image segmentation. Once these three steps are passed, the input images are ready to go through the machine learning or deep learning-based systems. In machine learning-based systems, first the visual features which are mostly based on color and texture information are extracted and then fed into supervised classifiers for final classification. But what if our dataset is not large enough? Can we go through the deep learning-based models to train them? Obviously not if we want to achieve efficient performance. Hence, in deep learning-based approaches, in addition to the three initial steps, there is one extra step as augmentation. Through this step, the dataset size and variations are enhanced, which causes

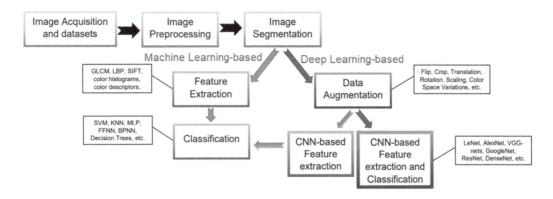

Figure 5.2 General process of skin disease classification and recognition based on machine learning and deep learning models.

the system to be more robust and perform accurately for future variant data. Moreover, generating additional training data through augmentation prevents the system from overfitting and boosts the performance. Deep learning-based approaches generally follows one of the two schemes shown in Figure 5.2: (a) Using the convolutional neural network (CNN)-based models only as the feature extractors and applying the extracted feature maps to machine learning-based classifiers for final classification and (b) Using CNN-based models as a single end-to-end network for both feature extraction and classification. The related details for each of these steps are presented in the following sections of this chapter.

To the authors' idea, learning theory and mathematics of the provided concepts solely cannot be useful for your future career or making whatever you learned as practical as possible. It is necessary to know how to design your own custom system through your codes and modify them. No need to worry about it! We also try to provide the necessary codes related to the basics of the skin disease recognition and classification so that you have more enjoyable journey in this field. The provided codes are based on OpenCV [12], Scikit-learn [31], Keras [15], and Tensorflow [7] libraries (presented in Figure 5.3) with *Python* programming language.

5.2 IMAGE ACQUISITION AND DATASETS

Getting access to the sufficient number of high-quality images is essential to evaluate the performance of skin disease recognition systems. Existence of the large amount of labeled data is even considered as the primary requirement for the training the reliable deep learning-based algorithms with high performance. Generally, skin disease images are acquired in two ways (shown in Figure 5.4): (a) Using dermoscope for dermoscopic images and (b) using normal digital cameras for clinical images. Each of these methods has their own advantages and disadvantages. Dermoscopy is a non-invasive imaging technique widely used in dermatology. In this method, the image of the skin surface with disease is acquired with a dermoscope (light magnifying device). The images acquired by the dermoscope usually have bright illumination

Libraries	Description
OpenCV	OpenCV, which stands for Open Source Computer Vision Library, is a library specially for the real-time computer vision. It aims to provide open and optimized codes for basic vision infrastructure.
Scikit-learn	Scikit-learn (also known as sklearn) is a machine learning library based on the Python programming language performing different classification, regression and clustering algorithms such as SVM, random forests, decision tree, KNN, k-means and DBSCAN.
Tensorflow	It supports different languages of Python, C++ and R. It can be easily used for designing and conducting the experiments of deep learning architectures with convenient formulation for data. Due to having flexible architecture, deep learning models can be easily run on multiple CPUs and GPUs.
Keras	It is also based on Python programming language and can run on top of TensorFlow appropriate for the beginners in the field of deep learning and can be run on multiple CPUs and GPUs.

Figure 5.3 Four common libraries with *Python* programming language along with their descriptions.

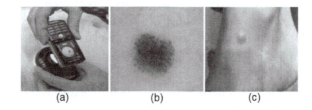

(a) (b) (c)

Figure 5.4 Skin disease image acquisition methods [19]: (a) Dermoscope in the process of image acquisition, (b) the acquired image from dermoscope, (c) clinical image acquired by a normal digital camera.

and more contrast so that the lesions are clear enough to be recognized. Having less noise and lighting variations (due to uniform illumination), the computer-aided methods perform better on these images. However, as the images are acquired by a non-invasive method, collecting a dataset with a sufficient number of images is time- and energy-consuming. Clinical images are acquired by taking photos from the affected body part of the patients using a normal digital camera. Consequently, the images may have variations in lighting conditions, resolution, and camera angle. However, as these images are usually captured by smartphone cameras, data can be collected more easily and the skin disease recognition systems based on these images can be used as the remote screening tool especially in underdeveloped countries. During the last decade, many publicly available datasets have been developed for skin disease diagnosis/recognition, classification, and severity grading based on these two image acquisition techniques. Some of these datasets including PH2 [29], ISIC [17], HAM10000 [1], Dermnet [4], Dermofit Image Library [5], Hallym [20], AtlasDerm [1], Danderm [2], Derm101 [11], 7-point criteria evaluation dataset [25], SD-198 [36], dermis [3], MoleMap [40], Asan dataset [20], Acne [33], and ACNE04 [39] are listed in Figure 5.5 with their details.

Dataset	NO of Images	Image Type	Disease Type
PH2	200	Dermoscopic	Common nevi, melanomas, atypicalnevi
ISIC	over 20000	Dermoscopic and clinical	Melanoma, seborrheic keratosis, benign nevi
HAM10000	10015	Dermoscopic	Pigmented lesions
Dermnet	23000	Dermoscopic, clinical, and pathology	23 categories
Dermofit Image Library	1300	Clinical	10 categories
Hallym	152	Clinical	Basal cell carcinoma
AtlasDerm	11057	Clinical	Almost all types
Danderm	over 3000	Clinical	Common skin diseases
Derm101	Thousands	Clinical	Almost all types
7-point criteria evaluation dataset	over 2000	Dermoscopic and clinical	Melanoma and non-melanoma
SD-198	6584	Clinical	198 categories
DermIS	Thousands	Clinical	Almost all types
MoleMap	102451	Dermoscopic and clinical	22 benign categories and 3 cancerous categories
Asan dataset	17125	Clinical	12 categories for Asian people
Acne	3000	Clinical	7 classes of acne types
ACNE04	1457	Clinical	4 classes of acne grading

Figure 5.5 Some of the publicly available datasets for skin disease diagnoses, classification, and severity grading.

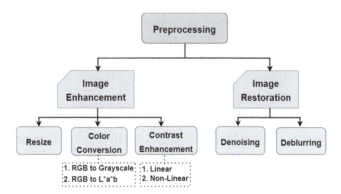

Figure 5.6 Common pre-processing techniques used in the field of skin disease recognition and classification systems.

5.3 PRE-PROCESSING

The quality of the input images has a significant impact on the performance of the whole system. Hence, applying pre-processing techniques to improve the image quality and clarity is an important step in designing the skin disease recognition and classification systems. Some of the common pre-processing techniques in this field are illustrated in Figure 5.6, which are explained in detail in the following sections.

5.3.1 Image Resizing and Normalization

As the input images for training the machine learning- and deep learning-based systems are required to be in the same sizes, the first step in pre-processing is applying

```
import cv2
import glob

for filename in glob.glob('images/*.jpg'): # path to your images folder
    print(filename)
    img=cv2.imread(filename)
    img_resize=cv2.resize(img, (360,360))
    cv2.imwrite(f'{filename}resized.jpg', img_resize)
```

Figure 5.7 *Python* code for resizing the images.

resizing. However, directly resizing the images into the desired sizes may result in distortion and loss in the images. So, it is usually useful in overcoming this issue to resize the images along their shortest side to a uniform scale (aspect ratio needs to be preserved). One sample *Python* code is illustrated in Figure 5.7 in which the OpenCV library is used as cv2. Additionally, the intensities of the images are usually normalized. It is performed on whole images before being fed into the system by subtracting the mean value from the intensities which is then divided by the standard deviation. This way of normalization based on subtracting a uniform value may not be able to efficiently normalize the images due to the large variations in the lighting condition, skin tones, and viewpoint in the skin disease images in the dataset, which results in a degradation on the performance of the system. This issue was addressed in [41] by performing the normalization based on subtracting it from the channel-wise mean intensity values obtained for each single image. As skin disease images may contain some other extra stuffs such as hair, they can be removed for more accurate analysis using thresholding, morphological, and deep learning techniques.

5.3.2 Color Space Conversion

Color space conversion is converting the depiction of a color from one graphical basis to another graphical basis by translating a picture from one color space to another to make the translated image seem as close as possible to the original one. As mentioned previously, the skin disease images are often captured using a digital camera or dermoscope. It is common to transform an RGB (red, green, blue) image to a scalar image because processing the scalar values (single channel) is easier, faster, more convenient, and even more accurate. For example, lesions are typically more visible in the blue channel, so it is common to keep only the blue channel, perform the luminance transformation, and preserve it. Skin lesions appear in a wide range of hues; however, absolute colors are not beneficial for picture segmentation. Color changes inside the lesion or in the surrounding skin, especially color changes at the edge of the lesion, are often used to make an initial assessment. As a result, converting RGB color-coordinated images into other color spaces is relatively popular and useful.

RGB to Grayscale. The RGB images based on three channels of Red (R), Green (G), and Blue (B), contain extra information which is mostly abundant for adopting them in the processing. It results in more complexity and storage space while the performance may improve slightly. Consequently, converting the RGB images into the Grayscale is one of the most popular pre-processing step in image processing and

```
import cv2
import glob

for filename in glob.glob('images/*.jpg'): # path to your images folder
    print(filename)
    input=cv2.imread(filename)

lab = cv2.cvtColor(input,cv2.COLOR_BGR2LAB) # lab color space conversion
cv2.imshow("l*a*b",lab)

L,A,B=cv2.split(lab) # split of lab color conversion

cv2.imwrite('aeL.jpg',L) # split of L
cv2.imwrite('aeA.jpg',A) # split of A
cv2.imwrite('aeB.jpg',B) # split of B
cv2.imwrite('aeLab.jpg',lab) # lab (together)

cv2.imshow("L_Channel",L) # For L Channel
cv2.imshow("A_Channel",A) # For A Channel (Here's what You need)
cv2.imshow("B_Channel",B) # For B Channel

cv2.waitKey(0)
cv2.destroyAllWindows()
```

Figure 5.8 *Python* code for color space conversion from RGB to $L^*a^*b^*$.

computer vision fields. RGB images are usually converted into grayscale images based on two methods: Average method and weighted method. The grayscale values based on the average method is computed as follows:

$$Grayscale = (R + G + B)/3 \qquad (5.1)$$

The values achieved from this are normalized between 0 and 255. Although this method is simple, it is not as accurate as expected from the human visual perception. Human eyes do not perform the same for all R, G, and B colors. This means that the sensitivity of our eyeballs to green, red, and blue lights are different (from the highest to lowest, respectively). So, in computing the grayscale values, these lights need to have different weights which leads to weighted method based on the following equation:

$$Grayscale = 0.299R + 0.587G + 0.114B \qquad (5.2)$$

RGB to L*a*b. To have a better visualization for skin disease regions, numerous approaches have applied RGB to $L^*a^*b^*$ color space conversion. In $L^*a^*b^*$ color space, L^*, and a^*, b^* stand for lightness, and color-opponent dimensions, respectively. The $L^*a^*b^*$ color space was developed to resemble human vision instead of the RGB and CMYK color models. By modifying and refining the output curves in the a and b components, the L component closely matches the human sense of lightness, allowing it to be used to change lightness contrast or make precise color balance adjustments. RGB or CMYK spaces, on the other hand, do not reflect the human visual experience and instead display the output of physical equipment which can be modified only by editing apps and using suitable mix modes. This issue can be overcome by changing the color space of the images from RGB to $L^*a^*b^*$. This conversion improves the final classification accuracy by providing a better representation of the input images and the skin disease areas. One sample *Python* code is illustrated in Figure 5.8 for color space conversion from RGB to $L^*a^*b^*$.

5.3.3   Contrast Enhancement

Contrast enhancement is usually necessary when an image has a small range of intensity values. It happens to an image which is either too dark or too bright. Contrast refers to the distinction between dark and bright regions. Increasing contrast involves making dark objects darker and bright objects brighter. Reducing contrast does the reverse, bringing the brightness and blackness extremes closer. The enhancement would imply an overall improvement in image quality.

Linear Contrast Enhancement. Contrast stretching methods are used in this form of contrast enhancement. By remapping or extending the gray-level values such that the histogram is stretched throughout the whole range, the contrast of the image is improved.

Nonlinear Contrast Enhancement. Nonlinear methods are extensively employed in medical image-based applications. This contrast enhancement is mainly concerned with histogram equalization and algorithmic methods of increasing contrast. The most significant flaw in such approaches is the loss of an object's exact brightness due to the many output image numbers compared to each value in the input image. One sample *Python* code is illustrated in Figure 5.9 for nonlinear contrast enhancement based on the histogram equalization.

```python
#Histogram Equlization Contrast Enhancement
def enhan(img):
    if(len(img.shape)==2):          #gray
        outImg = ex.equalize_hist(img[:,:])*255
    elif(len(img.shape)==3):        #RGB
        outImg = np.zeros((img.shape[0],img.shape[1],3))
        for channel in range(img.shape[2]):
            outImg[:, :, channel] = ex.equalize_hist(img[:, :, channel])*255

    outImg[outImg>255] = 255
    outImg[outImg<0] = 0
    return outImg.astype(np.uint8)

def main():
    img_name = sys.argv[1]
    img = imageio.imread(img_name)
    result = enhan(img)
    plt.imshow(result)
    plt.show()

if __name__ == '__main__':
    main()
```

Figure 5.9 *Python* code for nonlinear contrast enhancement based on the histogram equalization.

5.3.4   Denoising

Denoising an image may be done in a variety of ways. Spatial filtering and transform domain filtering are the two most fundamental image processing techniques. Neighborhood and a predetermined operation adjust the gray value of each pixel according to the pixel values of a square neighborhood centered at that pixel in spatial filtering (e.g. mean filters, median filters, wiener filters, lee filters, anisotropic diffusion filters, total variation filters, etc.).

Gaussian Filter. The blurring method based on the Gaussian function is known as the Gaussian blur or Gaussian smoothing and has often been used to minimize image noise. This method produces a smooth blur similar to that seen when viewing an image through a transparent screen, instead of the effect generated by an out-of-focus lens or the shadow cast by an object in standard lighting. Gaussian smoothing is also employed as a pre-processing step to improve picture structures of various sizes in computer vision methods. An example of *Python* code is illustrated in Figure 5.10 for Gaussian filter-based denoising.

```python
import cv2
import glob

for filename in glob.glob('images/*.jpg'): # path to your images folder
    print(filename)
    img=cv2.imread(filename)
    img_resize=cv2.resize(img, (360, 360)) #image scaling
    new_image_gauss = cv2.GaussianBlur(img_resize, (360, 360), 0) #Add Gaussian Blur
    new_image = cv2.medianBlur(img_resize, figure_size) #Add Meadian Blur

    #Outputs

    cv2.imwrite(f'{filename}resize.jpg', img_resize)
    cv2.imwrite(f'{filename}gaussian.jpg', new_image_gauss)
    cv2.imwrite(f'{filename}median.jpg', new_image)
```

Figure 5.10   *Python* code for denoising based on Gaussian and median filters.

5.3.5   Deblurring

Blurring occurs mostly during the image acquisition process. It is caused by a lack of focus or motion between the source picture and the camera which is required to be removed before using them in the further processes. Deblurring strategies include the Lucy-Richardson algorithm, the inverse filter, the Wiener filter deblurring method, and the neural network method. Wiener filter has been used in medical applications as one of the most powerful and popular deblurring techniques which also removes the noise.

5.4   SEGMENTATION

This procedure aims to differentiate between diseased and healthy parts of the skin. A highly accurate segmentation algorithm should be used to accurately identify the region of the illness. The color-based segmentation approach based on the K-means clustering method is the best method that produces a satisfactory outcome. Clustering is a method for dividing an image into K cluster groups based on the Euclidean distance between an image pixel and a chosen cluster centroid.

The procedure begins by calculating the image's histogram to obtain the number of the centroids as K which represents the number of the clusters. Then, if the difference between the pixel and centroid values is minimal, clustering is applied spatially or in color space. The desired number of clusters (i.e. K) is determined using the K-means or the Fuzzy C-means method. Comparing these two methods, the Fuzzy C-means method is more flexible because it identifies the items that interact with many clusters in the partition. Because the items in the image tend to have similar qualities, the clustering method is susceptible to them. As a result, the RGB picture was transformed to L^*a^*b color to explain the color's intensity better. Then, using a clustering method, the diseased areas are segmented.

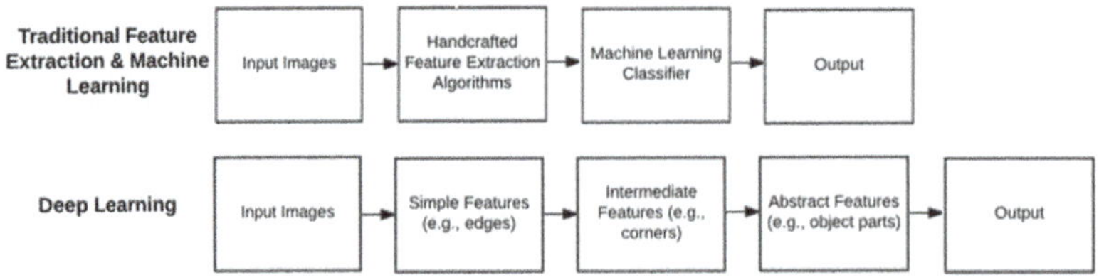

Figure 5.11 General comparison between the machine learning- and deep learning-based classification systems [8].

5.5 MACHINE LEARNING-BASED RECOGNITION

As shown in Figure 5.11 (top), the input images pass two main steps of hand-crafted feature extraction and machine learning-based classification once they are pre-processed and the skin disease area is segmented. These two steps are explained in detail in the following sections.

5.5.1 Feature Extraction

The information collected from the image is referred as features. The features are the information taken from pictures in numerical values, which are difficult for human beings to comprehend and relate to one another. The dimensions of the features retrieved from a picture are usually substantially lower than the original image.

In general, two characteristics may be derived from images depending on the application: Local and global features. The image retrieval, object detection, and classification are all done using global features, but skin disease diagnosis and identification are made with local features. Combining global and local features enhances identification accuracy while it increases the computation cost. This technique of discovering and extracting image features from vast textual data is known as feature extraction in machine learning.

GLCM Features. Texture characteristics are derived from the statistical distribution of intensity combinations at defined places relative to each other in the image in statistical texture analysis. Statistical features are classified into first-order, second-order, and higher-order statistics, based on the number of the intensity points in each combination. The GLCM approach is a technique for obtaining statistical texture characteristics of the second order. This method has been used in a wide range of applications. Third- and higher-order textures consider connections between three or more pixels. These are theoretically conceivable, but they are not widely used due to the time required to calculate them and the difficulty in interpreting the results.

Each row and column of the GLCM corresponds to the number of gray levels (G) in the image. The matrix element P (i, j |Δx, Δy) includes the second-order statistical probability values for changes in gray levels "i" and "j" at a certain distance (d) and specific angle (θ). The element of this matrix is the frequency with which two pixels,

```
import matplotlib.pyplot as plt
import cv2
from skimage.feature import graycomatrix, graycoprops

#segmented images
image=cv2.imread(filename)
PATCH_SIZE = 21
# select some patches from disease areas of the image
disease_locations = [(280, 454), (342, 223), (444, 192), (455, 455)]
disease_patches = []
for loc in disease_locations:
    disease_patches.append(image[loc[0]:loc[0] + PATCH_SIZE, loc[1]:loc[1] + PATCH_SIZE])
# select some patches from affected areas of the image
aff_locations = [(38, 34), (139, 28), (37, 437), (145, 379)]
aff_patches = []
for loc in aff_locations:
    aff_patches.append(image[loc[0]:loc[0] + PATCH_SIZE, loc[1]:loc[1] + PATCH_SIZE])
# compute some GLCM properties each patch
xs = []
ys = []
for patch in (disease_patches + aff_patches):
    glcm = graycomatrix(patch, distances=[5], angles=[0], levels=256,
                        symmetric=True, normed=True)
    xs.append(graycoprops(glcm, 'dissimilarity')[0, 0])
    ys.append(graycoprops(glcm, 'correlation')[0, 0])
```

Figure 5.12 *Python* code for GLCM feature extraction.

separated by pixel distance (Δx, Δy), appear in a specific neighborhood, one with intensity "i" and the other without, in that neighborhood.

The use of a significant number of intensity levels G necessitates storing a significant amount of temporary data, i.e. a G×G matrix for every combination of (Δx, Δy) or (d, θ). The GLCMs are quite sensitive to the number of texture samples on which they have been calculated because of their great dimensionality. One sample *Python* code for GLCM feature extraction is illustrated in Figure 5.12.

HOG. When extracting features from image data, the Histogram of Oriented Gradients (HOG) is often employed as feature extraction. It is commonly used in computer vision applications, such as skin disease detection and recognition.

Even without exact information of the associated gradient or edge coordinates, the distribution of local intensity gradients or edge orientations may frequently represent the look and structure of a local item rather effectively. This assertion leads to the definition of the HOG approach, which has been extensively employed in human identification and has been used in its mature version in Scale-Invariant Feature Transformation [18]. The feature representation is based on the accumulation of gradient orientations across a pixel of a tiny spatial area known as a "cell" and the subsequent building of a 1D histogram whose concatenation provides the feature vectors to be evaluated for future purposes. Let L be an intensity (grayscale) function that describes the image under consideration. The image is split into cells with the same size of N×N pixels, and the gradient's $\theta_{x,y}$ orientation in each pixel (Figure 5.13 a) is determined (Figure 5.13b,c) using the following equation:

$$\theta_{x,y} = \tan^{-1} \frac{L(x, y+1) - L(x, y-1)}{L(x+1, y) - L(x-1, y)} \tag{5.3}$$

Successively, the orientations $\theta_{j,i}$, i = 1...N_2, i.e. belonging to the same cell j, are quantized and accumulated into a M-bin histogram (Figure 5.13 d, e). Finally, all

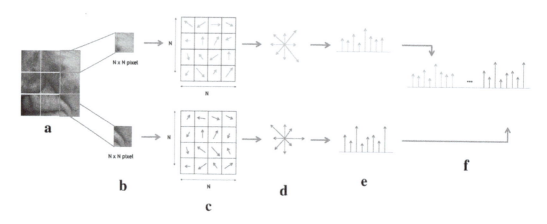

Figure 5.13 Extraction of HOG features: The segmented image's cell size N×N pixels. In an M-bin histogram of orientations, the orientation of every pixel is calculated and collected. In order to generate the final feature vector, all cell histograms are concatenated.

generated histograms are sorted and concatenated into a single HOG histogram (Figure 5.13f) as the feature vector to be used in the further processing. One sample *Python* code for HOG feature extraction is illustrated in Figure 5.14.

```
import matplotlib.pyplot as plt

from skimage.feature import hog
from skimage import data, exposure


image = data.astronaut()

fd, hog_image = hog(image, orientations=8, pixels_per_cell=(16, 16),
                    cells_per_block=(1, 1), visualize=True, channel_axis=-1)

fig, (ax1, ax2) = plt.subplots(1, 2, figsize=(8, 4), sharex=True, sharey=True)

ax1.axis('off')
ax1.imshow(image, cmap=plt.cm.gray)
ax1.set_title('Input image')

# Rescale histogram for better display
hog_image_rescaled = exposure.rescale_intensity(hog_image, in_range=(0, 10))

ax2.axis('off')
ax2.imshow(hog_image_rescaled, cmap=plt.cm.gray)
ax2.set_title('Histogram of Oriented Gradients')
plt.show()
```

Figure 5.14 *Python* code for HOG feature extraction.

5.5.2 Classifiers

K-Nearest Neighbors (KNN). A simple nearest neighbor classifier classifies each pixel into the same class as the training data with the highest intensity. If the single derived neighbor is an outlier of another class, a nearest neighbor classifier may make an incorrect judgment. The KNN classifier uses K patterns to prevent this. As it makes no underlying assumptions about the statistical structure of the data, it is considered as a nonparametric classifier.

KNN Algorithm:

- Training set includes classes.

- Examine K items near to the item desired to be classified.

- New item placed in class with the most number of close items.

The K-nearest neighbor classification is accomplished by utilizing a training dataset that includes both the input and the target variables and then comparing the test data, which only contains the input variables, to that reference set. One sample *Python* code for the KNN classifier is illustrated in Figure 5.15.

```
from sklearn.neighbors import KNeighborsClassifier

knn = OneVsRestClassifier(KNeighborsClassifier())

knn.fit(X_train,y_train)
# predict for one observation
knn.predict(X_test[0].reshape(1,-1))
# predict for multiple observation (images) at once
knn.predict(X_test[0:10])
# make prediction on entire test data
predictions = knn.predict(X_test)
```

Figure 5.15 *Python* code for the KNN classifier.

Support Vector Machine (SVM). A classification task often includes two sets of data: Training data and testing data, each of which contains a number of data samples. Each instance has one goal value and multiple features in the training set. The purpose of SVM is to create a model that predicts the target value of data instances in the testing set when just the features are supplied.

SVM is based on the supervised learning (a model is trained with labeled or annotated data which is employed for predicting the class label of the future unseen data). For the purpose of determining whether or not the system is performing correctly, known labels are employed, which can be used to aid the system in learning to behave appropriately by pointing to the intended response and evaluating the system's correctness. For classification, supervised learning algorithms are based on statistical learning theory and may generate reliable, accurate, and valuable results with fewer training samples. In general, a conventional binary classifier is trained using a collection of data that falls into two categories, and the SVM training method creates a training model that predicts the class for new data. However, multi-class classification has recently been addressed by breaking the multi-classes down into many classification tasks and creating several SVM classifiers.

SVM has been shown to outperform neural networks and RBF (Radial Basis Function) classifiers in several applications. SVM is based on two different mathematical procedures. In feature space, SVM separates two datasets using an optimal linear separating hyperplane. The minimal margin between the two sets is maximized to obtain this optimal hyperplane. Consequently, the final hyperplane will rely only on boundary training patterns known as support vectors.

An input vector is nonlinearly mapped onto a high-dimensional feature space concealed from the input and output. The best hyper-plane for differentiating the features is constructed. SVM transforms input vectors into a higher-dimensional vector space, from which an ideal hyper-plane is built that may be used to evaluate data with linear separability, whereas kernel functions like Gaussian RBF are used to study data that are not linearly separable.

In general, the output of an SVM is a concatenation of the training instances that has been projected onto a high-dimensional feature space via the application of kernel function training examples. SVM is based on structured risk minimization from the statistical learning theory, which is a kind of machine learning. Its primary goal is to reduce the margins between classes and the actual cost by controlling overfitting problems and classification ability. One sample *Python* code for SVM classifier is illustrated in Figure 5.16.

```
from sklearn.svm import SVC
svc = SVC(kernel='linear',gamma='auto')
svc.fit(X_train, y_train)
```

Figure 5.16 *Python* code for SVM classifier.

5.6 DEEP LEARNING-BASED RECOGNITION

In this section, we discuss the general concept of the CNNs along with details about their layers.

5.6.1 Data Augmentation

The performance of the deep learning-based models is highly dependent on the size of the training data. Large amount of training data prevents the model from overfitting, increases the generalizability of the model, and enhances its performance. However, collecting such a large number of data especially in medical image analysis is really challenging, costly, and energy-consuming due to patient privacy, rarity of diseases, and the necessity for labeling the images by a professional expert. On the other hand, as the future testing data seen by the network can contain several variations making them different from the training ones, it is important to train the model with various data and make it robust through learning more robust features. To address these issues, the original training data are artificially transformed by applying data augmentation. Data augmentation is composed of a wide range of techniques through which random jitters and perturbations are applied to generate new training samples from the original images keeping the class labels unchanged to improve the size and the quality of the original dataset. In testing phase, no data augmentation is applied. Applying different augmentations on the training data, the testing accuracy is increased as the model can learn more significant properties with different possible

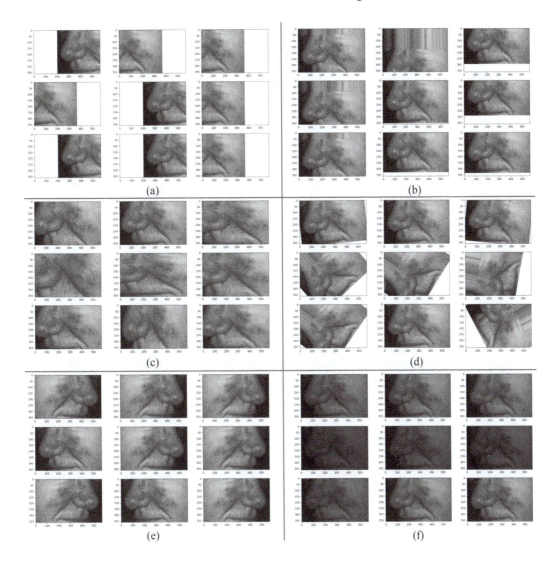

Figure 5.17 Visual representation for different augmentation techniques. Here, (a), (b), (c), (d), (e), and (f) represent the shifting, shearing, zooming, rotation, flipping, and shading augmentations.

variations. Common data augmentation techniques are the geometric transformations (e.g. translations, rotation, flip, shearing, scale changing and cropping), color space augmentation (e.g. shading), and Kernel filters. Several augmentation results are visually presented in Figure 5.17 for an image. One sample *Python* code for data augmentation is illustrated in Figure 5.18.

5.6.2   CNN-Based Models

As illustrated in Figure 5.11, in contrary to the machine learning-based classification systems, the deep learning-based models, e.g. CNNs, extract the features from the images automatically through the training process in a hierarchical way. In these

```
from keras.preprocessing.image import ImageDataGenerator, array_to_img, img_to_array, load_img

datagen = ImageDataGenerator(
    rotation_range=40,
    width_shift_range=0.2,
    height_shift_range=0.2,
    shear_range=0.2,
    zoom_range=0.2,
    horizontal_flip=True,
    fill_mode='nearest')
```

Figure 5.18 *Python* code for data augmentation.

models, some basic representations are encoded in the lower level layers of the network and later these basic layers are used by the higher level layers to generate more abstract concepts. This hierarchical learning makes the CNNs as end-to-end learners eliminating the need for extracting the hand-designed features. For a given image, raw pixel intensities are fed into the CNNs as inputs to extract the features using a series of hidden layers. As the feature extraction is carried out in a hierarchical fashion (the output of each layer in the network is used as the input of the next layer), only simple features such as edges are extracted in the lower level layers which are then used in the next layers to determine intermediate features such as corners and contours. These corners and contours are combined to form the object parts in the higher-level layers. The final layer as the output layer is utilized for classification and achieving the label of the output class.

One of the advantages of the CNNs over the traditional machine learning-based approaches is that the feature extraction step is skipped in these models focusing only on the process of training the model. Another main advantage is the outstanding performance accuracy achieved by the CNNs due to the emergence of highly optimized GPUs, fast computers, and large labeled datasets. Nowadays, researchers are able to design the deeper neural networks with huge number of training data which results in higher classification accuracy whereas this behavior cannot be accomplished using the traditional machine learning algorithms. As compared in Figure 5.19 [30], the performance of the machine learning algorithms plateau at a specific amount of the data and increasing the size of the dataset has no influence on their performance while the performance of deep learning models follows an increasing trend.

5.6.3 Neural Networks

Before discussing details of the CNNs, it is essential to first know the basics of the neural networks (NNs) which are the fundamental blocks of the deep learning systems. Artificial NN architectures have been inspired by our real-life biological NNs to mimic our brain capabilities in recognizing complex patterns. A simple example of a NN architecture is illustrated in Figure 5.20. In this architecture, a simple computation is carried out in each node and the output of each computation as a signal is carried by each connection from one node to another with specified weights. Some of these weights amplify the signals as they play an important role in the final classification and others diminish the strength of the signals as they have less importance for final classification. Here, X_1, X_2, and X_3 are the inputs as a single row from the design matrix. The value 1 is fed into the model as bias in the design matrix. Generally,

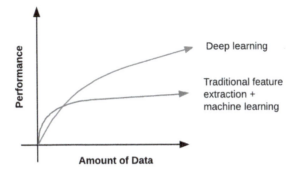

Figure 5.19 Comparison between the performance of the deep learning-based algorithms and the traditional machine learning-based methods showing that the accuracy of the deep learning algorithms are increased by enhancing the amount of the training data.

these inputs can be the vectors of the features such as color histograms, HOG, LBP, etc., but in the deep learning models, the inputs are usually the raw pixel intensities of the images. The overall process of the system can be formulated as $f(net)$ where $net = \sum_{i=1}^{n} W_i X_i$ (i.e. weighted sum of the inputs) on which an activation function of f is applied to determine whether the neuron fires or not.

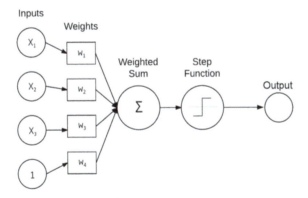

Figure 5.20 A simple neural network architecture with three inputs of X and one hidden layer with weights of W to compute the output class label. The weighted sum of the inputs are passed an activation function through which it is determined that whether the neuron fires or not.

Activation Functions. There are several popular activation functions broadly applied on deep learning models, six of which are illustrated in Figure 5.21. "Step function" is the most simple activation function as threshold function used by the Perceptron algorithm. It is defined as follows:

$$f(t) = \begin{cases} 1 & \text{if } net > 0 \\ 0 & \text{otherwise} \end{cases} \tag{5.4}$$

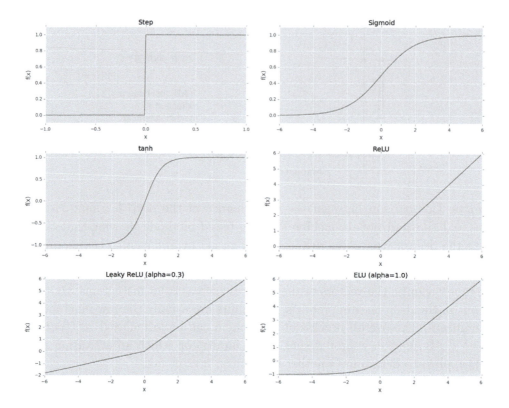

Figure 5.21 Six activation functions commonly used in deep learning architectures [8].

Although this function is easy to use, it suffers from challenges when the gradient descent is applied in training the network as the step function is not differentiable. Sigmoid is another common activation function which follows the equation:

$$t = \sum_{i=1}^{n} W_i X_i \qquad s(t) = \frac{1}{1 + e^{-1}} \qquad (5.5)$$

As the Sigmoid function is continuous, differentiable, and symmetric, it performs better than the step function for deep learning algorithms. However, it also suffers from some limitations: (i) Its outputs are not zero centered and (ii) the gradient will be degraded by saturated neurons due to extremely small delta for gradient. Having a similar shape to the Sigmoid, the hyperbolic tangent (tanh) is one of the extensively used activation functions as follows:

$$f(z) = tanh(z) = \frac{e^z - e^{-z}}{e^z + e^{-z}} \qquad (5.6)$$

In contrary to Sigmoid, it is zero centered, but still the saturated neurons kill the gradients. The better choice rather than the previously mentioned functions is the rectified linear unit (ReLU) as:

$$f(x) = max(0, x) \qquad (5.7)$$

Based on its shape, it is also called "ramp function." It is zero for negative values and increases linearly for positive values. It not only outperforms the activation functions of Sigmoid and tanh, but also is computationally efficient and not saturable. However, the problem arises only when the input value is zero because the gradient cannot be taken. To overcome this limitation, another form of ReLUs is used, namely Leaky ReLUs [28], in which the small, non-zero gradient is allowed when the unit is not active:

$$f(net) = \begin{cases} net & \text{if } net >= 0 \\ \alpha \times net & \text{otherwise} \end{cases} \tag{5.8}$$

The final activation function is the exponential linear units (ELUs) introduced [16] as follows:

$$f(net) = \begin{cases} net & \text{if } net >= 0 \\ \alpha \times (exp(net) - 1) & \text{otherwise} \end{cases} \tag{5.9}$$

α is a constant whose typical value is 1.

5.6.4 CNN-Based Classifiers

The introduced feedforward neural networks can be used in image processing. However, they are not efficient enough due to the many connections made between all nodes in one layer and those in the next layer. In this case, the efficiency can be improved by carefully pruning these connections according to the structure of the images which results in a better performance. A very common type of NN is CNN in which the spatial relationships in data are preserved through very few connections between layers. A typical CNN architecture applied in the field of skin disease classification and recognition is illustrated in Figure 5.22. CNNs are composed of different layers of Convolutional (CONV), Activation, Pooling (POOL), Fully connected (FC), Batch normalization (BN), and Dropout (DO), which are explained in details in the following paragraphs.

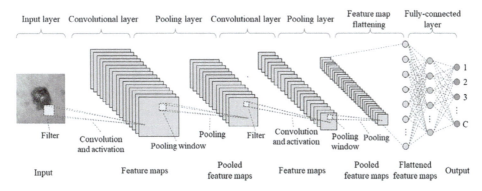

Figure 5.22 A typical CNN model for skin disease classification and recognition [27].

Convolutional Layers. The fundamental building layer of the CNNs is the CONV layer including a set of K learnable filters as "kernels" (nearly always as square matrices). These small filters extend throughout the full depth of the volume. As illustrated in Figure 5.23, the input of a CONV layer passes three main steps.

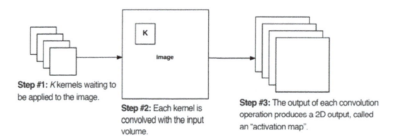

Step #1: K kernels waiting to be applied to the image.

Step #2: Each kernel is convolved with the input volume.

Step #3: The output of each convolution operation produces a 2D output, called an "activation map".

Figure 5.23   Three main steps that a volume (input image or a feature map) passes in a CONV layer [8].

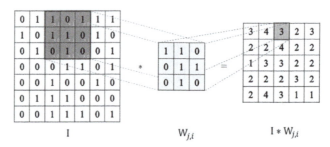

Figure 5.24   Visual presentation of the 2D convolution operation [27].

The initial depth of the inputs equals the number of the channels in the image (e.g. three when the images are RGB images). The depth in the deeper CONV layers of the network is the number of filters in the previous layer. In the first step, the number of filters and their kernel sizes are determined each of which is convolved with the input volume in the second step. One example for convolution operation is illustrated in Figure 5.24. Here, $W_{j,i}$ is the filter with the kernel size of 3×3, where j and i are the filter and layer number, respectively. This small window moves over a 2D grid of an image or a feature map from left to right and up to down. The step size according which the window is moving across the image is determined by *stride* step size which is normally either 1 (moving one pixel at a time) or 2 (skip two pixels at a time). In each movement, the elements of the window are multiplied to the corresponding elements of the grid. Convolution operation is completed by summing up these element-wise multiplication values to achieve a scalar value. All the obtained values from convolution result in another 2D grid called feature map/activation map as shown in the third step of Figure 5.23. Once all the K filters are applied to the input volume, K 2D activation maps are achieved which are stacked along the depth dimension as the final output volume (illustrated in Figure 5.25).

It should be mentioned that zero-padding is applied along the borders of the input to keep the original image size so that the output volume size matches that of the input (shown in Figure 5.26). Assume I, F, S, and P are the input square image size (one dimension), filter kernel size, stride value, and the amount of zero-padding, respectively. The following equation is required to be integer to have a valid CONV layer:

$$((I - F + 2P)/S) + 1 \tag{5.10}$$

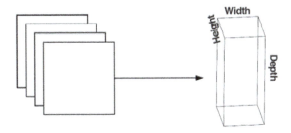

Figure 5.25 Stacking the K activation maps achieved from K filters which forms the input volume for the next layer of the CNN [8].

Input with zero-padding

Kernel

692	-315	-6
-680	-194	305
153	-59	-86

0	0	0	0	0	0	0
0	95	242	186	152	39	0
0	39	14	220	153	180	0
0	5	247	212	54	46	0
0	46	77	133	110	74	0
0	156	35	74	93	116	0
0	0	0	0	0	0	0

-99	-673	-130	-230	176
-42	692	-315	-6	-482
312	-680	-194	305	124
54	153	-59	-86	-24
-543	167	-35	-72	-297

Output of convolution

Figure 5.26 Applying zero-padding to the original input before convolution through which the spatial dimensions of the input is preserved and the output volume size matches the original input volume size.

Activation Layer. A nonlinear activation function such as ReLU, ELU, etc. (explained in Section 5.7.4) is applied after each CONV layer which aims to make the NN able to perform almost any nonlinear functions. As illustrated in Figure 5.27, the outputs of activation functions (here the ReLU) are new tensors as feature maps.

Pooling Layer. In addition to applying the CONV layers with a *stride* > 1, the size of an input volume can also be reduced by applying POOL layers between two consecutive CONV layers. Pooling can also be useful in handling overfitting. There are two common POOL layers of max pooling and average pooling. Typically, max pooling is applied in the middle of the CNN architecture with the purpose of spatial size reduction while average pooling is common as the final layer of the network to prevent entirely using FC layers. One example of applying max pooling on a sample input with strides of 1 and 2 is illustrated in Figure 5.28.

Batch Normalization. Batch normalization was first introduced in [21] which is placed after the activation layers to normalize the activations of the input volume

Input   **ReLU**

-249	-91	-37
250	-134	101
27	61	-153

0	0	0
250	0	101
27	61	0

Figure 5.27 An input going through a ReLU activation through which the negative values are considered as zero and the positive values remain unchanged.

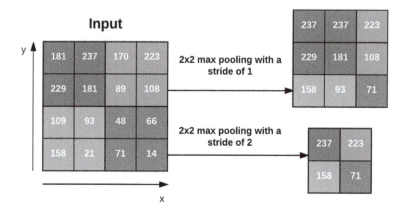

Figure 5.28 Applying a 2×2 max pooling filter with strides of 1 and 2 on a sample input [8].

before applying them to the next layer. Giving x as the mini-batch of the activations, its normalized value is computed as follows:

$$\hat{x}_i = \frac{x_i - \mu_\beta}{\sqrt{\sigma_\beta^2 + \varepsilon}} \qquad (5.11)$$

where μ_β and σ_β^2 are computed over each mini-batch of β during the training as the mean and variance values, respectively, and ε is defined as a very small value. Using batch normalization layer, the mean and standard deviation of the activations are forced to be kept as zero and unit. This layer also reduces the number of the epochs and makes it less dependent on initializing and tuning the parameters carefully.

Dropout. The dropout layer, first introduced in [35], is a form of regularization which prevents the model from overfitting by randomly eliminating the neurons in the training by removing the connection between the neurons of one layer and the next layer with the probability of p. This dropping process leads to slightly different networks so that the weights of the networks are tuned by optimizing various forms of the original networks. Figure 5.29 illustrates one example of applying dropout with 50% probability to the network through which half of the connections are removed.

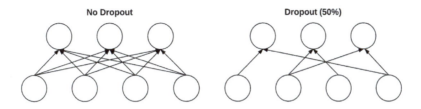

Figure 5.29 Two CNNs with and without applying a dropout layer. Here, half of the connections are removed by applying dropout with the probability of 0.5 or 50% [8].

Fully Connected Layers. The last layer we are explaining in this chapter is the FC layers which are named based on the fact that neurons in these layers are fully-connected to all the previous layer's activations. These layers are always designed as the last layers of the network, never in the middle of the CNN architecture. In each CNN architecture, one or two FC layers are commonly applied which is followed by the softmax classifier to compute the final output probabilities for each class (classification step). An example of *Python* code for a CNN-based end-to-end classifier is illustrated in Figure 5.30.

```
import numpy as np
from keras.models import Sequential
from keras.layers import Dropout, Dense, Flatten
from keras.optimizers import SGD
from keras.layers.convolutional import Conv2D, MaxPooling2D
from keras.utils import np_utils as u
from keras.datasets import cifar10

model = Sequential()
model.add(Conv2D(32, (3, 3), input_shape=(32, 32, 3), padding='same',
              activation='relu'))
model.add(Dropout(0.2)) #20% of the nodes are set to 0
#padding=valid this means that the output dimension can
model.add(Conv2D(32, (3, 3), activation='relu', padding='valid'))
model.add(BatchNormalization()) # This is works as regularization
model.add(MaxPooling2D(pool_size=(2, 2))) #maxpool with a kernet of 2x2
#In a convolution NN, we neet to flatten our data before input it into the ouput/dense layer
model.add(Flatten())
model.add(Dense(512, activation='relu')) #Dense layer with 512 hidden units
model.add(Dropout(0.3)) #this time we set 30% of the nodes to 0 to minimize overfitting
#Finally the output dense layer with 10 hidden units corresponding to class 3
model.add(Dense(3, activation='softmax'))
#Few simple configurations
model.compile(loss='categorical_crossentropy',
           optimizer=SGD(momentum=0.5, decay=0.0004), metrics=['accuracy'])
model.fit(X, y, validation_data=(X_test, y_test), epochs=25,
       batch_size=512)
#Finally print the accuracy of our model!
print("Accuracy: &2.f%%" %(model.evaluate(X_test, y_test)[1]*100))
```

Figure 5.30 *Python* code for a CNN-based end-to-end classifier.

5.6.5 CNNs as Feature Extractors

As explained in the previous section, CNNs are used as end-to-end image classifiers whose input is an image which propagates forward through the network and gives

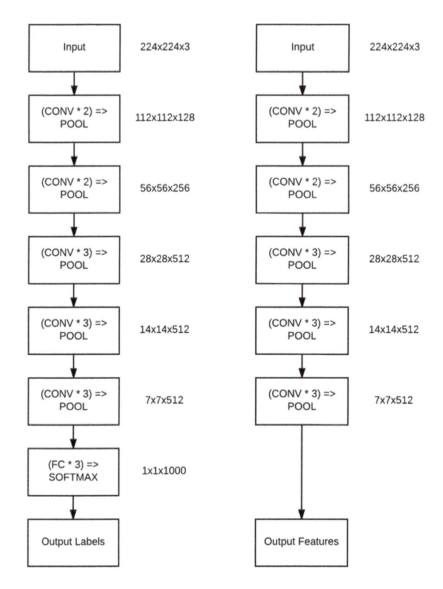

Figure 5.31 The left image is the original VGG16 network architecture as an end-to-end image classifier. Once the FC layers along with the softmax classifier are removed, the achieved feature maps from the previous layers are considered as the extracted features which are usually applied to a machine learning-based classifier for final classification [32].

the final classification probabilities as the output. However, as shown in Figure 5.31, CNNs can also be used as the only feature extractor by stopping the propagation at a specific layer and using the information extracted from that point of the network as feature vectors. In this example, the original VGG16 network [34] as the end-to-end classifier is illustrated at the left side. At the right side, the same network is treated as a feature extractor by removing the FC layers and the softmax classifier at the end of the network so that the last layer of the network is now a max pooling layer.

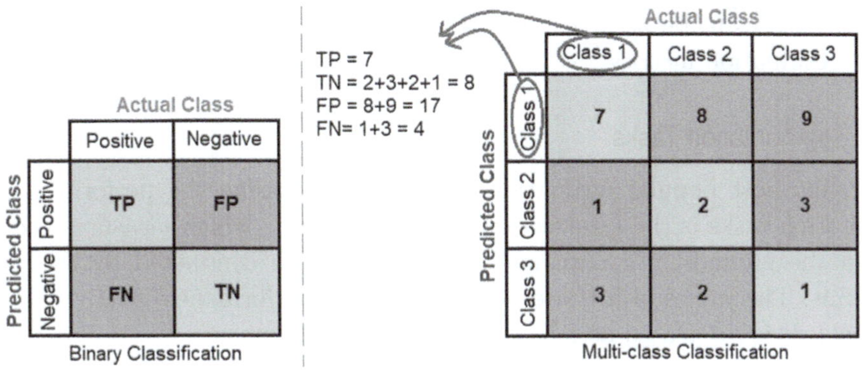

Figure 5.32 Confusion matrices for binary and multi-class classification.

In this case, the output shape is $7 \times 7 \times 512$ which shows that there are 512 filters with the kernel size of 7×7 (512, 7×7 activations which are considered as a feature vector with $7 \times 7 \times 512 = 25088$ values). Getting such a vector for each of N image in a dataset, you can obtain a design matrix of N images. These feature vectors can be used to train a machine learning-based classifier such as SVM to recognize the classes of the images.

5.7   EVALUATION METRICS

Most of the metrics for evaluating the performance of skin disease recognition and classification systems are based on the information achieved from the confusion matrix (the table for presenting the performance of the prediction model). In this matrix which is designed for both binary and multi-class classification cases, each entry indicates the number of predictions/classification made correctly or incorrectly by the model. As shown in Figure 5.32, for the binary classification problem, there are two classes of positive and negative. In this confusion matrix, TP, TN, FP, and FN represent True Positive, True Negative, False Positive, and False Negative, respectively. True Positive refers to the number of predictions made by the model through which the sample is correctly classified as positive class while the actual class of the sample is also positive. Similarly, if the actual class is negative and the model also predicts its class as negative, it is considered to be a True-Negative prediction. False Positive is the number of incorrect predictions by the classifier in which the actual class of negative is predicted as positive. In contrast, False Negative refers to the number of misclassifications where the model predicts the actual class of positive as negative.

Unlike binary classification, multi-class classification is based on the predictions made by the system for more than two classes. So, here, there are no positive or negative classes and we need to find TP, TN, FP and FN for each class individually based on the information we obtain from the whole table. One simple example of a three-class classification task with the calculation of the values of these four parameters for class 1 are presented in Figure 5.32. Similarly, they can be calculated for the other classes to be used in the further calculations for estimating the values of different evaluation metrics.

5.7.1 Segmentation Tasks

One of the most popular evaluation metrics for assessing the performance of the segmentation tasks is the Intersection-over-Union (IoU) which measures the overlap between the ground-truth data and the segmentation conducted by the proposed algorithms. The values of IoU are between 0 (poor performance) to 1 (the best performance). It is calculated as follows:

$$IoU = \frac{Area\,of\,overlap}{Area\,of\,union} = \frac{target \bigcap prediction}{target \bigcup prediction} \tag{5.12}$$

Here, *Area of overlap* is the value of overlapping between the ground-truth segmented area and the predicted one by the algorithms. *Area of union* is the union of the two items. Additionally, there are five more evaluation metrics for segmentation task which are calculated based on the TP, TN, FP, and FN obtained from the confusion matrix.

One of these metrics is *accuracy* which is defined as the fraction of the number of right predictions made by the model over the total number of predictions as follows:

$$Accuracy = \frac{TP + TN}{TP + TN + FP + FN} \tag{5.13}$$

Here, for the segmentation task, the pixel-level accuracy and the pixel values over and below 128 are considered as positive and negative, respectively. *Recall/Sensitivity* which indicates the True-Positive rate is calculated by dividing correctly predicted data as positive over all the positives:

$$Recall/Sensitivity = \frac{TP}{FN + TP} \tag{5.14}$$

Specificity as an evaluation metric is the proportion of correctly predicted negatives in relation to all the negatives as follows:

$$Specificity = \frac{TN}{FP + TN} \tag{5.15}$$

The *Jaccard index/Jaccard similarity coefficient* measures the similarity and differences between the sample sets and ranges from 0 to 1. It is calculated based on the following equation:

$$Jaccard = \frac{TP}{TP + FN + FP} \tag{5.16}$$

The most broadly used metric for validating skin disease segmentation algorithms is the *Dice similarity coefficient* through which the similarity between two sets of data can be measured as follows:

$$Dice = \frac{2TP}{2TP + FN + FP} \tag{5.17}$$

5.7.2 Classification Tasks

Common evaluation metrics for assessing the performance of the classification are accuracy, Precision, Recall/Sensitivity, F1-Score, Specificity, and Matthews Correlation Coefficient (MCC) Score, among which accuracy, Recall/Sensitivity, and Specificity are common metrics between the segmentation and classification as defined in the previous section with a difference that in classification task they are not in pixel level but they are calculated at the whole image level.

The *Precision* is determined by calculating the fraction of the positive class correct predictions over all the predictions as positive (correct or incorrect):

$$Precision = \frac{TP}{TP + FP} \tag{5.18}$$

The mean of the Precision and Recall is defined as the *F1-Score* which has a value between 0 (i.e. either the Precision or the Recall is zero) to 1 (i.e. perfect Precision and Recall). This is similar to Dice similarity coefficient and is calculated as follows:

$$F1\text{-}Score = \frac{2 * (Precision * Recall)}{Precision + Recall} \tag{5.19}$$

A more informative and valid score can be obtained by MCC in evaluating the classification performance which makes it more preferable rather than accuracy and F1-Score. Based on MCC, the accuracy of the classifier is measured through the comparison made between the observed and expected results as follows:

$$MCC = \frac{(TP * TN) - (FP * FN)}{\sqrt{(TP + FP) * (TP + FN) * (TN + FP) * (TN + FN)}} \tag{5.20}$$

These metrics are calculated for each of the classes in the classification task and the overall performance is evaluated by taking the average of all these values. As an example, the performance of multi-class classifier in Figure 5.32 is evaluated based on these evaluation metrics for the first class as follows:

$accuracy = \frac{7+8}{7+8+17+4} = 0.41 = 41\%$, $Recall/Sensitivity = \frac{7}{7+4} = 0.63 = 63\%$, $Specificity = \frac{8}{17+8} = 0.32 = 32\%$, $Precision = \frac{7}{7+17} = 0.29 = 29\%$, $F1\text{-}Score = \frac{2*0.29*0.63}{0.29+0.63} = 0.39 = 39\%$.

One sample *Python* code for computing the confusion matrix and evaluation metrics is illustrated in Figure 5.33.

In addition to the above-mentioned evaluation metrics which are calculated according to the values of the confusion matrix, the area under the curve (AUC) for the receiver operating characteristic (ROC) can perfectly evaluate the ability of the classifier. The ROC curve is formed by plotting the True Positive rate (i.e. Sensitivity/Recall) against the False-Positive rate (i.e. 1−Specificity) under different threshold values whose AUC is determined by taking its integral. The examples of ROC curves for different machine learning-based classifiers with their AUCs are illustrated in Figure 5.34. The AUC takes a value between 0 and 1 and the values close to 1 indicate the high capability of the classifier. Comparing the results, the highest value of AUC which is 0.90 belongs to SVM+Naïve Bayes and SVM classifiers which proves that those classifiers outperform the others.

```
# Confution Matrix and Classification Report
class_name = []
for i in validation_generator.class_indices:
    class_name.append(i)
# print(class_name)

num_of_test_samples = 1200
batch_size = 32
from sklearn.metrics import classification_report
from sklearn.metrics import jaccard_score

Y_pred = model.predict_generator(validation_generator, num_of_test_samples // batch_size+1)
y_pred = np.argmax(Y_pred, axis=1)
print('Confusion Matrix')
print(confusion_matrix(validation_generator.classes, y_pred)) #Output of confusion matrix
print('Classification Report')
target_names = class_name
#Output of precision, recall, f1 score, accuracy
print(classification_report(validation_generator.classes, y_pred, target_names=target_names))
print(jaccard_score(validation_generator.classes, y_pred, average=None)) #output of jaccard score
```

Figure 5.33 *Python* code for computing the confusion matrix and different evaluation metrics.

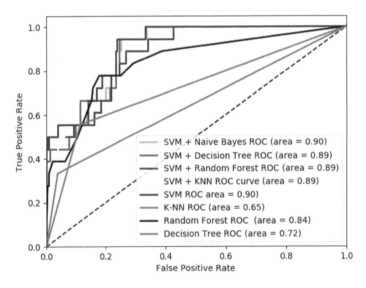

Figure 5.34 Comparison between the AUC of the ROC curves belonging to 8 different machine learning-based classifiers.

5.8 CONCLUSION

In this chapter, we reviewed the concept of skin disease recognition and classification based on computerized techniques, i.e. machine learning and deep learning-based methods. We started by discussing about the skin disease image acquisition methods and the available datasets as the main and primary requirement for computer-aided systems.

Categorizing the current skin disease recognition and classification algorithms into two main groups of machine learning and deep learning based, we first discussed two common steps of pre-processing and segmentation. By applying pre-processing, the input images are resized all in the same image size and their quality and clarity are improved by color space conversion, contranst enhancement, denoising, and deblurring. The exact area of the skin disease is separated from the healthy skin by applying different techniques of segmentation which improves the recognition performance. The segmented skin disease areas are employed for extracting the hand-crafted features. These features are utilized by a machine learning-based classifier such as SVM, KNN, etc. to carry out the recognition and classification task. The CNN-based models have achieved better performance as they are trained using a larger dataset. CNN-based models can be treated either as the end-to-end classifiers or only as the feature extractors. Then performance of the system for segmentation and classification tasks can be evaluated by different metrics usually obtained from the entries of the confusion matrices.

To understand all the steps and concepts related to skin disease recognition at a more intimate level, we implemented them by hand using different libraries of OpenCV, Scikit-learn, Keras, and TensorFlow in *Python*.

5.9 EXERCISES

1. For the given R, G, B matrices of a 3×3 image in Figure 5.35, calculate the grayscale values of the corresponding elements based on the weighted method.

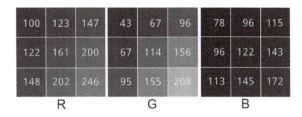

Figure 5.35 R, G, and B matrices for a 3×3 image.

2. In Figure 5.36, X is the input image matrix and h is the kernel of a convolutional layer. (a) Compute the pixel value of the output image by convolving the input image with the kernel with a stride of 1 and 2. (b) Repeat the process with considering zero-padding.

25	100	75	49	130
50	80	0	70	100
5	10	20	30	0
60	50	12	24	32
37	53	55	21	90
140	17	0	23	222

X

1	0	1
0	1	0
0	0	1

h

Figure 5.36 Input image of X to be convolved with the kernel of h.

3. For the confusion matrix of multi-class classification given in Figure 5.37, computes the average evaluation metrics of accuracy, sensitivity, specificity, precision, and F1-Score.

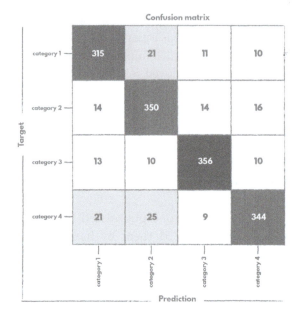

Figure 5.37 The confusion matrix for a 4-class classification task in which the *Target* and *prediction* indicate the actual classes and the predicted ones, respectively.

4. Adopt the dataset of ACNE04 [39] to apply the severity classification in *Python*. To this end, first split the data into 70% training and 30% test set. Resize the images into 240×240 pixels and then enhance their quality by applying different pre-processing techniques. Extracting GLCM features, train an SVM and a KNN classifier. Compare their performance based on five different evaluation metrics.

5. Apply the same classification scenario on the same dataset of ACNE04 [39] in *Python* with a CNN classifier. To this end, first split the data to 70% training and 30% test set. Resize the images into 240×240 pixels and then enhance the training set by applying five different augmentation techniques of rotation, horizontal flipping, translation, shearing, and zooming. Improve their quality by applying different pre-processing techniques. Design a CNN model with 5 convolutional layers, 5 max pooling layers, and 2 fully connected layers followed by a softmax classifier. Use zero-padding with stride of 1. (a) Optimize the number and kernel size of the filters in each convolutional layer. (b) Assess the performance of your system based on 5 different evaluation metrics. (c) How the performance of the model and the speed of the training can be improved by adding extra layers? Compare the results. (d) If your model faces the overfitting issue, how you can deal with it? Apply your solution to your model and compare the performance.

Bibliography

1. Atlasderm. https://www.atlasdermatologico.com.br, 2019. [Online: accessed 11-Sep.-2019].

2. Danderm. https://www.danderm.dk/, 2019. [Online: accessed 11-Sep.-2019].

3. Dermis. http://www.dermis.net/dermisroot/en/home/indexp.htm, 2019. [Online: accessed 11-Sep.-2019].

4. Dermnet. http://www.dermnet.com/, 2019. [Online: accessed 11-Sep.-2019].

5. Dermofit image library. https://licensing.edinburgh-innovations.ed.ac.uk/i/software/dermofit-image-library.html, 2019. [Online: accessed 11-Sep.-2019].

6. All about common skin disorders. https://www.healthline.com/health/skin-disorders, 2022. [Online: accessed 17-Feb.-2022].

7. Martín Abadi, Paul Barham, Jianmin Chen, Zhifeng Chen, Andy Davis, Jeffrey Dean, Matthieu Devin, Sanjay Ghemawat, Geoffrey Irving, Michael Isard, et al. TensorFlow: A system for large-scale machine learning. In *12th {USENIX} symposium on operating systems design and implementation ({OSDI} 16)*, pages 265–283, 2016.

8. Rosebrock Adrian. Deep learning for computer vision with *Python*, PyImageSearch 2017. ISBN:9781722487836.

9. Abder-Rahman A Ali and Thomas M Deserno. A systematic review of automated melanoma detection in dermatoscopic images and its ground truth data. In *Medical Imaging 2012: Image Perception, Observer Performance, and Technology Assessment*, volume 8318, page 83181I. International Society for Optics and Photonics, 2012.

10. Catarina Barata, Jorge S Marques, and M Emre Celebi. Deep attention model for the hierarchical diagnosis of skin lesions. In *Proceedings of the IEEE/CVF Conference on Computer Vision and Pattern Recognition Workshops*, pages 2757–2765, 2019.

11. Almut Boer and KC Nischal. Get set for the https://ijdvl.com/www-derm101-com-a-growing-online-resource-for-learning-dermatology-and-dermatopathology/. A growing online resource for learning dermatology and dermatopathology. 2007.

12. Gary Bradski and Adrian Kaehler. *Learning OpenCV: Computer vision with the OpenCV library*. O'Reilly Media, Inc., 2008.

13. Alexander Börve. How long do you have to wait for a dermatologist appointment? From *https://www.firstderm.com/appointment-wait-time-see-dermatologist/*, 2021. [Online: accessed 14-April-2021].

14. Shouvik Chakraborty, Kalyani Mali, Sankhadeep Chatterjee, Soumen Banerjee, Kaustav Guha Mazumdar, Mainak Debnath, Pikorab Basu, Soumyadip Bose, and Kyamelia Roy. Detection of skin disease using metaheuristic supported artificial neural networks. In *2017 8th Annual Industrial Automation and Electromechanical Engineering Conference (IEMECON)*, pages 224–229. IEEE, 2017.

15. Franffcois Chollet. 2015. Keras. Available at: https://github.com/fchollet/keras. Accessed 16 November, 2022.

16. Djork-Arnte Clevert, Thomas Unterthiner, and Sepp Hochreiter. Fast and accurate deep network learning by exponential linear units (ELUs). *arXiv preprint arXiv:1511.07289*, 2015.

17. Noel CF Codella, David Gutman, M Emre Celebi, Brian Helba, Michael A Marchetti, Stephen W Dusza, et al. Skin lesion analysis toward melanoma detection: A challenge at the 2017 International Symposium on Biomedical Imaging (ISBI), hosted by the International Skin Imaging Collaboration (ISIC). In *2018 IEEE 15th International Symposium on Biomedical Imaging (ISBI 2018)*, pages 168–172. IEEE, 2018.

18. Navneet Dalal and Bill Triggs. Histograms of oriented gradients for human detection. In *2005 IEEE Computer Society Conference on Computer Vision and Pattern Recognition (CVPR'05)*, volume 1, pages 886–893. IEEE, 2005.

19. Tanvi Goswami, Vipul K Dabhi, and Harshadkumar B Prajapati. Skin disease classification from image—A survey. In *2020 6th International Conference on Advanced Computing and Communication Systems (ICACCS)*, pages 599–605. IEEE, 2020.

20. Seung Seog Han, Myoung Shin Kim, Woohyung Lim, Gyeong Hun Park, Ilwoo Park, and Sung Eun Chang. Classification of the clinical images for benign and malignant cutaneous tumors using a deep learning algorithm. *Journal of Investigative Dermatology*, 138(7):1529–1538, 2018.

21. Sergey Ioffe and Christian Szegedy. Batch normalization: Accelerating deep network training by reducing internal covariate shift. In *International Conference on Machine Learning*, pages 448–456. PMLR, 2015.

22. Afsana Ahsan Jeny, Abu Noman Md Sakib, Masum Shah Junayed, Khadija Akter Lima, Ikhtiar Ahmed, and Md Baharul Islam. Sknet: A convolutional neural networks based classification approach for skin cancer classes. In *2020 23rd International Conference on Computer and Information Technology (ICCIT)*, pages 1–6. IEEE, 2020.

23. Masum Shah Junayed, Nipa Anjum, Abu Noman, and Baharul Islam. A deep CNN model for skin cancer detection and classification. 2021.

24. Masum Shah Junayed, Abu Noman Md Sakib, Nipa Anjum, Md Baharul Islam, and Afsana Ahsan Jeny. EczemaNet: A deep CNN-based eczema diseases

classification. In *2020 IEEE 4th International Conference on Image Processing, Applications and Systems (IPAS)*, pages 174–179. IEEE, 2020.

25. Jeremy Kawahara, Sara Daneshvar, Giuseppe Argenziano, and Ghassan Hamarneh. Seven-point checklist and skin lesion classification using multitask multimodal neural nets. *IEEE Journal of Biomedical and Health Informatics*, 23(2):538–546, 2018.

26. Seema Kolkur and DR Kalbande. Survey of texture based feature extraction for skin disease detection. In *2016 International Conference on ICT in Business Industry & Government (ICTBIG)*, pages 1–6. IEEE, 2016.

27. Hongfeng Li, Yini Pan, Jie Zhao, and Li Zhang. Skin disease diagnosis with deep learning: A review. *arXiv preprint arXiv:2011.05627*, 2020.

28. Andrew L Maas, Awni Y Hannun, Andrew Y Ng. Rectifier nonlinearities improve neural network acoustic models. In *Proc. ICML*, volume 30, page 3. Citeseer, 2013.

29. T Mendoncÿa, PM Ferreira, J Marques, ARS Marcÿal, and J Rozeira. A dermoscopic image database for research and benchmarking. *Presentation in Proceedings of PH2 IEEE EMBC*, 2013.

30. Andrew Ng. What data scientists should know about deep learning. https://www.slideshare.net/ExtractConf, 2015.

31. Fabian Pedregosa, Gaël Varoquaux, Alexandre Gramfort, Vincent Michel, Bertrand Thirion, Olivier Grisel, et al. Scikit-learn: Machine learning in *Python*. *Journal of Machine Learning Research*, 12:2825–2830, 2011.

32. Adrian Rosebrock. *Deep Learning for Computer Vision with Python: Practitioner Bundle*. PyImageSearch, 2017.

33. Xiaolei Shen, Jiachi Zhang, Chenjun Yan, and Hong Zhou. An automatic diagnosis method of facial acne vulgaris based on convolutional neural network. *Scientific Reports*, 8(1):1–10, 2018.

34. Karen Simonyan and Andrew Zisserman. Very deep convolutional networks for large-scale image recognition. *arXiv preprint arXiv:1409.1556*, 2014.

35. Nitish Srivastava, Geoffrey Hinton, Alex Krizhevsky, Ilya Sutskever, and Ruslan Salakhutdinov. Dropout: a simple way to prevent neural networks from overfitting. *Journal of Machine Learning Research*, 15(1):1929–1958, 2014.

36. Xiaoxiao Sun, Jufeng Yang, Ming Sun, and Kai Wang. A benchmark for automatic visual classification of clinical skin disease images. In *European Conference on Computer Vision*, pages 206–222. Springer, 2016.

37. Talicia Tarver. Cancer facts & figures 2012. American Cancer Society (ACS) Atlanta, GA: American Cancer Society, 2012. 66 p., pdf.

38. Philipp Tschandl, Cliff Rosendahl, and Harald Kittler. The HAM10000 dataset, a large collection of multi-source dermatoscopic images of common pigmented skin lesions. *Scientific Data*, 5(1):1–9, 2018.

39. Xiaoping Wu, Ni Wen, Jie Liang, Yu-Kun Lai, Dongyu She, Ming-Ming Cheng, and Jufeng Yang. Joint acne image grading and counting via label distribution learning. In *Proceedings of the IEEE/CVF International Conference on Computer Vision*, pages 10642–10651, 2019.

40. Xin Yi, Ekta Walia, and Paul Babyn. Unsupervised and semi-supervised learning with categorical generative adversarial networks assisted by wasserstein distance for dermoscopy image classification. *arXiv preprint arXiv:1804.03700*, 2018.

41. Zhen Yu, Xudong Jiang, Feng Zhou, Jing Qin, Dong Ni, Siping Chen, Baiying Lei, and Tianfu Wang. Melanoma recognition in dermoscopy images via aggregated deep convolutional features. *IEEE Transactions on Biomedical Engineering*, 66(4):1006–1016, 2018.

COVID-19 Diagnosis-Based Deep Learning Approaches for COVIDx Dataset: A Preliminary Survey

Esraa Hassan

Department of Machine Learning and Information Retrieval, Faculty of Artificial Intelligence, Kafrelsheikh University, Kafrelsheikh, Egypt

Mahmoud Y. Shams, Noha A. Hikal, and Samir Elmougy

Department of Information Technology, Faculty of Computers and Information, Mansoura University, Mansoura, Egypt

CONTENTS

COVID-19 IS A SERIOUS and widespread disease that puts the lives of many people at risk around the world because it directly affects the lungs. Chest X-ray (CXR) images and computed tomography (CT) scans are partially utilized for reliable screening analysis of patients infected with the virus. Recognizing COVID-19 from different medical images is very challenging because it takes longer time, as well as requires accurate visualization of batches that are prone to human error. As a result, artificial intelligence (AI) techniques are constantly used to achieve high performance. Deep learning (DL) architectures have bypassed the classic machine learning (ML) that are popular among the AI methodology. Furthermore, DL automatically acts as an extractor, selector, and feature classifier. In this chapter, a comprehensive review of the utilization of DL approaches to COVID-19 identification and lung segmentation is presented. Furthermore, we focus on studying both X-ray images and CT for feature extraction and classification. In addition, a review

DOI: 10.1201/9781003251903-6

of articles using DL approaches to classify the enrolled images using COVID$_X$ dataset is presented. This dataset is commonly used recently to classify infected or normal patients. Finally, we highlight the difficulties encountered in the process of detecting COVID-19-based DL approaches. The challenges of diagnosing recorded images are presented as well as suggesting recommendations to address these challenges.

6.1 INTRODUCTION

In December 2019, the COVID-19 virus was discovered in Wuhan (China) that immediately spread around the globe [1]. This virus enters into the lung cells and spread throughout the respiratory system, where it replicates and kills the cells [2, 3]. The virus is extremely dangerous and can cause death in persons who have weaker immune systems [4]. Around the world, infectious disease specialists and clinicians have attempted to find a cure for the disease [5]. COVID-19 is the highest cause of death in many countries in the world, and hence the detection is crucial in the early phases of its development. However, using the Polymerase Chain Reaction (PCR) test for detection is time-consuming which is a risk for patients with COVID-19. So, first and foremost, medical imaging applications are used to identify COVID-19, then PCR test is used for assisting clinicians in deciding finally. Two medical imaging procedures are used to diagnose COVID-19: X-ray and CT scan. The X-ray modality is the initial approach for diagnosing COVID-19. It has the advantage of being inexpensive and low-risk for human health radiation dangers [6]. Detection of COVID-19 based on X-rays is a difficult process. The radiologist must pay close attention to the white spots in these images that contain water and pus, as they are very lengthy and problematic. The initial technique for diagnosing COVID-19 is to use X-rays, which has the advantages of being inexpensive and is considered a low risk of human health radiation concerns. However, COVID-19 detection using X-ray is a difficult task. Because the white spots in these images, which contain water and pus, are exceedingly long and problematic, the radiologist must pay particular attention to them. Because X-rays have an important rate of inaccuracy, CT scans can be worked for more precise detection. Nonetheless, CT are meaningfully further costly than X-ray images [7]. Several slices from each patient supposed with COVID-19 are provided at the time of CT scan recording. To diagnose COVID-19 based on CT scan images, physicians and radiologists must deal with a huge number of images. Figure (14.3) [8] shows the taxonomy of the most recent approaches based on deep learning (DL) for both supervised and unsupervised learning are presented. In supervised learning, convolutional neural networks (CNNs) and recurrent neural networks (RNNs) are preferred to achieve precise diagnosis and classification in real time for medical images. On the other hand, unsupervised deep learning approaches based on autoencoder (AE), deep belief neural network (DBNN), and the combination of CNN and AE are presented for clustering and diagnosis of the applied medical images. In special cases, these approaches can be utilized to diagnose and scan brain tumors from MRI scans and several applications of AI in data processing by accomplishing the level of human precisions in many tasks including analysis of medical images. In this chapter, a summary of diagnosis of coronavirus using DL models is presented [7, 9]. Section

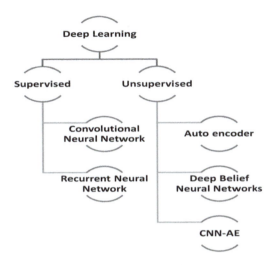

Figure 6.1   A general explanation for sub-deep learning methods [8].

6.3 explains the COVID$_X$ dataset, Section 6.4 illustrates the related works with DL models. Finally, an outline of future directions are presented in Section 6.5.

6.2   DATASET DESCRIPTIONS

Collecting datasets for COVID-19 cases is an important task for helping the research in support of their diagnosing tasks. The COVID$_X$ dataset [10] and COVID$_X$-CT [11] are presented in this survey to explain their roles in several recent types of research in the COVID-19 pandemic.

6.3   COVID$_X$ DATASET

COVID$_X$ is an open-access benchmark dataset comprising 13,975 CXR images from 13,870 patient cases, containing the most publicly available COVID-19 positive examples. A representative X-ray scan from the COVID$_X$ dataset [10] is shown in Figure 6.2.

6.4   COVID$_X$-CT DATASET

It was composed of 104,009 chest CT slices from 1489 patient instances, and it is called COVID$_X$-CT. This CT imaging data is noteworthy since it was produced from CT imaging data as shown in Figure 6.3. Their contribution includes cleaning and formatting raw data in a format appropriate for benchmarking, as well as giving bounding box annotations for body regions inside CT images [11, 12].

COVID$_X$ has several advantages that make it common and more usable in many types of research that are related to COVID-19 cases. The first advantage is rapid triaging that can be done in parallel with viral testing for helping relieve the high capacities of cases exclusively in zones most affected. Figure 6.4 shows the main categories for the COVID$_X$ dataset, and the needed libraries for implementation. The

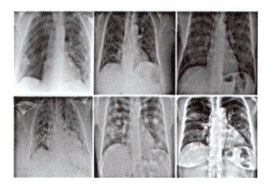

Figure 6.2 Examples of X-ray scans from the $COVID_X$ dataset [10].

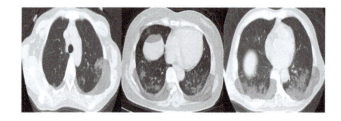

Figure 6.3 Examples of CT scans of COVID-19 cases [11, 12].

first library utilized in *Python* is Matplotlib, which visualize and create animated, static, and interactive plots of the applied medical images. Scikit Learn provides all the methodologies that assist the implementation of the applied images. Numpy presents comprehensive mathematical operations and functions, which generate a random number with linear algebra routines. All libraries are grouped in *Python* version 3.6. In this work, we classify the three well-known categories of COVID-19, which are infected COVID-19, Normal, and Pneumonia. Also, CXR imaging can be quite effective for triaging in geographic areas where patients are instructed to stay home until the onset of advanced symptoms [13]. The second challenge is availability and accessibility where it is considered standard equipment in most healthcare systems. In particular, CXR imaging is much more readily available than CT imaging, especially in developing countries where CT scanners are cost prohibitive due to high equipment and maintenance costs [11]. More specifically, Gunraj et al. [11] combined five data repositories to create the $COVID_X$ dataset by collecting the next categories of cases from each of the data repositories as: Non-COVID-19 pneumonia patient cases and COVID-19 patient cases from the COVID-19 Image Data Collection COVID-19 patient cases COVID-19 Chest X-ray Dataset.

6.5 RELATED WORK

This section presents a selection of related works of COVID-19 for classification, segmentation, and other tasks based on DL models. On July 11th, 2020, the most

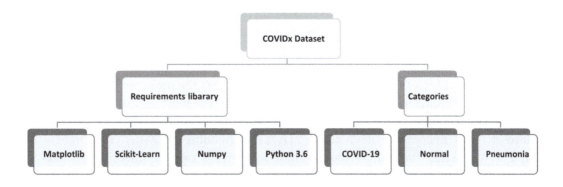

Figure 6.4 The main categories for the COVID$_X$ dataset and the requirement libraries required for implementation [10].

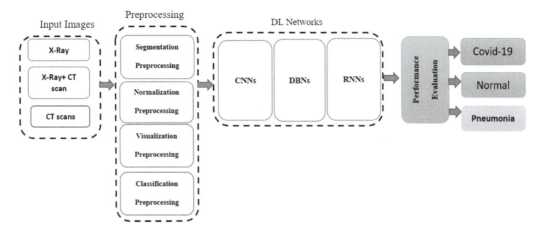

Figure 6.5 The general architecture of classification of the X-ray/CT images using DL approaches.

recent selection of papers is made using the keywords mentioned. The standard architecture for COVID-19 diagnosis employing DL methods using the COVID$_X$ database is shown in Figure 6.5 [14]. In response to decision-making difficulties, Gunraj et al. [12] introduced the COVID-Net CT-2 dataset that enhanced deep neural network (DNN) architectures for COVID-19 detection from chest CT images trained on the largest quantity and diversity of multinational cases. It consisted of 4501 patients from at least 15 countries, which they used to explain and observed the decision-making behavior of COVIDNet CT. Ebadi et al. [15] introduced COVID$_X$-US, an open-access dataset of COVID-19-related ultrasound imaging data. The dataset contained ultrasound imaging data. The dataset was collected from a variety of images. It contained 150 lung ultrasound recordings, 12,943 processed images of cases with COVID-19, non-COVID-19 infections, and other lung diseases. Chowdhury et al. [16] proposed an ensemble deep convolutional neural network (ECOVNet) to detect COVID-19 that is implemented on 15,959 CXR images from 15,854 cases, with normal, pneumonia, and COVID-19 patients. They described the predictions in a way that is understandable. The method can distinguish COVID-19 with a positive predictive value of precision 91.6 % for COVID-19 cases. Ambati et al. [17] proposed an AC-COVIDNet model by computing the sensitivity of the proposed method over the publicly available COVID-19 dataset. It is observed that the proposed AC-COVIDNet exhibits very promising performance as compared to the existing methods even with limited training data. Cherti et al. [18] introduce a pretraining model on a largescale dataset using ImageNet with a range 21K/1K depending on X-ray chest images as well as comparing the full transfer learning applied on various objective datasets in mutually physical and medical domains.

Qi et al.[19] incorporate the local phase of X-ray image features into a multi-CNN feature architecture. The mathematical evaluation is performed on a dataset containing 8851 normal, 6045 types of pneumonia, and 3795 COVID-19 X-ray scan images. Only 7.06% of the labeled data and 16.48% of unlabeled data for training are utilized. Moreover, 5.53% of the data were used for validation purposes. Their method reached 93.61% with a mean accuracy value applied on a large-scale dataset of the tested 70.93% data. Celaya et al. [25] showed that the PocketNet architecture produces results comparable to traditional NN when performing various segmentation and classification tasks, while reducing the number of parameters by several orders of magnitude, reducing the GPU memory used by up to 90%, and speeding up data processing. The training time is as high as 40%, enabling such models to be trained in resource-limited environments.

Yang et al. [26] explored this difficult topic by showing a simple but successful technique in which the server and client share only a part of the model. Extensive testing using real-world healthcare tasks and baseline data shows that their method provides equivalent or better results while reducing privacy and security issues. They tested the FLOP algorithm on the COVID-19 dataset and demonstrated that it enables different institutions to collaborate and effectively train partially shared models without the need to disclose local patient data for federal learning.

Ali et al. [27] achieved a 5% rise in the F1-Score and a gap twice as great for general classification using CNN architecture to learn the optimum embedding space. Dong et al. [20] proposed a deep model for robust COVID-19 detection that employs Mixed High-Order Moment Feature (MHMF) and Multi-Expert Uncertainty-Aware Learning to capture compact and disentangled representational characteristics. MHMF can fully explore the benefits of using high-order statistics and extracting discriminative features of complex distributions in medical imaging. Hao et al. [21] used the innovative combination to classify COVID-19 chest radiographs which is rich in data but extremely unbalanced. They showed that by training all available tags, only 10% of the labeled data is needed to achieve accuracy. Its use of COVID-19 data in a fully supervised classification scenario demonstrates its model, which uses a generic ResNet backbone. They explored the benefits of using high-order statistics and extracted discriminative features from the complex distribution of medical images. Zhang et al. [33] suggested a new unsupervised method for learning medical visual representations directly from naturally occurring image-textual data pairs. They used a domain-agnostic strategy of pretraining and encoding the medical image with related data with text bidirectional incompatible purpose between the two modes that involve no additional input. The technique on ImageNet-labeled data in order to achieve improved results.

Wong et al. [13] proposed a COVIDNet architecture that relies on a deep convolutional neural network (DCNN). They are custom designed to detect COVID-19 cases from chest X-ray images. Implemented using COVIDNet, this is the first source network plan to diagnose COVID-19 from CXR images from the time of initial release. COVIDNet uses explanatory power technology to construct predictions, which not only provides a deeper understanding of key aspects related to COVID cases but also helps clinicians perform better tests. The COVIDNet results of PARAM, MAC, and accuracy reached 11.75, 7.50, and 93.3 (%), respectively. Wang et al. [7] hypothesized that, compared to the basic reality of expert chest radiologists, computer-assisted deep learning algorithms can reliably predict the severity of lung disease on CXR associated with infection with SARSCoV2, and the experimental results of this study supported this prediction. Between the predicted score and the DNN of COVIDNet S, the R2 of the geographic range is 0.664 0.032 and 0.635 0.044.

Abbasi et al. [29] proposed an independent COVID-19 diagnosis and severity prediction method, which uses deep feature maps from CXR imaging to diagnose COVID-19 and predict its severity. They used different shallow supervised classification algorithms and proposed that the model adopts a three-phase classification method (health vs. sick, COVID-19 vs. pneumonia, and COVID-19 severity). They not only tested COVID$_X$ in a laboratory environment with 10 cross-validation and external validation datasets but also tested COVID$_X$ in a real environment with experienced radiologists. In all evaluation situations, COVID$_X$ outperforms all reliable myths for gravity prediction. Motamed et al. [23] Did not use tags or training data from images of unknown categories. They provide a RANDGAN (Random Bad Generation Network), which can identify photos of an unknown category (COVID-19) from known and labeled categories (normal and viral pneumonia) (COVID-19). In the COVID$_X$ dataset, they used transfer learning to distinguish the lungs. After that, they appear for anomaly detection. Qi et al. [19] proposed a new multi-feature architecture (CNN) to improve COVID-19 multi-class classification from CXR images. They provide a quantitative evaluation of the two datasets, as well as qualitative results for visual inspection. Quantitative analysis was performed on data from 8851 normal scans, 6045 types of pneumonia, and 3323 COVID-19 images. The model achieved an average accuracy of 95.57% for the three categories, an accuracy of 99%, a recovery rate, and the F1-Score of COVID-19 cases.

Huang et al. [28] integrated local phase CXR imaging features into a multi-feature CNN architecture. A total of 8851 normal (health) scans, 6045 types of pneumonia, and 3795 COVID-19 CXR scans were quantitatively analyzed. Only 7.06% of the data are labeled, while 16.48% are unlabeled, and 5.53% are used for verification. Type achieves an average accuracy of 93.61% on the large-scale test data (70.93%) of computer tomography (CT). Chen et al. [8] used a large chest X-ray dataset. This study suggested using a set of DCNN built-in EfficientNet, called Covent, to detect COVID-19. First, they supplemented the huge open-access chest X-ray collection, then transferred the pre-trained ImageNet weights to Efficient Net, and some custom fine-tuning of the top layers, and then a set of model snapshots to classify the chest X-rays corresponding to COVID-19, Normal, and Pneumonia.

Dev et al. [32] recommended using a hierarchical convolutional network (HCN) architecture to automatically enrich data with various attributes. To extract features, HCN uses the first convolutional layer of COVIDNet, followed by the previously trained convolutional layer of a known network. The use of the COVIDNet convolutional layer ensures that the representation related to the CXR modal is extracted. They also recommend using ECOC to encode multi-class problems into binary classifications to improve recognition performance. Compared with previous studies, the experimental results show that the HCN architecture can achieve better results. The suggested method uses CXR imaging to accurately classify potential COVID-19 patients so that the burden of proof can be shared, and testing capacity can be improved. In Shoeibi et al. [14], a comprehensive review of research using DL methods for COVID-19 diagnosis and automated lung segmentation was conducted, with a focus on research using X-ray and CT imaging. In addition, comments are provided on articles that use DL methods to predict the prevalence of coronavirus worldwide. Finally, the limitations of using the DL method to automatically identify COVID-19 and the future research prospects of computer tomography (CT) feature selection for medical diagnosis are emphasized.

For COVID-19 image classification, Nwosu et al. [24] offered ResNet, a two-path semi-supervised deep learning model based on Residual Neural Network (ResNet), where two routes refer to a supervised path and an unsupervised path, respectively. Furthermore, to address the data imbalance, they devise a weighted supervised loss in which the minority classes are given a higher weight in the training process. The suggested model can achieve promising performance even when trained on a small number of labeled training images, according to experimental results on the $COVID_X$ dataset. Ali et al. [27] achieved a 5% rise in the F1-Score and a gap twice as great for general classification. To learn the optimum embedding space, Zhang et al. [22] combined multi-similarity loss with a hard-mining sample technique and an attention mechanism and offered similar images to the query image. The model is trained and validated using a COVID-19 dataset obtained from three different sources from around the world. The results of COVID-19 image retrieval and diagnosis tasks show that the proposed model can serve as a robust solution for CXR analysis. To produce points in the convex hull of desired imbalanced class, Deepshikha et al. [34] imposed an extra requirement on generator network G. In addition, the adversarial game with classifier C pushes the conditional distribution learned by G to the peripheral of the respective class, accounting for the problem of class imbalance in anomaly detection. The goal of Sarosh et al. [35] was to examine the function of deep learning and artificial intelligence in many aspects of COVID-19 management in general and specifically in COVID-19 detection and categorization. The DL model is used to examine clinical patterns, such as CT scans and X-ray images, and predict the disease state of patients. The DL model is designed to detect COVID-19 pneumonia and classify and distinguish COVID-19, community-acquired pneumonia (CAP), viral and bacterial pneumonia.

Banerjee et al. [36] showed the results of two popular models, COVIDNet and CoroNet, which were evaluated on three publicly available datasets and an additional institutional dataset, which were collected between January and May 2020 at Emory Hospital, including patients that used RTPCR to test for COVID-19 infection. Zhang et al. [33] proposed using a minimal learning method called print weights, which uses many samples from known diseases (such as pneumonia) to improve COVID-19 detection performance. They suggest using a minimal learning method called weight imprinting, which uses many samples from known diseases (such as pneumonia) to improve COVID-19 detection performance. The general structure of the most common input image classifications, including X-rays and CT scans, is presented in Figure 6.5. Here, the combination of both modalities can be utilized, besides the use of unique modalities. In some cases, the pre-processing stages include segmentation and classification as an initial stage before matching and classification. The use of DL involves pre-processing and feature extraction. As we mentioned before, DL nets containing CNNs, DBN, and RNN are commonly utilized as feature extractors and classifiers for enrolled X-ray and CT images. The performance evaluation was determined based on the confusion matrix, including accuracy, F1-Score, recall, and precision. Most of the recent classifiers based on DL approaches utilized two classes, 0 and 1, that represented COVID-19 and normal cases as the output of the proposed architecture. Visualization is essential in the pre-processing stage because it enables the radiology to notice the batches and the spread of region of interest (ROI) that represent the COVID-19 or pneumonia cases. Table 6.1 summarizes the most recent approaches used to classify the CT/X-ray images, including the number of modalities, architecture, and performance obtained. For the following reasons, we selected those specific criteria in the table instead of others. Firstly, we studied all the diagnosis architectures using the COVID$_X$ dataset. Secondly, we highlighted the major tasks, including image classification and learning based on CT and X-ray images. Moreover, the survey focused on the utilization of deep learning approaches for COVID-19 classification and diagnosis. All the mentioned studies are currently being performed and presented to ensure the reliability of utilizing DL approaches to classify and predict the COVID-19 cases based on CT and X-ray images using a well-known COVID-x dataset. In Figure 6.5, the pipeline of the COVID-19 image classification major stages is shown. The first stage includes applying images that might be X-ray, CT, or a combination of both CT and X-ray images. Afterward, the pre-processing of the enrolled images, including segmentation, normalization, visualization, and classification, is performed before the classification and feature extraction process. Hence, the features are extracted to obtain the feature vector that is applied in the DL networks based on CNN, DBNN, and RNN to classify and match the CT and CXR images. The last stage is the performance evaluation matrices to obtain and classify three major class labels, which are COVID-19, Normal, and Pneumonia.

TABLE 6.1   Comparative Study of the COVID-19 Diagnosis Architecture Using the COVID$_X$ Dataset

Author	Year	Task	No. Cases	Pre-Processing	Architecture	Performance
Gunraj et al. [12]	2021	Geographic extent and Opacity extent scoring	396	Randomly translation Rotations Horizontal flips Zooms intensity shifts Cutout Gaussian noise.	DNN	Geographic Extent = 0.664 Opacity extent = 0.635 Mean = 0.032 STD = 0.044
Ebadi et al. [15]	2021	Image Classification	12,943	Cropping	DNN	N/A
Ambati et al. [17]	2021	Image Classification	13,870	N/A	CNN	Accuracy = 98.08%
Cherti et al. [18]	2021	Few-Shot learning Image classification transfer learning	13,870	Standard random resized crop data augmentation	Pre-trained model, Resent 50	Accuracy = 95.78%
Celaya, et al. [25]	2021	Medical image segmentation Semantic segmentation	13,870	Randomly reflecting	CNN	Accuracy = 0.9941%
Qi et al. [19]	2021	Computed tomography (CT) COVID-19 diagnosis Decision making	13,870	N/A	ResNet50	Accuracy = 86.46%
Yang et al. [26]	2021	Federated learning	13,870	N/A	CNN	Flop = 94.54 ± 0.21
Ali et al. [27]	2021	Classification	13,870	Geometric transformations and color space transformations	CNN	Accuracy = 95%
Li et al. [20]	2021	Computed tomography (CT)	19	Flipping, translation, rotation using random five different angles	Patch-based convolutional neural network	Accuracy = 97.89%
Hao et al. [21]	2021	Active learning General classification representation learning Unsupervised representation learning	13,870	Random crop Random flip Color distortion Gaussian blur Random grayscale	CNN-GP hybrid model	Accuracy = 96%
Zhang et al. [22]	2021	COVID-19 Diagnosis	13,870	N/A	CNN	Accuracy = 88%
Gunraj et al. [10]	2020	COVID-19 Diagnosis	13,870	Translation rotation horizontal flip Zoom intensity shift	CNN	N/A
Gunraj et al. [11]	2020	Decision-making	N/A	Cropping box Jitter Rotation Horizontal and vertical shear Horizontal flip Intensity	DNN	Sensitivity = 98.1% Positive Predictive Value = 98.1%

(*Continued*)

TABLE 6.1 *(Continued)* Comparative Study of the COVID-19 Diagnosis Architecture Using the COVIDₓ Dataset

Author	Year	Task	No. Cases	Pre-Processing	Architecture	Performance
Huang et al. [28]	2020	Image classification	217,000	Color jittering transformation	DNN	Accuracy = 92.4%
Abbasi et al. [29]	2020	Severity prediction	13,870	N/A	CNN	ROC= 0.99 ± 0.01 PR F1 = 0.99 ± 0.01 0.98
Hirano et al. [30]	2020	Anomaly detection COVID-19 diagnosis COVID-19 image segmentation transfer learning	13,870	Normalized images	CNN	ROC curve from 0.71 to 0.77
Motamed et al. [23]	2020	Anomaly detection COVID-19 diagnosis COVID-19 image segmentation transfer learning	13,870	Augmentation	CNN	AUC of 0.54 to 0.71
Li et al. [31]	2020	Image enhancement	13,870	Adaptive histogram equalization	CNN	Accuracy = 94.44%
Chowdhury et al[16]	2020	Image classification	14,914	Partial translation-invariant	CNN	Accuracy = 97%
Dev et al. [32]	2020	Image classification	13,870	Rotation Translation Flipping	CNN	Accuracy = 98.08%
Shoeibi et al. [14]	2020	Image classification	13,870	N/A	CNN	N/A
Zhang et al. [33]	2020	Image retrieval metric Learning	13,870	Horizontal flip Perspective Sharpness Blurring Grid distortion Elastic transformation	CNN	Accuracy = 94%
Deepshikha et al. [34]	2020	Anomaly detection2	14,200	GAN generator	Transfer learning	Accuracy = 70%
Sarosh et al. [35]	2020	Survival prediction transfer learning	13,870	GAN augmentation	CBB	Accuracy = 98%
Banerjee et al. [36]	2020	COVID-19 diagnosis	1786	Rotation, horizontal flip	CNN	Accuracy = 93%

6.6 CONCLUSION

COVID-19 is a developing pandemic disease that has put the health of many people throughout the world in danger in a very short time. It has a direct effect on lung cells and, if not detected early, can result in permanent damage, even death. X-ray or CT images, as well as PCR 15 results, are used by professionals to accurately detect the disease. Instead of COVID-19, the PCR results reveal another form of lung infection, such as respiratory tuberculosis. DL networks were utilized to conduct a wide-ranging analysis of studying COVID-19 identification. The public databases for diagnosing and predicting COVID-19 are discussed. The most up-to-date DL approaches were used to diagnose, segment, and forecast the spread of COVID-19. The availability of a large public database is one of the problems in developing a reliable and accurate COVID-19 diagnosis system. We believe that with more public databases, researchers will be able to construct better DL models to reliably detect and predict COVID-19. As a result, this will aid in the development of the finest-performing paradigm. We believe that the fusion of the data models can assist in diagnosis and prediction performance. To construct an accurate model, the extracted features from the DL architecture can be combined with the applied ML to enhance the obtained classification results. Furthermore, constructing an appropriate segmentation model is difficult because it requires professionals to delineate the lungs. Therefore, the visualization process of the chest X-ray or CT is utilized to detect the most significant features that are essential to the accurate diagnosis of these cases. In the future, we plan to use feature selection algorithms applied to $COVID_X$ and other updated collected datasets to investigate and visualize the images with reliable and efficient accuracy.

Bibliography

1. J. Wu, J. Liu, S. Li, Z. Peng, Z. Xiao, X. Wang, R. Yan, J. Luo. "Detection and analysis of nucleic acid in various biological samples of COVID-19 patients." Travel Medicine and Infectious Disease, 37 (2020): 101673?

2. T. Corse, L. Dayan, S. Kersten, F. Battaglia, S. R. Terlecky, and Z. Han, "Clinical outcomes of COVID-19 patients with pre-existing, compromised immune systems: A review of case reports," Int. J. Med. Sci., vol. 17, no. 18, pp. 2974–2986, 2020, doi: 10.7150/ijms.50537.

3. D. Wu, T. Wu, Q. Liu, and Z. Yang, "The SARS-CoV-2 outbreak: What we know," Int. J. Infect. Dis., vol. 94, pp. 44–48, 2020, doi: 10.1016/j.ijid.2020.03.004.

4. S. Elmuogy, N. A. Hikal, and E. Hassan, "An efficient technique for CT scan images classification of COVID-19," vol. 40, pp. 5225–5238, 2021, doi: 10.3233/JIFS-201985.

5. E. Hassan, M. Shams, N. A. Hikal, and S. Elmougy, "Plant seedlings classification using transfer," 2021 International Conference on Electronic Engineering (ICEEM), pp. 1–7, 2021.

6. O. M. Elzeki, M. Shams, S. Sarhan, M. A. Elfattah, and A. E. Hassanien, "COVID-19: A new deep learning computer-aided model for classification," PeerJ Comput. Sci., vol. 7, pp. 1–33, 2021, doi: 10.7717/peerj-cs.358.

7. A. Sedik et al., "Deploying machine and deep learning models for efficient data-augmented detection of COVID-19 infections," Viruses, vol. 12, no. 7, pp. 1–29, 2020, doi: 10.3390/v12070769.

8. P. Ghamisi et al., "New frontiers in spectral-spatial hyperspectral image classification: The latest advances based on mathematical morphology, Markov random fields, segmentation, sparse representation, and deep learning," in IEEE Geoscience and Remote Sensing Magazine, vol. 6, no. 3, pp. 10–43, Sept. 2018.

9. K. M. Hosny, M. A. Kassem, and M. M. Foaud, "Skin cancer classification using deep learning and transfer learning," 2018 9th Cairo Int. Biomed. Eng. Conf. CIBEC 2018 - Proc., pp. 90–93, 2019, doi: 10.1109/CIBEC.2018.8641762.

10. H. Gunraj, L. Wang, and A. Wong, "Chest X-ray images," Front. Med., vol. 7, pp. 1–12, 2020, doi: 10.3389/fmed.2020.608525.

11. H. Gunraj, L. Wang, and A. Wong, "COVIDNet-CT: A tailored deep convolutional neural network design for detection of COVID-19 cases from chest CT images," Front. Med., vol. 7, pp. 1–12, 2020, doi: 10.3389/fmed.2020.608525.

12. H. Gunraj, A. Sabri, D. Koff, and A. Wong, "COVID-Net CT-2: Enhanced deep neural networks for detection of COVID-19 from chest CT images through bigger, more diverse learning," pp. 1–15, 2021, [Online]. Available: http://arxiv.org/abs/2101.07433.

13. A. Wong et al., "COVID-Net S: Towards computer-aided severity assessment via training and validation of deep neural networks for geographic extent and opacity extent scoring of chest X-rays for SARS-CoV-2 lung disease severity," 2020, [Online]. Available: http://arxiv.org/abs/2005.12855.

14. Shoeibi, A., Khodatars, M., Alizadehsani, R., Ghassemi, N., Jafari, M., Moridian, et al. Automated detection and forecasting of COVID-19 using deep learning techniques: A review. arXiv preprint arXiv:2007.10785.

15. A. Ebadi, P. Xi, A. Maclean, S. Tremblay, S. Kohli, and A. Wong, "COVID$_X$-US - An open-access benchmark dataset of ultrasound imaging data for AI-driven COVID-19 analytics," pp. 1–12, 2021.

16. N. K. Chowdhury, M. A. Kabir, M. M. Rahman, and N. Rezoana, "ECOVNet: an ensemble of deep convolutional neural networks based on EfficientNet to detect COVID-19 from chest X-rays," 2020, doi: 10.7717/peerj-cs.551.

17. A. Ambati and S. R. Dubey, "AC-CovidNet: attention guided contrastive CNN for recognition of COVID-19 in chest X-ray images," 2021, [Online]. Available: http://arxiv.org/abs/2105.10239.

18. M. Cherti and J. Jitsev, "Effect of large-scale pre-training on full and few-shot transfer learning for natural and medical images," pp. 1–18, 2021, [Online]. Available: http://arxiv.org/abs/2106.00116.

19. X. Qi, J. L. Nosher, D. J. Foran, and I. Hacihaliloglu, "Multi-Feature semi-supervised learning for COVID-19 diagnosis from chest X-ray images," vol. 2019, pp. 1–11, 2021, [Online]. Available: http://arxiv.org/abs/2104.01617.

20. Q. Li, B. Dong, D. Wang, and S. Wang, "Identification of secreted proteins from malaria protozoa with few features," IEEE Access, vol. 8, pp. 89793–89801, 2020, doi: 10.1109/ACCESS.2020.2994206.

21. Heng Hao, P. Bangert, "Highly efficient representation and active learning framework for imbalanced data and its application to COVID-19 X-ray classification," pp. 1–16, 2021.

22. J. Zhang, P. Xi, A. Ebadi, H. Azimi, S. Tremblay, and A. Wong, "COVID-19 detection from chest X-ray images using imprinted weights approach," pp. 1–5, 2021, [Online]. Available: http://arxiv.org/abs/2105.01710.

23. S. Motamed, P. Rogalla, and F. Khalvati, "RANDGAN: Randomized generative adversarial network for detection of COVID-19 in chest X-ray," Sci. Rep., vol. 11, no. 1, pp. 1–10, 2021, doi: 10.1038/s41598-021-87994-2.

24. L. Nwosu, X. Li, L. Qian, S. Member, and S. Kim, "Semi-supervised learning for COVID-19 image classification via ResNet." EAI Endorsed Transactions on Bioengineering and Bioinformatics. 1. 170754. 10.4108/eai.25-8-2021.170754.

25. A. Celaya, J. A. Actor, R. Muthusivarajan, E. Gates, C. Chung, D. Schellingerhout, B. Riviere, et. al. "PocketNet: A smaller neural network for medical image analysis," pp. 1–9, 2021 [Online]. Available: http://arxiv.org/abs/2104.10745.

26. Y. Yang and S. Soatto, "FDA: Fourier domain adaptation for semantic segmentation." pp. 4084–4094.

27. A. Ali, T. Shaharabany, and L. Wolf, "Explainability guided multi-site COVID-19 CT classification," 2021, [Online]. Available: http://arxiv.org/abs/2103.13677.

28. Huang, C., Wang, Y., Li, X., Ren, L., Zhao, J., Hu, Y., et al. "Clinical features of patients infected with 2019 novel coronavirus in Wuhan, China." Lancet, 2020, 395(10223), pp. 497–506.

29. W. A. Abbasi, S. A. Abbas, and S. Andleeb, "COVID$_X$: Computer-aided diagnosis of COVID-19 and its severity prediction with raw digital chest X-ray images," pp. 1–19, 2020 [Online]. Available: http://arxiv.org/abs/2012.13605.

30. H. Hirano, K. Koga, and K. Takemoto, "Vulnerability of deep neural networks for detecting COVID-19 cases from chest X-ray images to universal adversarial attacks," PLoS One, vol. 15, no. 12, 2020, doi: 10.1371/journal.pone.0243963.

31. Li, L., Qin, L., Xu, Z., Yin, Y., Wang, X., Kong, B., et al. Using artificial intelligence to detect COVID-19 and community-acquired pneumonia based on pulmonary CT: Evaluation of the diagnostic accuracy. Radiology, 2020, 296(14.4), E65–E71.

32. K. Dev, S. A. Khowaja, A. S. Bist, V. Saini, and S. Bhatia, "Triage of potential COVID-19 patients from chest X-ray images using hierarchical convolutional networks," Neural Comput. Appl., November, 2021, doi: 10.1007/s00521-020-05641-9.

33. Zhang, Hang, H. Zhang et al., "ResNeSt: Split-attention networks," 2022 IEEE/CVF Conference on Computer Vision and Pattern Recognition Workshops (CVPRW), 2022, pp. 2735–2745.

34. K. Deepshikha and A. Naman, "Removing class imbalance using Polarity-GAN: An uncertainty sampling approach," 2020 [Online]. Available: http://arxiv.org/abs/2012.04937.

35. P. Sarosh, S. A. Parah, R. F. Mansur, and G. M. Bhat, "Artificial intelligence for COVID-19 detection—A state-of-the-art review," Arxiv, vol. 2, November 2020.

36. I. Banerjee P. Sinha, S. Purkayastha, N. Mashhaditafreshi, A. Tariq, J. J. Jeong, H. Trivedi, and J. W. Gichoya, "Was there COVID-19 back in 2012? Challenge for AI in diagnosis with similar indications," vol. 19, 2020 [Online]. Available: http://arxiv.org/abs/2006.13262.

Automatic Grading of Invasive Breast Cancer Patients for the Decision of Therapeutic Plan

Hossain Shakhawat, Matthew Hanna, Kareem Ibrahim, Rene Serrette, Peter Ntiamoah, Marcia Edelweiss, Edi Brogi, Meera Hameed, Masahiro Yamaguchi, Dara Ross, and Yukako Yagi

CONTENTS

WHOLE SLIDE IMAGING (WSI) with artificial intelligence (AI) has become a spearhead for smart healthcare in diagnostic pathology. This chapter discusses an automatic system designed using WSI and AI technology to grade invasive breast cancer patients enabling therapeutic decisions. Cancer is one of the top five leading causes of death in the United States and invasive breast cancer is the most

DOI: 10.1201/9781003251903-7

worrisome form of cancer as its treatment is complicated and requires considering numerous factors. This chapter addresses the shortcomings of current practices for selecting patients for therapy, introduces WSI technology for automated analysis, and describes how the proposed system utilizes WSI and AI to bridle the challenges of grading patients for therapy. The result of the validation experiments which ensures the potential of the system for clinical implementation is also included.

7.1   INTRODUCTION

Breast Cancer (BC) is one of the most frequent forms of cancer. BC becomes invasive when cancer-affected cells travel from their origin to the surrounding breast tissue through the bloodstream or lymphoid system. According to the most recent study of the National Cancer Institute (NCI), 12.9% of women born in the United States have the chance to develop invasive BC [1]. Meaning, 1 in 8 women will be diagnosed with invasive BC at some point during her lifetime and this rate is increasing over time, despite all the research for decades. In 2021, an estimated 330,840 new cases of invasive BC were expected to be diagnosed in women in the United States [2], while this count was 268,600 in 2019 [3]. This risk factor is lower for men, 1 in 833 men has the risk of being diagnosed with invasive BC.

Besides lung cancer, invasive BC has the highest death rate for women. Human epidermal receptor growth factor receptor 2 (HER2) positive is one of the most aggressive forms of BC that decreases survival rates and high-recurrence. Approximately 25% of BC patients are HER2 positive [4,5]. HER2 (also known as ERBB2) is a member of the human epidermal growth factor receptor family. The HER2 receptor is responsible for the growth, division, and repair of a normal cell. In an HER2-positive BC patient, the HER2 gene makes too many copies of itself which leads to uncontrolled growth and division of breast cells. Therefore, HER2 is used as a predictive and prognostic biomarker not only for BC but also for gastric and stomach cancer as well. HER2-positive tumors respond differently to the conventional treatment plans of BC than the HER2-negative tumors. In clinical practice, HER2 assessment is performed routinely for all breast cancer patients to determine the HER2 amplification status and to decide on the treatment plan.

The treatment of invasive BC patients includes surgery, chemotherapy, radiation, hormone therapy, and targeted therapy. These treatments can be given independently or in a combination depending on the size of the tumor, location of the tumor, type of cancer, stage of cancer, HER2 status, and some other factors. In 1998, the United States Food and Drug Administration approved trastuzumab for the treatment of HER2-positive patients of BC and gastric cancer. Later in 2010, it was approved for HER2-positive stomach cancer patients. HER2-positive patients benefit from anti-HER2-targeted therapy such as trastuzumab or lapatinib. Trastuzumab attaches itself to the HER2 protein, thus blocking cancer cells from receiving growth signals. Such targeted therapy helps increase the survival time if given appropriately [6]. But if this treatment is given to an HER2-negative patient it may cause cardiac toxicity and other serious side effects [7]. Expense is another consideration [8, 9]. A single false HER2-positive decision may cost $50,000 plus the cost of mitigating the side effects

[10]. However, the health and economic damage are far greater in reality. Therefore, the selection of patients for HER2-targeted therapy is a crucial step while deciding on the treatment plan for a BC patient.

HER2 positivity can be determined by assessing the HER2 gene amplification or subsequent HER2 protein overexpression. An immunohistochemistry (IHC) test is used for assessing HER2 protein expression and in situ hybridization (ISH) test for assessing HER2 gene amplification. In clinical practice, the IHC test is utilized initially to determine the HER2-positive status. However, IHC test results are not conclusive for the cases. Therefore, a reflex ISH test is required and performed using Fluorescence ISH (FISH) or Chromogenic ISH (CISH) to confirm the result, according to the American Society of Clinical Oncology and College of American Pathologists (ASCO/CAP) guideline [11]. Most laboratories use FISH as the reflex test. However, the FISH test requires special training, special setup, and a lot of time for preparing the specimen and quantifying the signals. On the other hand, the CISH test requires the pathologists to count myriads of signals manually which is laborious and time consuming. Some commercial tools are available for assistance, but they are not reliable and not user friendly. The limitations of current tests signify the need for a more practical and efficient system for HER2 assessment. Moreover, in a study of 1787 patients, it was found that 9.7% of the HER2 negative patients benefited from anti-HER2 therapy [12] which might be due to the false-negative results of the HER2 test. Therefore, the need for a more practical and accurate system is undeniable to select patients for therapy and to advance patient care.

In this chapter, we present an automatic HER2 grading system for selecting invasive BC patients for giving HER2 therapy. This system is developed based on the HER2 quantification method proposed by our group in a previous study [13], and validated by Memorial Sloan Kettering Cancer Center. The system is expected to determine HER2 status automatically and provide useful information to decide on the therapeutic plan, thus improve patient care with optimal therapies.

7.2 CLINICAL PRACTICES FOR HER2 QUANTIFICATION

HER2 positivity can be determined by counting the HER2 gene inside the nucleus or the subsequent HER2 protein receptors outside the nucleus. Figure 7.1 shows the example of a normal cell and HER2-positive cell. Based on the biomarker used, HER2 tests can be divided into two categories: Gene-based test and protein-based test. Both types of tests are intended to detect HER2 positivity but they quantify different biological targets for this purpose. IHC assesses the HER2 protein expression while ISH tests such as FISH or CISH detect the HER2 gene amplification. IHC, FISH, and CISH are the FDA approved tests for HER2 assessment.

IHC detects proteins in the cell membrane (cell boundary). If a HER2 protein is in the cell membrane, the membrane will be stained in IHC. Pathologists then score the staining strength in the cell membrane as 0, 1+, 2+, or 3+, according to the criteria given in Table 7.1. After that, the HER2 protein expression is classified as negative (scores 0 and 1+), equivocal (score 2+), or positive (score 3+). IHC is a qualitative test, and the score is subject to inter and intraobserver variation. Moreover, the

TABLE 7.1 IHC Scoring Method

Scoring Criteria	IHC Score	HER2 Protein Expression
Membrane staining in < 10% of invasive tumor cells	0	Negative
Weak incomplete membrane staining in ≥ 10% of invasive tumor cells	+1	Negative
Weak to moderate complete membrane staining in ≥ 10% of invasive tumor cells or strong complete membrane staining in < 30% cells	2+	Equivocal
Strong complete membrane staining in ≥ 30% of invasive carcinoma cells	3+	Positive

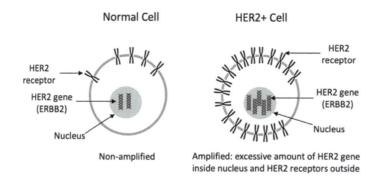

Figure 7.1 Example of a normal (left) and an HER2-positive (right) cell.

staining strength varies depending on the fixation time and the antibody used for staining. In routine clinical practice, HER2 status is determined using the IHC test in the first stage. After that, if the IHC test result is equivocal (not clearly positive or negative) for a case then a reflex ISH test is performed to determine the HER2 status, according to the ASCO/CAP 2018 guidelines. However, the reliability of IHC test scores is a concern due to the subjective nature of the test. The percentage of HER2 flex test and HER2 positive patients varies significantly if the threshold for detecting equivocal cases is changed slightly. Figure 7.2 illustrates the current HER2 assessment approach.

FISH and CISH are the popular options for the reflex ISH test. Both are quantitative assays; however, FISH is used in most of the laboratories. Both FISH and CISH count HER2 genes and Chromosome Enumeration Probe 17 (CEP17) per nucleus. The ASCO/CAP 2018 guidelines requires counting the signals for at least 20 nuclei. Then the HER2-to-CEP17 ratio and average HER2 copy number per nucleus are calculated for the 20 nuclei to determine the HER2 status. FISH tests utilize fluorescence microscopy and requires a special darkroom setup to observe the signals. In a FISH slide, the nucleus appears blue while the HER2 and CEP17 signals appear as red and green dots, accordingly, shown in Figure 7.3. The test requires special training and only board-certified pathologists can perform the test who can differentiate the blue nuclei of cancer cells from the blue nuclei of benign cells. Cost is another

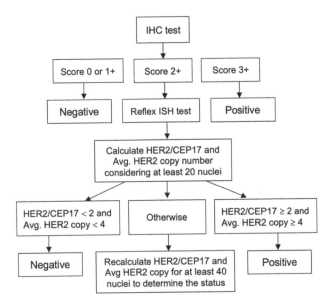

Figure 7.2 Current clinical approach for HER2 assessment.

Figure 7.3 (a) IHC membrane stain, (b) FISH, and (c) CISH.

issue for FISH. An IHC usually costs $100 to $150 per case while it is three times higher for a FISH test [10]. Slide preparation time is long for FISH and the signals fade over time, making it difficult to archive the slides for future use. Automated and semi-automated quantification methods as available for FISH [14,15]. However, the turnaround time and cost of the test are not available for comparison.

On the other hand, the CISH test utilized bright field imaging and no special setup is required to observe the biomarker signals. In a CISH slide, the nucleus is blue, HER2 is black, and CEP17 is magenta, shown in Figure 7.3. No special training is required for the CISH test and can be performed by a technician at a usual computer desk. Slide preparation time and the cost is cheaper for CISH than FISH, more comparable to IHC slides. Archiving CISH slides is convenient as the CISH dyes fade at a slower rate. The tumor heterogeneity can be identified easily in a CISH slide, even using a low magnification image. Moreover, the tissue morphology can be observed in a CISH slide. Therefore, the insitu carcinoma and non-tumor areas can be differentiated easily from tumor areas which makes the signal counting easy in tumors and increases the reliability of the test. It can be concluded that CISH is more suitable for routine clinical applications considering its cost, preparation time, and other usability. However, there is no reliable tool available to count signals from CISH slides for automated HER2 quantification. Counting the signals manually from CISH slides is an arduous task and time consuming as well. Some commercial tools are available for CISH, but they could not differentiate singular nucleus and quantified overlapped and partially missing nuclei. Moreover, they included false-positive biomarker signals in the quantification.

Another major challenge of HER2 quantification is tumor heterogeneity. HER2 tumor heterogeneity indicates a situation where cancer has different tumor cells expressing different levels of HER2 amplification in the same patient [16]. It includes obvious HER2-positive tumor cells with HER2 negative or equivocal cells for the same case. It was found that heterogeneity affects the accuracy of 11%–13% of the HER2 tests [17,18]. The chemotherapy can also increase the HER2 gene copy number [19], making the HER2 assessment more challenging for a patient who has received chemotherapy.

7.3 WHOLE SLIDE IMAGING AND AUTOMATED IMAGE ANALYSIS

WSI system provides the platform for automated image analysis and diagnosis utilizing AI for digital pathology. WSI scanner converts the entire pathological specimen into a high-resolution digital image, mostly know as whole slide image. In digital pathology, the analysis and diagnosis are performed using this image. In traditional pathology, the diagnosis is performed based on the microscopic view of some selected areas from the specimen. The analysis depends on the pathologist and his selected areas. Consequently, the analytic result sometimes differs among pathologists. While the WSI allows scrutinization of the entire specimen using automated tools on a computer screen which improves analysis and helps reduce the bias among pathologists. Figure 7.4 shows the system of monitor-based analysis using WSI. WSI images are very large in dimensions, usually in the order of gigapixels. A pyramid shaped model is used to store the multi-resolution WSI, where each layer of the pyramid represents a different magnification or resolution of the image. The top layer provides the lowest-resolution image, and the bottom layer is for the highest-resolution. This type of system is suitable for developing an automated analysis system for digital pathology. For example, some pathological tasks can be performed using a low-resolution image like tissue segmentation while some tasks like analyzing tissue structures require a high-resolution image. A higher-resolution image provides finer details of the specimen, but it requires high time to process this huge amount of data.

Image processing and machine learning techniques can be applied on WSI to extract features, correlate, and interpret them for diagnostic decisions, which is implausible in an observation dependent diagnosis. Machine learning can be applied to WSI to automate pathological operations such as counting mitosis [20], detecting tissue artifacts [21], detecting infectious agents like asbestos, identifying cancer [22, 23], classifying cancers [24–26], and others.

However, the accuracy of the analysis largely depends on the quality of the image. WSI quality is sometimes insufficient for analysis due to scanning failure or poor specimen, mainly. Poor quality images could lead to inaccurate analysis and wrong

Figure 7.4 Analysis using WSI on a computer screen.

diagnosis. Figure 7.5 shows the impact of the focus problem on the analysis. Other quality issues include color variation, noise, and artifacts. The color of a pathology specimen can be affected by multiple sources. However, the staining protocol and the image capturing device being the most important for automated analysis. The elements of staining protocol such as reagent, temperature and staining time decides the color of the glass specimen. Then the color of the WSI is affected by the scanner characteristics. The color of the WSI can affect the automated analysis tool, especially if it relies on the color-based features of the image.

Selecting the appropriate regions is another key consideration for automated analysis tools. Most of the tools rely on the operator to select the regions while others detect the regions automatically for analysis. A group of the tools utilizes the entire WSI which is time consuming. For an automated HER2 quantification method, the region selection is crucial, especially if the tumor heterogeneity is present, illustrated in Figure 7.7.

7.4 ARCHITECTURE OF THE AUTOMATED CANCER GRADING SYSTEM

The proposed automatic HER2 grading system utilizes the bright field dual-color CISH WSI for HER2 quantification as CISH offers more user friendly and practical features. The automated cancer grading system can be described in three parts: Specimen preparation, WSI acquisition, and automated HER2 scoring as illustrated in Figure 7.8. Automated HER2 cancer grading can be affected by multiple sources such as the quality of the glass slide specimen, quality of the WSI, the color of the WSI, and location of the tumor used for the quantification.

Accurate HER2 quantification is conditional upon the quality of the glass slide preparation. Major factors that decide the quality of the specimen include the sectioning technique, thickness of the tissue, tissue digestion, washing steps, detection kit, probe, and hybridization time. Therefore, we optimized the specimen preparation protocol to keep the specimen standard which is useful for automated analysis. We prepared 4-micron thick serial section slides in the order CISH-H&E-FISH for the proposed system. We used the VENTANA Ultra Red ISH DIG detection kit and INFORM HER2 Dual ISH DNA Probe Cocktail to estimate the number of copies of the HER2 gene in CISH slides. The hybridization time was 6 hours. Failure in the

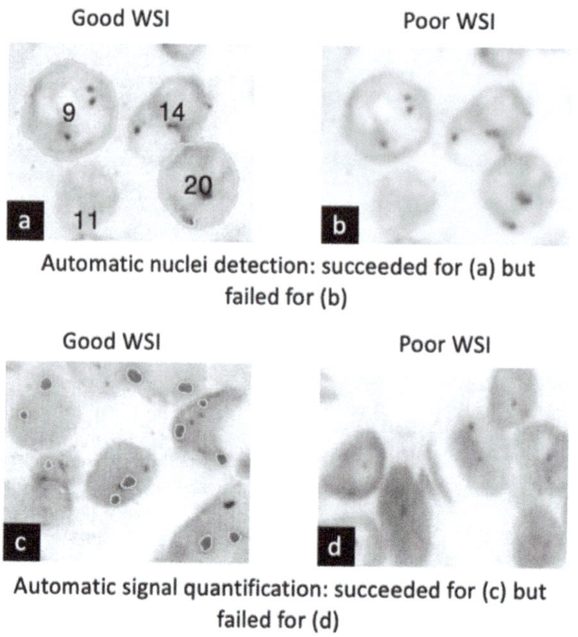

Figure 7.5 Impact of image quality on automated tools.

specimen preparation step results in a deviant glass slide which leads to inaccurate analysis or failure.

When the glass slide is prepared, it is scanned using the WSI scanner. The proposed system requires a high-resolution image to segment the biomarkers. The CISH slides were scanned using a brightfield WSI scanner at 40× magnification (numerical aperture 0.95) which provided an image resolution of 0.13 $\mu m/pixel$. However, the quality of the image is not up to the mark for analysis sometimes. A good quality image is a precondition for efficient image analysis, especially if the analysis is based on counting biological signals. Therefore, we developed a practical WSI quality evaluation method for clinical applications [27, 28]. The proposed system evaluates the quality of WSI prior to analysis to ensure only sufficient quality images are used for HER2 quantification. When the quality of WSI is confirmed, the invasive tumor regions were selected for HER2 quantification using the serial sectioned H&E WSI to exclude in situ carcinoma and non-tumor regions. The proposed system relies on the pathologist for annotating the invasive tumor regions on the H&E slide, which is a routine practice in breast pathology. We copied the annotation using an image registration application.

After that, the automatic HER2 quantification method detected the singular nuclei, HER2 gene, and CEP17 signals from the regions. The proposed system calibrates the color of the regions for reliable HER2, and CEP17 detection. The method counted HER2 and CEP17 signals for each nucleus to calculate the values of two parameters: HER2-to-CEP17 ratio and average HER2 copy number per nucleus. After that, the HER2 grade is determined based on these parameters following the ASCO/CAP 2018 guideline. The proposed system quantified at least three distinct invasive

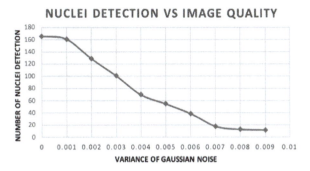

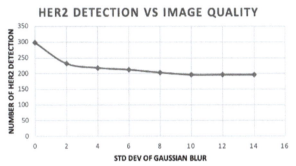

Figure 7.6 Impact of noise on nuclei detection method (top) and focus blur on HER2 detection (bottom).

regions to derive the HER2 status. The number of distinct regions was higher for the cases with tumor heterogeneity. Visualizing the quantified data (e.g., nuclei detection, HER2, and CEP17 detection) is an important property of the system, it allows the pathologist to observe and verify the score. The system also generates a diagnostic report and exports the results into the laboratory information system (LIS) when the operator approves.

The proposed system uses a CISH WSI image for HER2 quantification [13]. However, the system was also validated for HER2 quantification using FISH WSI by optimizing few parameters. The FISH slides were scanned using a confocal WSI scanner at 40× (NA 1.25) with water immersion which gave an image resolution of 0.165 $\mu m/pixel$. The system used the extended focus layer image for the FISH HER2 quantification.

7.5 QUALITY EVALUATION OF WSI

7.5.1 Image Quality Issues for Automated Analysis

The quality of automated analysis highly depends on the quality of the image used. With the advent of many robust and fast WSI scanners, the WSI is getting more popular among pathologists and practitioners for routine clinical works. The achievable limit of quality has improved significantly over the years for WSI. However, the

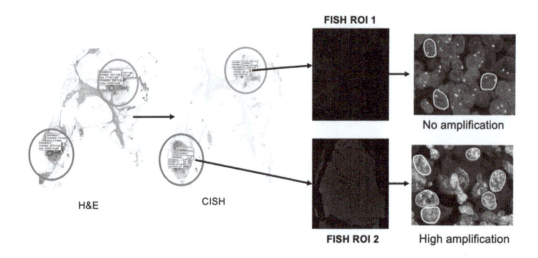

Figure 7.7 Different regions show different levels of HER2 amplification.

occasional failures in quality remain an issue up to the present time. Scanning failure and faulty specimens are the principal sources of quality failures in WSI.

Scanning failures introduce out of focus regions and noises in the image which trouble the automated image processing or machine learning algorithms. The WSI scanners utilize an optimal number of focus points, selected manually or automatically from the tissue surface to convert the specimen into a digital image. However, the tissue surface is uneven and focus depth varies from region to region. If a focus point is selected from a region having a different depth compared to its neighboring regions, the neighboring regions could suffer from focus blur problems. Therefore, the WSI sometimes has out of focus regions. Focus error can also be introduced due to errors in stage alignment and optical alignment.

The faulty specimen contains tissue artifacts such as tissue folds, tissue wrinkles, tissue tears, air bubbles, tissue dirt, pen marks, and others. Tissue artifacts are produced due to the errors that happened while preparing the specimen. Tissue folds and air bubbles are the most common artifacts found in pathology. Tissue fold is the overlapping of tissue which is usually caused by faulty bleed or fatty tissue. The air bubble is air trapped under the tissue cover, produced during flotation, and mounting. Such artifacts hide or alter significant features which may lead to inaccurate analysis. For example, tissue-fold regions stock more dye than normal regions, making them look like tumors.

Both scanning failures and failures in specimen preparation degrade the image quality. However, if the quality degradation is caused by scanning failures such as focus problem or noise it can be resolved by rescanning the specimen. But rescanning could not help improve the quality of artifact-affected regions. An intelligent approach is to detect such artifacts and eliminate them from analysis for a risk-free diagnosis. We proposed a practical quality evaluation method for automated analysis using WSI. This method detects useless regions of the image (i.e., tissue artifacts and

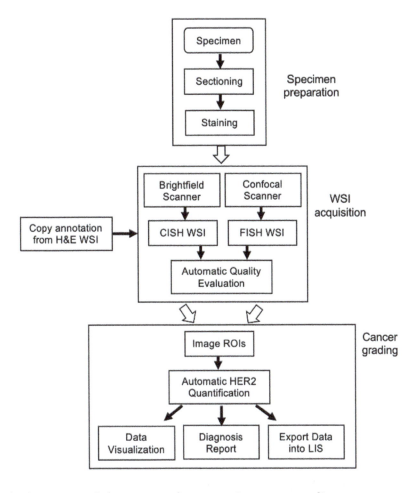

Figure 7.8 Architecture of the proposed automatic cancer grading system.

glass areas) and then evaluates the image quality of the WSI excluding the useless regions. If the image quality is sufficient then it is used for analysis, otherwise, the slide is recommended for rescanning. The quality evaluation method communicates with the analysis method to exclude the useless regions from the analysis. Existing image quality evaluation methods treat artifacts affected regions in the same way as normal tissue. Thus, they could not be utilized for selecting slides for analysis or rescanning. The proposed method enables reference-free evaluation as practically it is not feasible to use a reference image in pathology. This method utilizes the low-resolution WSI for detecting tissue artifacts while the high-resolution image for quality evaluation and HER2 quantification. Using a low-resolution WSI for artifact detection enables faster detection with optimal accuracy. While the quality evaluation and HER2 quantification require a high-resolution image. Utilizing multi-resolution WSI for different tasks according to the need makes the system efficient for practical use.

7.5.2   Artifact Detection for Quality Evaluation

We detected tissue fold and air bubbles as these are the most common artifacts in pathology. The proposed method utilized low-resolution WSI for detecting artifacts using support vector machines (SVMs). Two binary SVM classifiers were trained using a set of selected features: One for detecting tissue fold and the other for air bubbles. A set of 9 suboptimal features were selected for each classifier from a bank of 126 features which includes conventional GLCM features, rotation invariant GLCM features, and other features derived based on the physical properties of artifacts. We utilized the sequential feature selection (SFS) method which provides a suboptimal feature set by evaluating its performance for the classification. The features were ranked, and a set of features selected based on the average classification error in a 10-fold cross-validation experiment. We relied on the SFS as it is computationally faster than the optimal feature selection method such as the branch and bound method, especially when the feature space is large. The orientation of artifacts may rotate at an arbitrary angle from slide to slide, thus we trained the SVM classifiers to achieve rotation invariant classification. In our experiment, we have found that using rotated images for augmenting the training data results in better accuracy for the classifier than using rotation-invariant features. Thus, we trained the classifiers using the selected features derived from original images and their rotated versions. Each image was rotated in 15-different angles starting from 15 to 345 degrees.

We experimented with 1× and 5× magnification and analyzed the effects of resolution on artifact detection accuracy. Using 5× magnification instead of the 1× image improves the artifact detection accuracy but the detection time increases significantly. In the 4-fold cross validation, the average classification accuracy was 99.40% for 5× magnification WSI compared to 98.11% for 1× magnification. The detection time varied from 4 to 10 seconds for 1× magnification depending on the WSI dimensions while it ranged from 14 to 50 sections. For this experiment, we used 5 WSIs and a personal notebook with a 2.6 GHz Intel Core i5 processor (without any external graphics card) for detecting the artifacts using the proposed method.

We also experimented with traditional machine learning methods such as SVM and deep learning methods such as CNN for detecting artifacts when a limited number of samples is available. The accuracy was 98% for detecting tissue fold and 97% for air bubbles using SVM. While using CNN the best accuracy was 87.33% and 92.17% for tissue-fold and air bubble detection, respectively. In the CNN experiment, we fine-tuned the final layers of a VGG16 model which was pre-trained on the ImageNet dataset. We utilized data augmentation for training the model and then investigated the model's performance for different epochs, batch sizes, learning rates, dropouts, and optimizers. The result of the study suggests that the SVM classifier with optimally selected features yields more accurate and stable classification than CNN-based model when the amount of training data is limited. Moreover, SVM is simple to implement and computationally fast.

The proposed method used 1× magnification image and divide the WSI into non-overlapped images blocks of 100×100 pixels. Then two different SVM classifiers detected tissue-fold and air bubble artifacts for these blocks. Image blocks containing either of the artifacts were eliminated from quality evaluation and analysis. The coordinate location of blocks at 1× magnification was manipulated to trace artifacts at higher magnification such as 20× for excluding them from quality evaluation and analysis.

7.5.3 Scoring WSI Quality

The proposed method was designed to evaluate WSI quality at higher magnification (i.e., 20× or 40×). The proposed method was demonstrated using a 20× WSI and then the procedure to evaluate 40× is explained in this chapter. At first, this method divides the entire WSI into non-overlapped and fixed-size blocks. Then, it eliminates the useless blocks such as artifact-affected blocks and glass blocks. An image block with more than 75% white pixels is considered a glass block and eliminated for further work. A pixel is considered white if its intensity is higher than 200 in grayscale. After that, the method scores each image block except the artifact-affected blocks and glass blocks. The proposed method utilizes the localization information of artifacts and glass blocks of low magnification to trace them at higher magnification. In the experiment, 20× magnification WSIs were used for quality evaluation and the block size was 2000 × 2000 pixels which made it easier to integrate the artifact detection results at higher magnification to eliminate the detected blocks. For the rest of the blocks, the quality was scored to estimate the quality of the WSI.

The proposed method utilizes two metrics for judging the quality of image blocks which are blur index and noise index. The blur index is obtained by measuring the distance between the local maximum and local minimum of edge pixels. The local maximum gradient is detected for each edge's direction. If this gradient is higher than a predefined threshold, then it is considered as an edge pixel and the width is estimated for that edge; illustrated in Figure 1.9. Then, the total width of all edges is divided by the number of edges which gives the blur index. Out of focus regions have small gradients on edge pixels compared to the sharp regions. Thus, a smaller value of the blur index indicates that the image is sharper. This method considers a pixel as noise if it is independent of its surrounding pixels and not belongs to an edge. Noises and edges of the image are detected by filtering. Then, the minimum difference between the center pixel and surrounding pixels in a 3 × 3 pixel window is calculated at all pixels to filter the edge pixels. After that, the average value of these minimum differences is used to quantify noise in the image. The higher value means that the image seems to have more noise or excess sharp pixels.

After that, the quality degradation index of a block is derived using a linear regression model

$$q = \alpha + \beta s + \gamma n \qquad (7.1)$$

where q is the quality degradation index for image blocks, s is sharpness degradation or blur index, n is noise index, and α, β, and γ are coefficients of prediction. The coefficient of prediction was derived by training regression model where the mean square error (MSE) between the original image (20×) and its degraded version was in place of the quality degradation in Equation (7.1). The magnification of the training images should be the same as the magnification of WSI, we want to evaluate. If we want to evaluate the WSI at 40×, the coefficients should be derived using 40× training images. After that, each image block is classified as a good, average, or poor block by applying thresholds on the quality degradation index. Finally, the percentage of poor blocks in the WSI is used to select a slide for analysis or rescan. This method shows high concordance with the subjective assessment of pathologists and objective assessment using MSE for pathological images.

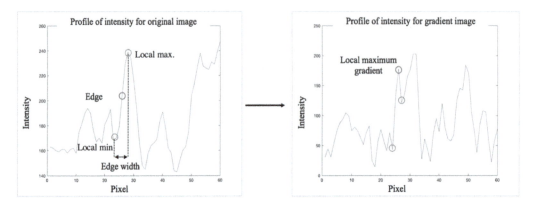

Figure 7.9 Edge width calculation for deriving the sharpness degradation angle.

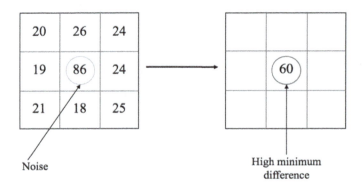

Figure 7.10 Identifying a typical noise pixel.

7.6 AUTOMATED HER2 SCORING FOR CANCER GRADING

7.6.1 Singular Nuclei Detection

The proposed method utilizes machine learning for detecting singular nucleus. A non-singular nucleus contains overlapped regions or partially missing regions. Including non-singular nuclei in the quantification compromises the reliability of the HER2 score as it misleads localizing and counting signals. Therefore, the proposed method selected the nuclei for quantification in two stages: The candidate detection stage and the selection stage. The candidate detection stage includes all possible singular nuclei which are then scored in the selection stage. Only high-score nuclei are used to make the quantification robust and reliable. In the earlier version, the proposed system utilized SVM for nuclei detection which is replaced by U-net based segmentation in the updated system. In a CISH specimen, three dyes decide the color of the nucleus, HER2 and CEP17 signals. In nuclei areas, the blue dye is prominent while it is black for HER2 and magenta for CEP17 signals. This method utilizes the color unmixing technique to separate the dye channels from the RGB image of CISH. The U-net was trained to differentiate background, boundary, and inner nucleus pixels from a color unmixed blue channel. The output of the U-Net is the probabilities of each pixel being background, boundary, or inside the nuclei.

In the FISH specimen, the crosstalk of dyes was not significant to differentiate them. Therefore, the method utilized grayscale extended layer image of nuclei channel for nuclei detection. After the U-net segmentation, each candidate is scored based on its shape and size feature. Scrutinizing cancerous cells with larger nuclei sizes is a common practice in cytopathology and the shape is useful to differentiate a complete singular nucleus from partially missing or jointed nuclei. The non-singular nucleus as shown in Figure 7.11b,c is not suitable for HER2 quantification.

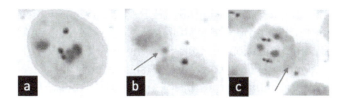

Figure 7.11 (a) Singular nucleus has no overlapped or missing parts, (b) partially missing, and (c) overlapped nuclei.

7.6.2 HER2 and CEP17 Detection

The HER2 and CEP17 signals are detected in four steps: Background separation, noise reduction, splitting jointed signals, and scoring the signals. For CISH quantification the color unmixed dye channels are used for signal detection. In the first stage, the black HER2 dye and magenta CEP17 dye channels are obtained using the color unmixing method which serves as the background for segmenting the HER2 and CEP17 signals, accordingly. In the second stage, noise is removed. Two types of noises were encountered for detecting the signals: Color noise and white noise. For example, in the CEP17 dye image, the magenta CEP17-dye is overlapped by blue nuclei dye or black HER2 dye. These are the color noise for CEP17 detection and are excluded from quantification as they are weak signals. In the HER2 dye image, the color noise is produced by overlapping the black dye with magenta or blue dye. White noise includes the pixels having high value in all channels of RGB images. When the noise is removed, the partly jointed signals are separated utilizing distance transformation and watershed algorithm. After that, each candidate signal is scored based on the likelihood of its corresponding dye amount. A signal with a high likelihood score is selected for quantification. In this way, the proposed method ensures that only prominent signals are used for quantification and weak as well as false-positive signals are rejected.

In the case of FISH, the extended focus image of CEP17 and HER2 channels is used as the background. The interference between dyes is negligible in FISH, therefore the nature of the noise is different here than in CISH. The noise is low intensity pixels which are removed by applying a threshold, derived based on the distribution of pixel intensities of the grayscale image. After that, partly jointed signals are separated. Then each signal is scored based on its pixels' intensity.

In the case of HER2 detection. The third step which separates the jointed signals were ignored as the HER2 signal may form a cluster of multiple signals. Instead, we developed a strategy that identifies HER2 clusters and then counts the number of signals in the cluster. The average size of HER2 signals is utilized to identify a cluster and then count the number of signals in it.

7.6.3   Signal Quantification for HER2 Scoring and Cancer Grading

In 2007, ASCO/CAP provided a guideline to determine the HER2 amplification grade, realizing the impact of inaccurate HER2 assessment which was updated in 2013 and later in 2018. The proposed system followed the ASCO/CAP 2018 guidelines to score the HER2 amplification and classify the HER2 cancer grade. At first, the proposed method counts HER2 and CEP17 signals for the selected nuclei. Then sorted the nuclei in descending order based on the HER2-CEP17 difference, nuclei with less than 2 CEP17 or HER2 signals were not included. Then calculated the HER2-to-CEP17 ratio and average HER2 copy number per nucleus for at least 20 nuclei taking from the top of the sorted list. After that, the HER2 grade is determined based on the ASCO/CAP chart, given in Table 7.2. For the ISH equivocal cases, an additional IHC test is recommended. If the reflex IHC test score is 2+, then ISH quantification is repeated using a higher number of nuclei such as 40 or 60. After that, the ISH result is considered final. If the reflex IHC score is 0/1+/3+, the result of the ISH score is considered resolved by IHC. The decision for providing HER2-targeted therapy is based on the HER2 grade of the patient.

TABLE 7.2   HER2 Grading Criteria

Parameter Scores	HER2 Quantification Report Based on ASCO/CAP Guide [2018]	HER2 Cancer Grade
HER2/CEP17 ≥ 2.0 Avg. HER2 copy ≥ 4.0	ISH Positive	Group 1
HER2/CEP17 ≥ 2.0 Avg. HER2 copy < 4.0	ISH equivocal	Group 2
HER2/CEP17 < 2.0 Avg. HER2 copy ≥ 6.0	ISH equivocal	Group 3
HER2/CEP17 < 2.0 Avg. HER2 copy ≥ 4.0 and < 6.0	ISH equivocal	Group 4
HER2/CEP17 < 2.0 Avg. HER2 copy < 4.0	ISH Negative	Group 5

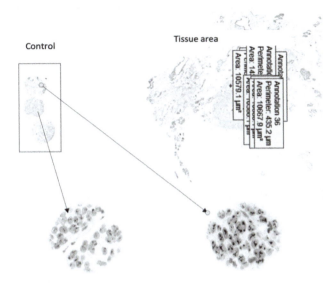

Figure 7.12 Color correction of HER2 and CEP17 signals using a control in the slide.

7.6.4 Color Correction for HER2 Scoring

Color is an important property for automated image analysis using WSI and variation in color can cause trouble for the analysis tools. In the WSI system, the color variation can be caused by the thickness of the specimen, staining concentration, scanner settings, WSI visualization software, or the display device. The proposed system utilizes the RGB values of pixels to score the signals. Therefore, it is necessary to apply color correction prior to signal detection and scoring. Automatic color correction is a common feature of automated image analysis systems. Two most common approaches of color correction in WSI are color calibration using a color chart and color transformation based on a reference image. The color chart-based method is intended to compensate for the color variations introduced by the scanners. While the other method is focused on the scanning issues and utilizes the shape of color distribution of a reference image. In the proposed system, slide controls are utilized to calibrate the color of the slide. Controls are circular patches of the specimen where the staining is standard. The proposed system calibrated the color of signals utilizing the color correction method proposed by Murakami et al. [29]. However, instead of applying the color correction to the entire WSI, the system estimates the color corrected values for pixels using the loop-up table and the corrected pixels are utilized scoring and selecting the signals. This approach saves time for estimating the corrected values.

7.7 PROCEDURES AND RESULTS OF THE VALIDATION EXPERIMENT

In our study, we demonstrated the proposed system on 36 invasive BC patients. The study was approved by Institutional Review Board (IRB#17-287). These patients were selected randomly and diagnosed with the IHC test, and it scored 5 patients

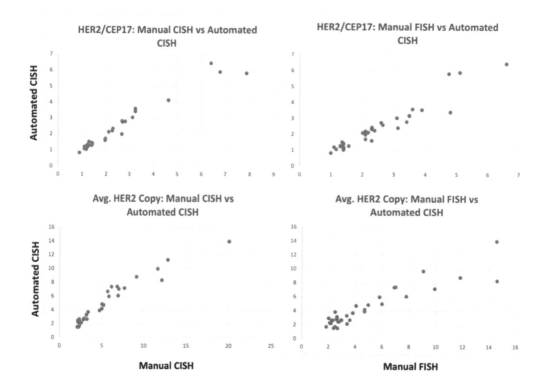

Figure 7.13 The correlation between automated HER2 quantification and pathologists' manual counts.

as negative, 8 as positive, and 23 as equivocal. Patients with all IHC classes were included in the validation study to ensure that the proposed system can grade HER2 amplification for all types of BC patients. We prepared the CISH and FISH slides following the optimized protocol. Then two expert breast pathologists manually counted the signals from CISH and FISH slides to score the HER2 amplification. Then the slides were scanned to obtain the WSI for automated quantification using the proposed system. The proposed system was validated for automated HER2 grading using CISH WSI by comparing the results with the pathologist's manual CISH and FISH counts, shown in Figure 7.13.

The HER2 quantification was performed using CISH WSI. The CISH HER2 quantification scores were compared with the pathologist's manual CISH counts. The correlation for HER2-to-CEP17 ratio between automated CISH quantification and manual CISH count was 0.97 for 29 patients. Manual CISH count was not available for 7 patients. The correlation was 0.96 for the average HER2 copy number. Then, we compared the automated CISH scores with the manual FISH counts. The correlation was 0.96 and 0.93 for the HER2-to-CEP17 ratio and average HER2 copy number, accordingly for 36 patients. The high positive correlation of the proposed method with manual FISH and CISH counts ensures the effectiveness of the proposed system for automated HER2 quantification using the CISH slide.

The proposed system was also demonstrated for automated HER2 quantification using FISH WSI. The automated FISH quantification was performed for 7 BC patients which were diagnosed as equivocal using prior IHC test. After that, the automated FISH quantification results were compared with the pathologist's CISH and FISH counts. The correlation for HER2-to-CEP17 ratio and average HER2 copy number between the automated FISH and manual FISH were 0.97 and 0.96, accordingly. The correlation for HER2-to-CEP17 ratio and average HER2 copy number were 0.98 and 0.99, accordingly in the comparison between automated FISH and manual CISH comparison. The findings of this study indicate that the proposed system can be utilized also for automated FISH quantification.

The proposed system had 100% agreement in Cohen's kappa statistic in terms of HER2 grading for both automated CISH and FISH quantification while comparing with manual CISH and FISH counts. However, we relied on the Bland-Altman statistic to confirm the efficacy of the proposed HER2 quantification system for replacing the existing system in the hospitals. The Bland-Altman plot, also known as the difference plot, measures the agreement between two quantitative assays and it is a popular choice for validating a new clinical technique by comparing it against the standard practice. The proposed system has low bias and a narrow limit of agreement with manual FISH counts which is the standard clinical practice for HER2 assessment. The proposed automated CISH quantification had a low bias of 0.05 with a 95% confidence interval -0.07 to 0.17 and a narrow limit of agreement from -0.48 to 0.58 with manual FISH counts. For automated FISH quantification, the bias was 0.01 and the limit of agreement was from -0.69 to 0.72. The result of the study demonstrates that the proposed system determines the HER2 grade efficiently for a BC patient regardless of its prior IHC score.

The time requirements of the proposed HER2 quantification method were investigated. The turnaround time for the current version of the application ranged from 1 to 7 minutes for HER2 quantification using the CISH WSI. The turnaround time was less than a minute if the quantification is performed using a FISH slide. The time was estimated using a personal notebook with a 2.6 GHz Intel Core i5 processor without any external graphics card. The quantification time depends on several factors which include the area of the quantified regions, nucleus density in image, signal density, and the image quality.

7.8   APPLICATION DEVELOPMENT AND SYSTEM INTEGRATION

The proposed system firstly evaluates the quality of the WSI. WSI of inadequate quality is selected for rescanning and good quality WSIs are stored in the image server for further use. The proposed system communicates with the image server to access a WSI for automated quantification. The proposed system quantifies the selected regions of WSI after excluding the useless artifacts and glass blocks. Currently, the system relies on the pathologist's annotation for selecting suitable regions for HER2 quantification.

An in-house application was developed using *Python* and OpenCV to implement the proposed system for clinical operation. The graphical user interface (GUI) of the

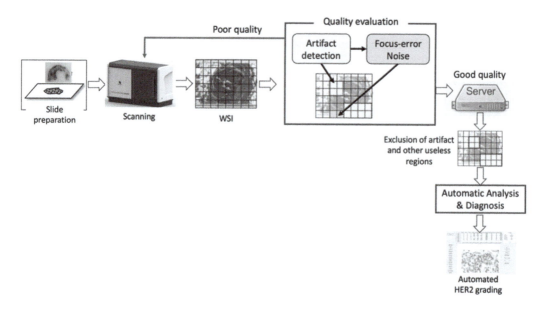

Figure 7.14 Workflow of proposed system.

software was developed using *Python* Tkinter. The application utilizes the server-client model and is integrated with Hospital's Laboratory information system (LIS). It generates a diagnostic report and visualization data for the quantification. The quantification results and data are stored in the LIS system and available for consultation.

Feedback from the pathologists is crucial for implementing the system for routine work. The application was developed and updated based on their comments to make it convenient for practical use. The application allows to pause current quantification and resume it later using a unique identifier of the case. It allows to search and group cases based on the HER2 result such as positive or negative, quantification date, and quantification types such as FISH or CISH. Moreover, it allows pathologists to include or exclude regions for quantification, to remove any nucleus for the result or manually include any nucleus to the quantification. It also provides warnings or notifications for possible tumor heterogeneity, HER2 clusters, HER2 dye noise, and other situations. A user can also provide emoji-based feedback and comments about the system using the application GUI. Figure 7.14 shows the structure of the system. This structure of the proposed system can be utilized for other automated image analysis and diagnosis applications. For example, mitosis detection or other analysis applications can be added under the automated analysis and diagnosis module by updating the system accordingly. A suitable magnification can be selected depending on the requirement of a particular analysis. The definition of useless components could change beyond tissue artifacts and glass blocks depending on the analysis. For example, some analyses may require detecting and excluding broken tissue blocks.

7.9   CONCLUSION

Existing commercial tools designed to quantify the CISH slides fail to detect singular nucleus and include non-singular nucleus in the quantification. They also failed to detect reliable CEP17 and HER2 signals and quantified false positives in the primary experiment. The proposed includes a limited number of singular nuclei which are suitable for quantification and prominent signals with high likelihood scores for accurate analysis. The nuclei and signal detection algorithms of the proposed system were validated by comparing it with pathologists' manual nuclei and signal selection.

Currently, the IHC test is performed followed by the manual FISH test as the reflex assessment to determine the HER2 grade of a patient. However, the FISH test is not performed by the breast pathologist as it requires a long time and more effort. This chapter presents an automated HER2 grading system using CISH WSI which requires less effort and has a significantly low turnaround time compared to the manual FISH counting. Moreover, this quantification does not require any training or special work environment making the system suitable for a breast pathologist. The system was validated for HER2 quantification using CISH slides by comparing it with the pathologist's manual CISH count and FISH counts. This ensures the reliability of the system. In addition, the system was demonstrated for FISH slides to confirm its validity. Thus, the proposed system allows the laboratories to select the quantification option freely depending on their convenience. Clinically, the FISH test is performed only for equivocal cases (IHC 2+), and one major reason is its cost. The proposed system can be utilized for routine HER2 assessment to verify the IHC results for all invasive BC patients as the IHC test result is not conclusive for all cases and proposed quantification is cost and time efficient, comparable to the IHC test.

However, there remain some scopes of improvement in the system. Sometimes, the CISH WSI is affected by the HER2 background noise which has similar properties to the HER2 signals. The proposed method can detect the presence of the noise in the slide but could not separate them from quantification if it occurs inside a nucleus. The proposed system was utilized for the cell block CISH images having higher thickness than 4 microns. The cell block images exhibit more nuclei overlapping than usual. Finding singular nuclei is a challenge for cell block images. Multinucleation is another issue faced while quantifying the cell block images. The nuclei detection accuracy falls for cell block CISH images. Currently, the system relies on the pathologist to select the regions on the H&E slide for quantification. In the future, automatic invasive cancer detection and representative region selection methods can be integrated with the system to make it more practical. The turnaround time could be optimized further to quantify a CISH case within a minute.

Bibliography

1. Howlader N, Noone AM, Krapcho M, et al. (eds). SEER Cancer Statistics Review, 1975–2017, National Cancer Institute. Bethesda, MD, https://seer.cancer.gov/csr/1975_2017/, based on November 2019 SEER data submission, posted to the SEER website, April 2020.

2. American Cancer Society. How Common Is Breast Cancer? Jan. 2021. Available at: https://www.cancer.org/cancer/breast-cancer/about/how-common-is-breast-cancer.html.

3. American Cancer Society. Breast Cancer Facts & Figures 2019–2020. Atlanta: American Cancer Society, Inc. 2019

4. Owens MA, Horten BC, Da Silva MM. HER2 amplification ratios by fluorescence in situ hybridization and correlation with immunohistochemistry in a cohort of 6556 breast cancer tissues. Clin Breast Cancer 2004; 5:63–69. PMID: 15140287

5. Yaziji H, Goldstein LC, Barry TS, Werling R, Hwang H, Ellis GK, et al. HER-2 testing in breast cancer using parallel tissue-based methods. JAMA 2004; 291:1972–1977. PMID: 15113815

6. Slamon DJ, Leyland-Jones B, Shak S, et al. Use of chemotherapy plus a monoclonal antibody against HER2 for metastatic breast cancer that overexpresses HER2. N Engl J Med 2001; 344:783–92.

7. Tan-Chiu E, Yothers G, Romond E, et al. Assessment of cardiac dysfunction in a randomized trial comparing doxorubicin and cyclophosphamide followed by paclitaxel, with or without trastuzumab as adjuvant therapy in node-positive, human epidermal growth factor receptor 2-overexpressing breast cancer: NSABP B-31. J Clin Oncol. 2005; 23(31):7811–7819.

8. Liberato NL, Marchetti M, Barosi G. Cost effectiveness of adjuvant trastuzumab in human epidermal growth factor receptor 2-positive breast cancer. J Clin Oncol. 2007; 25:625–633 [PubMed: 17308267]

9. Kurian AW, Thompson RN, Gaw AF, Arai S, Ortiz R, Garber AM. A cost-effectiveness analysis of adjuvant trastuzumab regimens in early HER2/neu-positive breast cancer. J Clin Oncol. 2007; 25:634–641. [PubMed: 17308268]

10. Carlson B. HER2 Tests: How do we choose? Biotechnol Healthc. 2008 Sep; 5(3):23–27.

11. A. C. Wolff, M. E. H. Hammond, K. H. Allison, B. E. Harvey, P. B. Mangu, J. M. S. Bartlett, et.al. Human epidermal growth factor receptor 2 testing in breast cancer: American Society of Clinical Oncology/College of American Pathologists clinical practice guideline focused update, Arch Pathol Lab Med. 2018 Nov; 142(11):1364–1382.

12. Paik S, Kim C, Wolmark N. HER2 status and benefit from adjuvant trastuzumab in breast cancer. N Engl J Med 2008; 358: 1409–11.

13. Md Shakhawat Hossain, Matthew G. Hanna, Naohiro Uraoka, Tomoya Nakamura, Marcia Edelweiss, Edi Brogi, et al. Automatic quantification of HER2 gene amplification in invasive breast cancer from chromogenic in situ hybridization whole slide images. J Med Imaging. 2019 Oct;6(4): 047501

14. J. Konsti et al., A public-domain image processing tool for automated quantification of fluorescence in situ hybridization signals. J. Clin. Pathol. 2008; 61(14.5), 278–282.

15. D. Furrer, S. Jacob, C. Caron, et al., Validation of a new classifier for the automated analysis of the human epidermal growth factor receptor 2 (HER2) gene amplification in breast cancer specimens. Diagn. Pathol. 2013; 8, 17.

16. HannaWM, Rüschoff J, Bilous M., et al. HER2 in situ hybridization in breast cancer: clinical implications of polysomy 17 and genetic heterogeneity. Mod Pathol 2014; 27: 4–18

17. Brunelli M, Manfrin E, Martignoni G, et al. Genotypic intratumoral heterogeneity in breast carcinoma with HER2/NEU amplification: evaluation according to ASCO/CAP criteria. Am J Clin Pathol 2009; 131: 678–82.

18. Seol H, Lee HJ, Choi Y, et al. Intratumoral heterogeneity of HER2 gene amplification in breast cancer: its clinicopathological significance. Mod Pathol 2012; 25: 938–48

19. Valent A, Penault-Llorca F, Cayre A, Kroemer G. Change in HER2 (ERBB2) gene status after taxane-based chemotherapy for breast cancer: polyploidization can lead to diagnostic pitfalls with potential impact for clinical management. Cancer Genet 2013; 206: 37–41.

20. Tellez D, Balkenhol M, Otte-Holler I, et al. Whole-slide mitosis detection in H&E breast histology using PHH3 as a reference to train distilled stain-invariant convolutional networks. IEEE Trans Med Imaging 2018; 37: 2126–2136.

21. Bautista, Pinky & Yagi, Yukako. Detection of tissue folds in whole slide images. Conference Proceedings: Annual International Conference of the IEEE Engineering in Medicine and Biology Society. IEEE Engineering in Medicine and Biology Society. Conference 2009; 3669–3672. 10.1109/IEMBS.2009.5334529.

22. Alanazi SA, Kamruzzaman MM, Islam Sarker MN, Alruwaili M, Alhwaiti Y, Alshammari N, Siddiqi MH. Boosting breast cancer detection using convolutional neural network. J Healthc Eng. 2021 Apr 3; 2021:5528622. doi: 10.1155/2021/5528622. PMID: 33884157; PMCID: PMC8041556.

23. Cruz-Roa, A., Basavanhally, A., González, F., Gilmore, H., Feldman, M., Ganesan, S., Shih, N., Tomaszewski, J., & Madabhushi, A. (2014). Automatic detection of invasive ductal carcinoma in whole slide images with convolutional neural networks. Medical Imaging 2014: Digital Pathology. https://doi.org/10.1117/12.2043872

24. Iizuka, O., Kanavati, F., Kato, K, et al. Deep learning models for histopathological classification of gastric and colonic epithelial tumours. Sci Rep 2020; 10:1504. https://doi.org/10.1038/s41598-020-58467-9

25. J. Ye, Y. Luo, C. Zhu, F. Liu and Y. Zhang, Breast cancer image classification on WSI with spatial correlations. ICASSP 2019–2019 IEEE International Conference on Acoustics, Speech and Signal Processing (ICASSP), 2019, 1219–1223, doi: 10.1109/ICASSP.2019.8682560.

26. Baris Gecer, Selim Aksoy, Ezgi Mercan, Linda G. Shapiro, Donald L. Weaver, Joann G. Elmore. Detection and classification of cancer in whole slide breast histopathology images using deep convolutional networks, Pattern Recognition, 2018; 84:345–356, ISSN 0031-3203,

27. Hossain M. Shakhawat, Tomoya Nakamura, Fumikazu Kimura, Yukako Yagi, Masahiro Yamaguchi. Automatic quality evaluation of whole slide images for the practical use of whole slide imaging scanner, ITE Trans. on MTA, 2020; 8(4):252–268.

28. Hossain M. Shakhawat, Tomoya Nakamura, Fumikazu Kimura, Yukako Yagi, Masahiro Yamaguchi. Practical image quality evaluation for whole slide imaging scanner, The 4th Biomedical Imaging and Sensing Conference 2018 (BISC2018), Proceedings of SPIE, 10711, 107111S, Apr. 2018.

29. Murakami Y, Abe T, Hashiguchi A, Yamaguchi M, Saito A, Sakamoto M. (2013). Color correction for automatic fibrosis quantification in liver biopsy specimens. J Pathol Inform 2013; 4:36

## CHAPTER 8

Prognostic Role of CALD1 in Brain Cancer: A Data-Driven Review

S. M. Riazul Islam

Department of Computer Science, University of Huddersfield, Huddersfield, UK

Subbroto Kumar Saha

Department of Biochemistry and Molecular Medicine, University of California, Davis, USA

Afsana Nishat

Microbiology and Cell Science, University of Florida, USA

Shaker El-Sappagh

Faculty of Computer Science and Engineering, Galala University, Suez, Egypt

CONTENTS

T HE CALDESMON GENE (CALD1) is involved in many cellular functions and has recently attracted attention in cancer due to its roles in cell migration, invasion and proliferation. Some researchers have found the correlation between CALD1 expression and prognosis of Glioblastoma. However, the application of CALD1-mediated treatment in a clinical setting and the use of their expression levels as prognostic markers of human cancers is still in its infancy. In this context, this chapter reviews the significance of CALD1 expression in brain cancer using a number of online bioinformatics platforms and tools. The chapter investigates the expression patterns, functions, and prognostic values of CALD1 in brain cancer by accessing and analyzing all currently available gene expression data. This systemic review will eventually be helpful to gain some insights into whether CALD1 expression can be used as a biomarker for the prognosis of brain cancer.

8.1 INTRODUCTION

In 2020, nearly 24,000 people received a brain cancer diagnosis in the United States alone and this cancer is the ninth leading cause of cancer-related mortality. Treatment of malignant tumors of the brain has remained a challenge for a long time. Detection of brain cancer at an early stage can be highly useful in attempting to cure them. Molecular-targeted treatments have recently transformed the therapeutic approach for different tumors. To adopt a targeted therapy for the treatment of brain cancer patients, it is important to better understand the status of various molecular processes including gene expression and methylation of the related genes. Various genes such as Fibronectin 1 (FN1), Nicotinamide phosphoribosyltransferase (NAMPT), and CALD1 are mutated in brain cancers [1–3]. Whereas FN1 is a well-known biomarker for brain cancer diagnosis, NAMPT is not expressed in all the studies. In contrast, although CALD1 is overexpressed in all the studies, it is not studied comprehensively from a data-driven perspective. Systematic reviews are helpful to understand the prognostic role of genes in human cancers and can help enhance our realization in targeted therapy [4,5]. In this chapter, we will review the roles of CALD1 in brain cancer.

CALD1 performs as a cytoskeleton-associated protein and regulates cell morphology and motility via actin filaments modulation [6]. The transcriptional variance of the CALD1 gene consists of 15 exons and is characterized by the recombination of alternative splicing modes. Different isoforms of CALD1 are responsible for the distinct functions of cell types. Earlier it was reported that CALD1 expression is restricted to vessel architectures in glioma [7]. Another study reveals that CALD1 transcriptomic level in glioma patient-derived tumor cells is associated with glioma grades progression [3]. Liu et al. [8] showed that CALD1 can be considered a prognostic biomarker and is correlated with immune infiltrates in gastric cancers. The association of CALD1 is also evident in bladder cancer [9].

Despite a good volume of related research, the application of CALD1-assisted targeted therapy is still in the early stages, and the use of its expression level as a prognostic marker in brain cancer is an area of active investigation. In this chapter, we thoroughly reviewed the biomarker utility and prognostic significance of CALD1 in human brain cancer using multiomics analysis. We comprehensively analyzed CALD1 expression pattern, various functions, and different prognostic impacts on brain cancer using all currently available gene expression data. This multiomics analysis eventually demonstrated that CALD1 expression can be embraced as a biomarker for the prognosis of BC patients.

The rest of this chapter is organized as follows. Section 8.2 provides the materials and methods. Whereas the multiomics analysis and results are presented in Section 8.3, the final section concludes the chapter.

8.2   MATERIALS AND METHODS

In this section, we provide the details of online bioinformatics tools used in this study.

8.2.1   ULCAN Web

CALD1 mRNA and protein expression were examined with TCGA level 3 RNA-seq and clinical proteomic tumor analysis consortium (CPTAC) data, respectively, using ULCAN web (ualcan.path.uab.edu) [10]. A student's t-test was used to analyze the data, and a p-value of 0.05 was considered significant. Positively co-expressed genes of CALD1 in GBM were retrieved from the ULCAN web. Co-expression coefficient r value > 0.30 and $P \leq 0.01$ was considered to extract co-expressed genes in this analysis.

8.2.2   GEPIA2 Web

CALD1 mRNA expression in GBM and normal tissue was examined with TCGA level 3 and genotype tissue expression (GTEx) RNA-seq data using GEPIA2 web (gepia2.cancer-pku.cn) [11]. A student's t-test was used to analyze the data, and a p-value of 0.05 was considered significant.

8.2.3   Human Protein Atlas

The expression of CALD1 protein in GBM tissues and normal tissues was examined using immunohistochemistry (IHC) staining data from the Human Protein Atlas (proteinatlas.org) [12]. IHC staining was performed using the HPA017330 antibody targeting CALD1.

8.2.4   cBioPortal Web

Mutations and Copy number alterations (CNAs) of CALD1 in GBM were examined using TCGA datasets via the cBioPortal web (cbioportal.org) [13]. Also, the co-relationship between CALD1 mRNA expression and CNAs was analyzed using the cBioPortal web.

8.2.5   Tumor IMmune Estimation Resource (TIMER) Web

Using the TIMER web (cistrome.shinyapps.io/timer) [14], a prognostic relationship was examined between CALD1 mRNA expression and Glioblastoma (GBM) patient survival. Survival curves were analyzed using a threshold log-rank test using scan mode expression values. The significance level was set at $p < 0.05$.

8.2.6   Enrichr Web

Based on 1575 positively co-expressed genes of CALD1 in GBM identified from UL-CAN, Enrichr (amp.pharm.mssm.edu/Enrichr) was used to determine Gene Ontology (GO) and pathway enrichment [15]. The bar graphs retrieved from Enrichr web were ranked by p-value from several databases, including the Kyoto Encyclopedia of Genes and Genomes (KEGG)-2021, Wiki pathway-2021, and GO (biological process and molecular function)-2021.

8.3   MULTIOMICS ANALYSIS AND RESULTS

In this section, we present the results of our data-driven review.

8.3.1   Gene and Protein Expression Analysis

To investigate the expression level of CALD1 in brain cancer and their normal counterparts, we first determined the mRNA expression pattern of CALD1 using TCGA data through UALCAN web. A significant high mRNA expression levels of CALD1 in GBM were found (Figure 8.1a). Since the number of normal samples is relatively lower, we compared the expression pattern using GTEx data via GEPIA2 web. The result (Figure 8.1b) also shows the overexpression of CALD1 in patients with GBM compared to the normal brain tissues. Also, CALD1 is significantly overexpressed in GBM tissues compared to their normal counterparts in terms of protein expression (Figure 8.1c), which was analyzed using CPTAC data through UALCAN web. To crosscheck CALD1 protein expression in normal brain and brain cancer tissues, we showed the immunohistochemistry results (Figure 8.1d) derived from Human Protein Atlas web. These results also confirm the overexpression of CALD1 at the protein levels in brain cancer samples relative to normal brain tissue at both the early and advanced stage of GBM.

8.3.2   Mutation and Genetic Alteration Analysis

We analyzed the alteration frequency of CALD1 copy number in brain cancer using the cBioPortal web. The results show that the mutation and amplification are responsible for the copy number alterations (CNA) (Figure 8.2a). The cBioPortal was queried with CALD1 gene over 7 studies covering 2178 GBM patients/2198 samples. It resulted in 10 mutations, including 3 duplicate mutations in patients with multiple samples. Most of the mutations are of missense type. We also noticed that mRNA expression levels and CNA are correlated in general, and diploid and gain are primary sources of CNA that contribute to the correlation in particular (Figure 8.2b).

8.3.3   Survival Analysis

We next investigated whether CALD1 mRNA expression has any potential role on GBM prognosis. To find the prognostic relevance of CALD1 in brain cancer, we performed survival analysis using TCGA data via TIMER web. Here, we used a Kaplan–Meier curve for the survival analysis, which is used to measure the fraction of

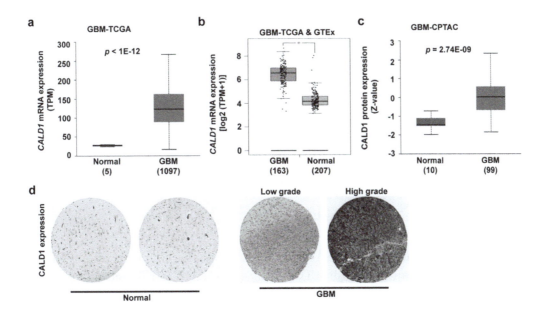

Figure 8.1 Expression of CALD1 in GBM. (a) mRNA expression of CALD1 in normal and GBM tissues was analyzed using TCGA data through UALCAN web. (b) mRNA expression of CALD1 in normal and GBM tissues was analyzed using TCGA and GTEx data through GEPIA2 web. (d) Protein expression of CALD1 in normal and GBM tissues was analyzed using CPTAC data through UALCAN web. (d) Protein expression of CALD1 in normal and GBM tissues by immunohistochemistry (IHC) was derived from Human Protein Atlas web. $P < 0.01$.

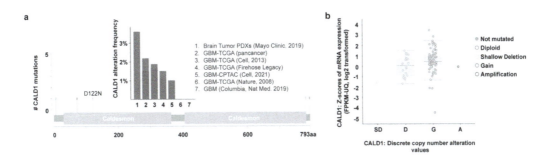

Figure 8.2 Mutation and genetic alterations of CALD1 in GBM. (a) Alteration frequency of CALD1 mutation and copy number in GBM was analyzed by using cBio-Portal web. (b) Correlation between CALD1 mRNA expression and copy number alteration. *Abbreviations:* SD: Shallow Deletion, D: Diploid, G: Gain, and A: Amplification.

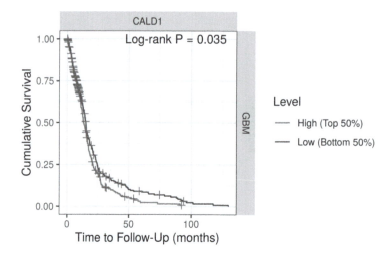

Figure 8.3 Relationship between CALD1 mRNA expression and clinical outcomes in GBM patients were analyzed using TCGA data through TIMER web.

patients with brain cancer living for a certain amount of time after getting treatment. The analysis shows that high levels of CALD1 expression correlated with poor survival, whereas low levels of CALD1 expression were associated with high survival rates (Figure 8.3).

8.3.4 Ontology and Pathway Analysis

Instead of a single gene, multiple genes collectively take part in a signaling pathway. These multiple genes are called correlated genes. We first identified genes that correlate with CALD1 in brain cancer using the UALCAN web. We then performed gene ontology (GO) and pathway enrichment analysis on the CALD1 along with the positively correlated genes via Enrichr web. The analysis shows that whereas the extracellular matrix organization is the primary biological process, cadherin binding is the main molecular function in the GO (Figure 8.4a,b). It also reveals that VEGFA-VEGFR2 signaling pathway is the key is the Wiki pathway, and protein processing in endoplasmic reticulum is the significant KEGG pathway (Figure 8.4c,d).

8.4 DISCUSSION AND CONCLUSION

In this chapter, we presented our review results of a multiomics analysis of CALD1 mRNA expression and clinical outcome data to understand the impact of CALD1 on human brain cancer. Whereas the gene expression analysis and immunostaining results demonstrate that CALD1 is overexpressed in glioblastoma, the survival analysis shows that the overexpression of CALD1 is associated with poor survival. The results are in agreement with several previous findings [16–18]. The GO analysis shows that cadherin binding is the major representative molecular function for CALD1 and its correlated genes, which is in agreement with a previous study that identified

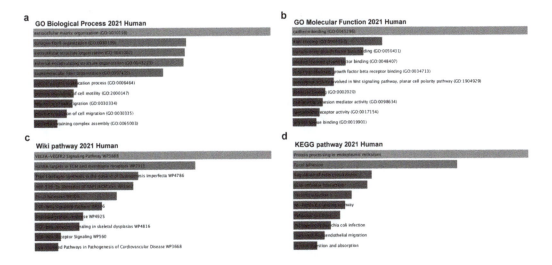

Figure 8.4 Gene ontology (GO) and pathway enrichment analysis for genes positively co-expressed with CALD1 in GBM through Enrichr web. (a and b) GO (biological process 2021 and molecular function 2021) enrichment analysis was performed using Enrichr web for the 1575 genes positively co-expressed with CALD1 in GBM. (c and d) pathway (Wiki 2021 and KEGG 2021) enrichment analysis was performed using Enrichr web for the 1575 genes positively co-expressed with CALD1 in GBM.

N-cadherin expression level is a critical indicator of invasion in non-epithelial tumors such as GBM [19]. We also noticed that VEGFA-VEGFR2 is one of the major pathways associated with GBM, which is also in agreement with earlier findings [20, 21]. In sum, this review concludes that CALD1 can be targeted to treat patients with brain cancer and can be considered a prognostic marker of GBM. However, appropriate experimental and clinical studies are required to validate the data mining-based results.

ACKNOWLEDGEMENT

S. M. Riazul Islam and Subbroto Kumar Saha contributed equally to this chapter.

Bibliography

1. E. Serres, F. Debarbieux, F. Stanchi, L. Maggiorella, D. Grall, L. Turchi, F. Burel-Vandenbos, *et al.*, "Fibronectin expression in glioblastomas promotes cell cohesion, collective invasion of basement membrane in vitro and orthotopic tumor growth in mice," *Oncogene*, vol. 33, no. 26, pp. 3451–3462, 2014.

2. A. Lucena-Cacace, M. Umeda, L. E. Navas, and A. Carnero, "NAMPT as a dedifferentiation-inducer gene: Nad+ as core axis for glioma cancer stem-like cells maintenance," *Frontiers in Oncology*, vol. 9, p. 292, 2019.

3. Q. Cheng, A. Tang, Z. Wang, N. Fang, Z. Zhang, L. Zhang, C. Li, and Y. Zeng, "CALD1 modulates gliomas progression via facilitating tumor angiogenesis," *Cancers*, vol. 13, no. 11, p. 2705, 2021.

4. S. K. Saha, S. M. R. Islam, K.-S. Kwak, M. Rahman, S.-G. Cho, *et al.*, "PROM1 and PROM2 expression differentially modulates clinical prognosis of cancer: a multiomics analysis," *Cancer Gene Therapy*, vol. 27, no. 3, pp. 147–167, 2020.

5. S. K. Saha, S. M. R. Islam, T. Saha, A. Nishat, P. K. Biswas, M. Gil, L. Nkenyer-eye, et al, "Prognostic role of EGR1 in breast cancer: a systematic review," *BMB Reports*, vol. 54, no. 10, p. 497, 2021.

6. S. Nie, Y. Kee, and M. Bronner-Fraser, "Caldesmon regulates actin dynamics to influence cranial neural crest migration in Xenopus," *Molecular Biology of the Cell*, vol. 22, no. 18, pp. 3355–3365, 2011.

7. P.-P. Zheng, T. M. Luider, R. Pieters, C. J. Avezaat, M. J. van den Bent, P. A. Sillevis Smitt, and J. M. Kros, "Identification of tumor-related proteins by proteomic analysis of cerebrospinal fluid from patients with primary brain tumors," *Journal of Neuropathology & Experimental Neurology*, vol. 62, no. 8, pp. 855–862, 2003.

8. Y. Liu, S. Xie, K. Zhu, X. Guan, L. Guo, and R. Lu, "CALD1 is a prognostic biomarker and correlated with immune infiltrates in gastric cancers," *Heliyon*, vol. 7, no. 6, p. e07257, 2021.

9. Y. Du, X. Jiang, B. Wang, J. Cao, Y. Wang, J. Yu, X. Wang, and H. Liu, "The cancer-associated fibroblasts related gene CALD1 is a prognostic biomarker and correlated with immune infiltration in bladder cancer," *Cancer Cell International*, vol. 21, no. 1, pp. 1–15, 2021.

10. D. S. Chandrashekar, B. Bashel, S. A. H. Balasubramanya, C. J. Creighton, I. Ponce-Rodriguez, B. V. Chakravarthi, and S. Varambally, "UALCAN: a portal for facilitating tumor subgroup gene expression and survival analyses," *Neoplasia*, vol. 19, no. 8, pp. 649–658, 2017.

11. Z. Tang, B. Kang, C. Li, T. Chen, and Z. Zhang, "Gepia2: an enhanced web server for large-scale expression profiling and interactive analysis," *Nucleic Acids Research*, vol. 47, no. W1, pp. W556–W560, 2019.

12. M. Uhlén, L. Fagerberg, B. M. Hallström, C. Lindskog, P. Oksvold, A. Mardinoglu, Å. Sivertsson, C. Kampf, E. Sjöstedt, A. Asplund, *et al.*, "Tissue-based map of the human proteome," *Science*, vol. 347, no. 6220, p. 1260419, 2015.

13. J. Gao, B. A. Aksoy, U. Dogrusoz, G. Dresdner, B. Gross, S. O. Sumer, Y. Sun, A. Jacobsen, R. Sinha, E. Larsson, *et al.*, "Integrative analysis of complex cancer genomics and clinical profiles using the cBioPortal," *Science Signaling*, vol. 6, no. 269, pp. pl1–pl1, 2013.

14. T. Li, J. Fan, B. Wang, N. Traugh, Q. Chen, J. S. Liu, B. Li, and X. S. Liu, "Timer: a web server for comprehensive analysis of tumor-infiltrating immune cells," *Cancer Research*, vol. 77, no. 21, pp. e108–e110, 2017.

15. E. Y. Chen, C. M. Tan, Y. Kou, Q. Duan, Z. Wang, G. V. Meirelles, N. R. Clark, and A. Ma'ayan, "Enrichr: interactive and collaborative html5 gene list enrichment analysis tool," *BMC Bioinformatics*, vol. 14, no. 1, pp. 1–14, 2013.

16. P.-P. Zheng, A. M. Sieuwerts, T. M. Luider, M. van der Weiden, P. A. Sillevis-Smitt, and J. M. Kros, "Differential expression of splicing variants of the human caldesmon gene (CALD1) in glioma neovascularization versus normal brain microvasculature," *American Journal of Pathology*, vol. 164, no. 6, pp. 2217–2228, 2004.

17. P.-P. Zheng, W. C. Hop, P. A. S. Smitt, M. J. van den Bent, C. J. Avezaat, T. M. Luider, and J. M. Kros, "Low-molecular weight caldesmon as a potential serum marker for glioma," *Clinical Cancer Research*, vol. 11, no. 12, pp. 4388–4392, 2005.

18. J. Meola, G. d. S. Hidalgo, J. C. R. e. Silva, L. E. C. M. Silva, C. C. P. Paz, and R. A. Ferriani, "Caldesmon: new insights for diagnosing endometriosis," *Biology of Reproduction*, vol. 88, no. 5, pp. 122–1, 2013.

19. J. S. Haug, X. C. He, J. C. Grindley, J. P. Wunderlich, K. Gaudenz, J. T. Ross, A. Paulson, K. P. Wagner, Y. Xie, R. Zhu, *et al.*, "N-Cadherin expression level distinguishes reserved versus primed states of hematopoietic stem cells," *Cell Stem Cell*, vol. 2, no. 4, pp. 367–379, 2008.

20. S. R. Michaelsen, M. Staberg, H. Pedersen, K. E. Jensen, W. Majewski, H. Broholm, M. K. Nedergaard, C. Meulengracht, T. Urup, M. Villingshøj, *et al.*, "VEGFA sustains VEGFR2 activation under bevacizumab therapy and promotes glioblastoma maintenance," *Neuro-Oncology*, vol. 20, no. 11, pp. 1462–1474, 2018.

21. C. Xu, X. Wu, and J. Zhu, "VEGF promotes proliferation of human glioblastoma multiforme stem-like cells through VEGF receptor 2," *Scientific World Journal*, 2013.

## CHAPTER 9

Artificial Intelligence for Parkinson's Disease Diagnosis: A Review

Moradul Siddique Yeasir, Arefin Tusher, Humaun Kabir, and Syed Galib

Department of Computer Science and Engineering, Jashore University of Science and Technology, Jashore, Bangladesh

Mohammad Farhad Bulbul

Department of Mathematics Jashore University of Science and Technology, Jashore, Bangladesh and
Department of Computer Science and Engineering, Pohang University of Science and Technology, Pohang, South Korea

CONTENTS

PARKINSON'S DISEASE (PD) is an illness of the neurological system that involves tremors, muscle stiffening, impairment of posture and movement, state of imbalance, changes in speech and writing, and reduction of concentration. Considering proper care, early and quick PD detection is essential for every probable patient. For detecting PD, however, there are no specific medical tests except the assessment of clinical symptoms. For a conclusive outcome, several physical issues must be noticeable in order to diagnose PD. Artificial Intelligence (AI)-driven approaches such as machine learning (ML), deep learning (DL), computer vision, and natural language processing (NLP) have been used for the classification and assessment of the severity of PD patients. However, a comprehensive and technical overview of AI methods through various data like non-motor signs and symptoms of PD are missing. Therefore, in this chapter, we present a systematic review of the literature based on the application of AI that serves to identify PD that have been recently published to diagnose PD from the medical and technical database. Hence, the review covers numerous

DOI: 10.1201/9781003251903-9

articles which retrieve, investigate, and encapsulate their objectives, collection of data with sources as well as the achievements according to different approaches. As a result, utilizing integrated AI methods could be a useful decision tool for early PD diagnosis which will help prevent other changes in a patient's health condition.

9.1 INTRODUCTION

Parkinson's disease (PD) is the world's second-most frequent neurological disorder after Alzheimer's disease. Moreover, PD is a lifelong and degenerative disease. PD was originally discovered by the English physician James Parkinson [1]. Generally, the disturbance of dopamine neuronal activity in the basal ganglia causes PD. It mostly affects around 1 million or about 1% of adults over the age of 50 in the United States [2]. Men are much more likely than women to be afflicted. The symptoms appear gradually and intensify over time. There are two main types of PD symptoms: Motor and non-motor. Movement issues, trembling, stiffening, walking difficulties, and postural imbalance are all instances of motor symptoms [3, 4]. In addition, vascular dementia, mental issue, melancholy, and severe depression are some non-motor symptoms. Although immunotherapy can help with the signs of PD and enhance the quality of life in patients, therapy alone may not able to cure the disease fully. In the following decades, the prevalence of PD is only anticipated to rise. Resting tremor affects more than 75% of Parkinson's sufferers at a certain time during their illness and symptomatic tremor occurs in 60% of sufferers during motion or action [5]. In addition numerous sufferers with PD in the community remain undiagnosed due to poverty, remote location, or other reasons. Thus, these patients become much worse day by day. It is reported that the number of individuals affected by PD will rise at a quicker pace until 2050 [6]. Therefore, most developed countries are becoming highly concerned.

Actually, early detection of PD is critical for minimizing consequences and complications. The symptoms and severity vary among individuals. Several motor symptoms have been identified that are either actively or passively connected to PD. The following are the signs and symptoms: Tremor, Dysarthria, Akinesia, Bradykinesia Dysphagia, Dystonia, Fall, Hypokinesia, Rigidity, Hypomimia, Kinesia Paradoxica, Hypophonia, Postural instability, and so on [7]. Moreover, specific medical tests which can definitively confirm PD in its early stages are absent [8, 9]. In a conventional clinical context, a specialist may ask the patient to undertake various cognitive and physical tests such as postural instability, sleep difficulties, speech problems, constipation, and so on, in order to diagnose PD in any stage [10]. Currently, computer-assisted prognostic systems are being used to aid in the detection of PD. A brain scan can be done by various image-based screening tests such as Magnetic Resonance Imaging (MRI), Functional Magnetic Resonance Imaging (fMRI), Single Photon Emission Computed Tomography (SPECT), brain ultrasound scans and Positron Emission Tomography (PET)/Computerized Tomography (CT) scans [11–13]. In addition, a specialist can obtain precise images of the brain's dopaminergic system using the dopamine transporters (DaT) scan technology [14, 15]. The above-mentioned screening technology cannot prove the existence of PD. However, it can assist the specialist

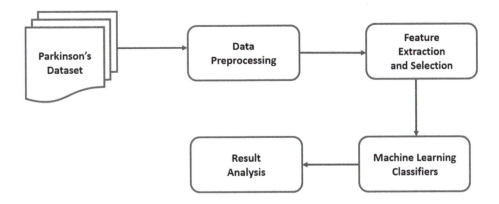

Figure 9.1 Generic methodology of PD detection using ML.

to confirm a prognosis as well as to identify the stage of such a Parkinson's imitation. However, identifying PD from several other neurological illnesses can be challenging, and it is entirely dependent on the skill of the radiologist.

9.2 BACKGROUND

Many studies have been conducted on the use of machine learning and speech pattern recognition in PD patients. In case of the PD detection using the different types of machine learning model such as Support Vector Machine (SVM), K-Nearest Neighbour (KNN), Decision Tree (DT), Naïve Bayes (NB), Gaussian Naïve Bayes (GNB), Logistic Regression (LR), Random Forest (RF), Linear Discriminant Analysis (LDA), Gradient Boosting (GB), and Fuzzy Logic (FL) have been used. Furthermore, the generic system of the ML model is shown in Figure 9.1.

In comparison to the standard principal component analysis (PCA), statistical compression techniques have provided much stronger classification of the data [16]. In a higher dimensional space sense, PCA can only minimize the dimension of the data.

Deep learning approaches have recently looked promising in clinical image analyses and disease identification applications. Later, features can be different types based on datasets like motion [17], voice [18], MRI [19], gait and tremor [20], handwriting [21], and Archimedes spiral [22]. Acoustic features [23], time frequency features [24], recursive feature elimination, and feature importance methods [25] were used for deriving the most significant features. The authors [26] conducted a feature selection approach called maximum information gain minimum correlation (MIGMC) using a similarity network model. Likewise, a computer vision-based study has been applied to recognize PD using micrography [27]. In the meantime, Peker et al. [28] integrated voice features and complex-valued neural networks to diagnosis of PD as well.

Patients with PD can suffer from speech impairment or vocal cord issues. Therefore, the assessment of the voice and speech has received a lot of attention as a way to detect and measure the severity of PD symptoms. According to studies [29,30], around 90% of people with Parkinson's (PWP) have some form of speech and voice impairment. Voice impairment could potentially be one of the first signs of the disorder. Little et al. [31] established a novel dysphonia scale called Pitch Period Entropy (PPE) which is unaffected from a number of uncontrolled influencing variables, such as noisy acoustic surroundings and natural, normal changes in vocal frequencies. Radha et al. [32] investigated PD detection from a voice signal using a Convolutional Neural Network (CNN), Artificial Neural Network (ANN), and Hidden Markov Model (HMM). Comparing to HMM and CNN-based PD systems, ANN-based PD detection perform better than others. When handling with unbalanced data, there are two approaches: Resampling and algorithmic [33]. The boosting procedures are also equivalent in terms of performance. Using a 10-fold cross-validation procedure, the authors demonstrated that ensemble methods achieved 100% accuracy through optimized by Particle Swarm Optimization (PSO) [34].

Motor disorder is also a symptom of PD where hand tremor can affect handwriting of PD patients. There are three tests to detect in handwriting impairments such as static spiral test (SST), dynamic spiral test (DST), and stability test on a specific spot (STCP) [35]. The input data are taken from the handwriting of a variety of people including ordinary and those with Parkinson's disease. Due to the small data (62 Parkinson's and 15 healthy) one of the studies [36] achieved 100% accuracy which applied the k-NN and ANN methods. The alterations of output in handwriting are the most common symptom of PD. In case, Kinematic and pressure spatio-temporal handwriting measurement factors in handwriting can be examined for the evaluation of PD [37]. The tasks are like Archimedean spiral, writing a small sentence, phonetic symbols, words, vowels, and so on [15].

Palmerini et al. presented a tremor-related assessment such as postural movement measure and a frequency-domain postural measure [38]. An audio recording (from the PC-GITA database) spectrum was computed and used as converted spectrogram images from recordings that applied in the Residual Neural Network (ResNet) framework which was pre-trained through ImageNet and Singular Value Decomposition (SVD) [39]. W. U. Wang et al. [40] provided a selection of significant features and identify PD using the Boosting method. Several patients have difficulty with syntax, verbal ability, semantics as well as language impairments. The studies investigated used a variety of Natural Language Processing (NLP) methods such as Bag of Words (BoW) and Term Frequency-Inverse Document Frequency (TF-IDF), and word embedding-based methods like word2vec (W2V) [41]. In this context, there are various identification state-of-the-art approaches proposed for PD that are shown in Table 9.1.

In Table 9.1, most of the proposed methods for PD diagnosis were evaluated using public datasets from UCI, which are comprised inductive and deductive reasoning [42]. On the contrary, several are personal data [43]. Most of all cases, the data collection system occurred through non-invasive way that can be conveniently obtained from sufferers and sent into the clinical support system.

TABLE 9.1   Related Work on PD Diagnosis Using ML and DL

Authors	Techniques											
	SVM	KNN	DT	NB	RF	LDA	DF	LR	GNB	DNN	GB	NN
[44]	•	•		•	•			•				
[40]	•	•	•		•			•		•		
[45]	•	•	•			•		•	•			
[17]	•			•	•							
[46]	•				•			•			•	
[47]	•	•	•	•	•			•		•		
[16]	•	•	•		•			•				
[48]	•	•	•	•	•			•				•
[49]	•	•	•	•		•						
[24]	•	•		•		•						
[50]	•		•				•	•				•
[51]	•			•	•			•				
[52]	•		•	•	•							

Abbreviations: SVM: Support vector machine; KNN: K-nearest neighbors; DT: Decision Tree; NB: Gaussian NB; RF: Random Forest; LDA: Linear Discriminant Analysis; DF: Decision Forest; LR: Linear Regression; GNB: Gaussian Naïve Bayes; DNN: Deep neural network; GB: Gradient boosting; NN: Nearest Neighbor.

9.3   PROBLEM ANALYSIS

Due to aging, detecting PD can create numerous challenges including loss of balance, poor movements, muscular rigidity, and slumped posture. Later, PD affects the nerve cells, where cells deliver messages to muscles that allows it to move. Neurons which are sick str unable to regulate muscles normally. PD diagnosis is not straightforward, which means a particular clinical test such as a blood test or an electrocardiogram (ECG) cannot indicate whether a person has the disorder. Moreover, the proper cause of the disease remains unknown [53]. In some scenarios, PD may be induced by inherited factors [54]. As there is no conclusive test for PD, there is a strong possibility of misdiagnosis. Conversely, it is challenging for a large number of people to make the proper physical appointments at the clinic for screening and therapy. As a sense, the majority of persons with PD are highly dependent on medical interventions. Conceptually, it has been claimed that early diagnosis may effectively control or reduce PD symptoms. Hence, an automated system for early diagnosis of PD is a major consideration, as the condition could become incurable if detected at an advanced stage.

Definitely, a sufferer can receive proper treatment if early and reliable diagnosis is feasible. Hereby, the DL and ML methods can provide radiological experts with information that help make better and more reliable decisions for a sufferer.

9.4   POSSIBLE SOLUTION

There is no known treatment for PD, the medical records and personal observations of the patient are used to evaluate the disease's severity. However, clinical symptoms are used to diagnose PD. Dopamine is a feature to assess as a suspected PD patient which is measured by Fluorodopa positron emission tomography (PET). Further, computed tomography (CT) and MRI can be used to determine a structural issue in a suspected PD patient [55]. Therefore, the most frequently used rating scales for the diagnosis of PD are the Unified Parkinson's Disease Rating Scale (UPDRS), Hoehn and Yahr scale, Schwab and England Activities of Daily Living (ADL) scale, PDQ-39, PD Non-Motor Symptoms (NMS) questionnaire and survey. Specialists prefer the UPDRS level (a four-point level) to grade PD severity owing to its clinometric features that are superior to other systems [56]. As a matter of fact, all of these scales may still not provide reliable and accurate results [57].

To diagnose PD, new approaches are required. As a result, less economical, simple, and reliable methods for diagnosing the disease and ensuring treatment should be adopted. Nonsurgical PD diagnosis procedures should be researched. DL and ML algorithms are utilized to differentiate between people with PD and those without PD. In this way, handwriting difficulties can be relevant to both the disorder and also its severity, therefore alterations in handwriting can also be used as a biomarker [58]. In current history, healthcare informatics methods have been frequently used in the rapid screening of major diseases. Therefore, numerous research has dealt with DL and ML techniques for solving the PD diagnosis issues. In addition, various types of datasets are widely available based on different PD features which are used to detect the disease. In the field, the rapid finger-tapping test (RFT) is a useful technique for assessing PD movement disorders [59]. Finally, the usage of enriched data with more patient medical history would be effective for training ML or DL models. Currently, the severity of the disorder was determined by the gait and tremor features [20]. A. Ul Haq et al. [60] show that deep neural network comparatively has better accuracy (98%), sensitivity (99%), specificity (95%) than ML techniques such as LR, SVM, KNN. The pc-Gita dataset achieved 99.7% accuracy as well as deep feature-based strategy, which showed better performance than ordinary acoustic features and transfer learning methods [61]. A ML technique, Boosted Logistic Regression provides impressive accuracy (97%) based on nonmotor symptoms, Rapid Eye Movement (REM), and olfactory loss [10]. Changes of dysphonia, PD patients suffer speech problems [45]. Hence, PD has been detected using a variety of speech data like words, numbers, short sentences, etc., along with the gait data analysis [6] and sensor data [52]. The authors revealed the outperformance of U-Net, an adaptive convolutional structure, which has better capability and performance than the Visual Geometry Group (VGG) [62].

9.5 CONCLUSION

Neurodegenerative disease detection and diagnosis has now become a promising research area. Whereas, PD is a long-term disorder that worsens with time. As a result, early identification of PD should receive a lot of attention. In this study, we discussed from the state-of the-art techniques to some common DL and ML classification techniques that have been applied to PD detection. Various feature extension and reduction methods have been suggested in order to improve classification performance. As a result, the methods described above can provide a reliable solution for detecting PD in its early stages.

In this chapter, we have suggested a method that acts as a bridge between ML and DL techniques as well as unweighted symptoms that can be averted by the dataset.

Compared to existing techniques, the integrated system can categorize with less parameters and achieve maximum prediction accuracy. This strategy can lower complexity of the prediction methods as well as minimize computational load. To achieve higher classification rates, most weighted features as well as innovative techniques can be used. In conclusion, our preliminary reports suggest the use of accelerometers as an acceptable diagnostic tool in PD. In the end, the hybrid methodology supports early PD detection, this one will benefit all of us.

Bibliography

1. G. K. York, "The History of James Parkinson and His Disease," J. Neurol. Sci., vol. 381, no. 2017, p. 35, 2017, doi: 10.1016/j.jns.2017.08.147.

2. L. Ali, C. Zhu, Z. Zhang, and Y. Liu, "Automated Detection of Parkinson's Disease Based on Multiple Types of Sustained Phonations Using Linear Discriminant Analysis and Genetically Optimized Neural Network," IEEE J. Transl. Eng. Heal. Med., vol. 7, no. September, pp. 1–10, 2019, doi: 10.1109/JTEHM.2019.2940900.

3. A. J. Hughes, S. E. Daniel, L. Kilford, and A. J. Lees, "Accuracy of Clinical Diagnosis of Idiopathic Parkinson's Disease: A Clinico-Pathological Study of 100 Cases," pp. 181–184, 1992.

4. B. R. Brewer, S. Pradhan, G. Carvell, and A. Delitto, "Application of Modified Regression Techniques to a Quantitative Assessment for the Motor Signs of Parkinson's Disease," vol. 17, no. 6, pp. 568–575, 2009.

5. L. E. Heusinkveld, M. L. Hacker, M. Turchan, T. L. Davis, and D. Charles, "Impact of Tremor on Patients with Early Stage Parkinson's Disease," Front. Neurol., vol. 9, no. AUG, pp. 1–5, 2018, doi: 10.3389/fneur.2018.00628.

6. S. Aich, H. C. Kim, K. Younga, K. L. Hui, A. A. Al-Absi, and M. Sain, "A Supervised Machine Learning Approach Using Different Feature Selection Techniques on Voice Datasets for Prediction of Parkinson's Disease," Int. Conf. Adv. Commun. Technol. ICACT, vol. 2019 February, no. 3, pp. 1116–1121, 2019, doi: 10.23919/ICACT.2019.8701961.

7. V. Majhi, S. Paul, G. Saha, and J. K. Verma, "Sensor-Based Detection of Parkinson's Disease Motor Symptoms," 2020 Int. Conf. Comput. Perform. Eval. ComPE 2020, pp. 553–557, 2020, doi: 10.1109/ComPE49325.2020.9200051.

8. R. Das, "Expert Systems with Applications: a Comparison of Multiple Classification Methods for Diagnosis of Parkinson Disease," Expert Syst. Appl., vol. 37, no. 2, pp. 1568–1572, 2010, doi: 10.1016/j.eswa.2009.06.040.

9. N. Singh, V. Pillay, and Y. E. Choonara, "Advances in the Treatment of Parkinson's Disease," Prog. Neurobiol., vol. 81, no. 1, pp. 29–44, 2007, doi: 10.1016/j.pneurobio.2006.11.009.

10. K. N. R. Challa, V. S. Pagolu, G. Panda, and B. Majhi, "An Improved Approach for Prediction of Parkinson's Disease Using Machine Learning Techniques," Int. Conf. Signal Process. Commun. Power Embed. Syst. SCOPES 2016 - Proc., pp. 1446–1451, 2017, doi: 10.1109/SCOPES.2016.7955679.

11. F. M. Skidmore et al., "Reliability Analysis of the Resting State can Sensitively and Specifically Identify the Presence of Parkinson Disease," Neuroimage, vol. 75, pp. 249–261, 2013, doi: 10.1016/j.neuroimage.2011.06.056.

12. M. Rumman, A. N. Tasneem, S. Farzana, M. I. Pavel, and A. Alam, "Early Detection of Parkinson's Disease Using Image Processing and Artificial Neural Network," 2018 Jt. 7th Int. Conf. Informatics, Electron. Vis. 2018 2nd Int. Conf. Imaging, Vis. Pattern Recognit., pp. 256–261, 2018, doi: 10.1109/ICIEV.2018.8641081.

13. A. Boutet et al., "Predicting Optimal Deep Brain Stimulation Parameters for Parkinson's Disease Using Functional MRI and Machine Learning," Nat. Commun., pp. 1–13, 2021, doi: 10.1038/s41467-021-23311-9.

14. S. J. Son, M. Kim, and H. Park, "Imaging Analysis of Parkinson's Disease Patients Using SPECT and Tractography," Sci. Rep., vol. 6, pp. 1–11, 2016, doi: 10.1038/srep38070.

15. P. Drotár, J. Mekyska, I. Rektorová, L. Masarová, Z. Smékal, and M. Faundez-Zanuy, "Evaluation of Handwriting Kinematics and Pressure for Differential Diagnosis of Parkinson's Disease," Artif. Intell. Med., vol. 67, pp. 39–46, 2016, doi: 10.1016/j.artmed.2016.01.004.

16. Y. Mittra and V. Rustagi, "Classification of Subjects with Parkinson's Disease Using Gait Data Analysis," 2018 Int. Conf. Autom. Comput. Eng. ICACE 2018, pp. 84–89, 2018, doi: 10.1109/ICACE.2018.8687022.

17. F. Cavallo, A. Moschetti, D. Esposito, C. Maremmani, and E. Rovini, "Parkinsonism and Related Disorders Upper Limb Motor Pre-Clinical Assessment in Parkinson's Disease Using Machine Learning," Park. Relat. Disord., vol. 63, no. September 2018, pp. 111–116, 2019, doi: 10.1016/j.parkreldis.2019.02.028.

18. J. S. Almeida, P. P. Rebouças, T. Carneiro, W. Wei, V. Hugo, and C. De Albuquerque, "Detecting Parkinson's Disease with Sustained Phonation and Speech Signals Using Machine Learning Techniques? Robertas Damaısevic," vol. 125, pp. 55–62, 2019, doi: 10.1016/j.patrec.2019.04.005.

19. G. Singh, M. Vadera, L. Samavedham, and E. C. Lim, "ScienceDirect Machine Framework for Machine Framework for Machine Framework for Multi-Class Diagnosis of Neurodegenerative Machine Learning-Based Framework for Diagnosis of Neurodegenerative Diagnosis of Neurodegenerative Diseases: Study on Parkinson's," IFAC-PapersOnLine, vol. 49, no. 7, pp. 990–995, doi: 10.1016/j.ifacol.2016.07.331.

20. E. Abdulhay, N. Arunkumar, K. Narasimhan, and E. Vellaiappan, "Gait and Tremor Investigation Using Machine Learning Techniques for the Diagnosis of Parkinson Disease," Futur. Gener. Comput. Syst., 2018, doi: 10.1016/j.future.2018.02.009.

21. C. Kotsavasiloglou, N. Kostikis, D. Hristu-Varsakelis, and M. Arnaoutoglou, "Biomedical Signal Processing and Control Machine Learning-Based Classification of Simple Drawing Movements in Parkinson's Disease," Biomed. Signal Process. Control, vol. 31, pp. 174–180, 2017, doi: 10.1016/j.bspc.2016.08.003.

22. R. Saunders-Pullman et al., "Validity of Spiral Analysis in Early Parkinson's Disease," vol. 23, no. 4, pp. 531–537, 2008, doi: 10.1002/mds.21874.

23. O. Yaman, F. Ertam, and T. Tuncer, "Automated Parkinson's Disease Recognition Based on Statistical Pooling Method Using Acoustic Features," Med. Hypotheses, vol. 135, p. 109483, 2020, doi: 10.1016/j.mehy.2019.109483.

24. Y. N. Zhang, "Can a Smartphone Diagnose Parkinson Disease? A Deep Neural Network Method and Telediagnosis System Implementation," Parkinsons. Dis., vol. 2017, no. 1, 2017, doi: 10.1155/2017/6209703.

25. Z. K. Senturk, "Early Diagnosis of Parkinson's Disease Using Machine Learning Algorithms," Med. Hypotheses, vol. 138, no. January, p. 109603, 2020, doi: 10.1016/j.mehy.2020.109603.

26. E. Rastegari, S. Azizian, and H. Ali, "Machine Learning and Similarity Network Approaches to Support Automatic Classification of Parkinson's Disease Using Accelerometer-Based Gait Analysis," Proc. Annu. Hawaii Int. Conf. Syst. Sci., vol. 2019-Janua, pp. 4231–4242, 2019, doi: 10.24251/hicss.2019.511.

27. C. R. Pereira et al., "A New Computer Vision-Based Approach to Aid the Diagnosis of Parkinson's Disease," Comput. Methods Programs Biomed., vol. 136, pp. 79–88, 2016, doi: 10.1016/j.cmpb.2016.08.005.

28. M. Peker, B. Şen, and D. Delen, "Computer-Aided Diagnosis of Parkinson's Disease Using Complex-Valued Neural Networks and mRMR Feature Selection Algorithm," J. Healthc. Eng., vol. 6, no. 3, pp. 281–302, 2015, doi: 10.1260/2040-2295.6.3.281.

29. S. Skodda, H. Rinsche, and U. Schlegel, "Progression of dysprosody in Parkinson's disease over time—A longitudinal study," Mov. Disord., vol. 24, no. 5, pp. 716–722, 2009, doi: 10.1002/mds.22430.

30. J. A. Logemann, H. B. Fisher, B. Boshes, and E. R. Blonsky, "Jeri A. Logemann," pp. 47–57, 2016.

31. M. A. Little et al., "NIH Public Access," vol. 56, no. 4, 2011, doi: 10.1109/TBME.2008.2005954.Suitability.

32. N. Radha, S. M. Rm, and S. Sameera, "Parkinson's Disease Detection using Machine Learning Techniques," vol. XXX, pp. 543–552, 2021, doi: 10.24205/03276716.2020.4055.

33. H. He and E. A. Garcia, "Learning from Imbalanced Data," vol. 21, no. 9, pp. 1263–1284, 2009.

34. P. Das, S. Nanda, and G. Panda, "Automated Improved Detection of Parkinson's Disease Using Ensemble Modeling," Proc.—2020 IEEE Int. Symp. Sustain. Energy, Signal Process. Cyber Secur. iSSSC 2020, 2020, doi: 10.1109/iSSSC50941.2020.9358898.

35. M. E. Isenkul, B. E. Sakar, and O. Kursun, "Improved Spiral Test Using Digitized Graphics Tablet for Monitoring Parkinson's Disease," 2014, pp. 171–175, 2014, doi: 10.13140/RG.2.1.1898.6005.

36. A. Ranjan, "An Intelligent Computing Based Approach for Parkinson Disease Detection," 2018 Second Int. Conf. Adv. Electron. Comput. Commun., pp. 1–3, 2018.

37. P. Drot, I. Rektorov, and L. Masarov, "Decision Support Framework for Parkinson's Disease Based on Novel Handwriting Markers," vol. 11, no. 4, 2014, doi: 10.1109/TNSRE.2014.2359997.

38. L. Palmerini, L. Rocchi, S. Mellone, F. Valzania, and L. Chiari, "Feature Selection for Accelerometer-Based Posture Analysis in Parkinson's Disease," IEEE Trans. Inf. Technol. Biomed., vol. 15, no. 3, pp. 481–490, 2011, doi: 10.1109/TITB.2011.2107916.

39. M. Wodzinski, A. Skalski, D. Hemmerling, J. R. Orozco-Arroyave, and E. Noth, "Deep Learning Approach to Parkinson's Disease Detection Using Voice Recordings and Convolutional Neural Network Dedicated to Image Classification," Proc. Annu. Int. Conf. IEEE Eng. Med. Biol. Soc. EMBS, pp. 717–720, 2019, doi: 10.1109/EMBC.2019.8856972.

40. W. U. Wang, J. Lee, F. Harrou, and Y. Sun, "Early Detection of Parkinson's Disease Using Deep Learning and Machine Learning," pp. 147635–147646, 2020, doi: 10.1109/ACCESS.2020.3016062.

41. K. E. Ed, R. Goebel, Y. Tanaka, and W. Wahlster, Text, Speech and Dialogue. 2019.

42. M. A. Little*, P. E. McSharry, E. J. Hunter, J. Spielman and L. O. Ramig, "Suitability of Dysphonia Measurements for Telemonitoring of Parkinson's Disease," in IEEE Transactions on Biomedical Engineering, vol. 56, no. 4, pp. 1015–1022, April 2009, doi: 10.1109/TBME.2008.2005954.

43. M. Su and K. S. Chuang, "Dynamic Feature Selection for Detecting Parkinson's Disease through Voice Signal," 2015 IEEE MTT-S Int. Microw. Work. Ser. RF

Wirel. Technol. Biomed. Healthc. Appl. IMWS-BIO 2015 - Proc., no. 2, pp. 148–149, 2015, doi: 10.1109/IMWS-BIO.2015.7303822.

44. G. Solana-lavalle and R. Rosas-Romero, "Computer Methods and Programs in Biomedicine Classification of PPMI MRI Scans with Voxel-Based Morphometry and Machine Learning to Assist in the Diagnosis of Parkinson's Disease," Comput. Methods Programs Biomed., vol. 198, p. 105793, 2021, doi: 10.1016/j.cmpb.2020.105793.

45. L. Ali, S. U. Khan, M. Arshad, S. Ali, and M. Anwar, "A Multi-model Framework for Evaluating Type of Speech Samples Having Complementary Information about Parkinson's Disease," 1st Int. Conf. Electr. Commun. Comput. Eng. ICECCE 2019, no. July, pp. 1–5, 2019, doi: 10.1109/ICECCE47252.2019.8940696.

46. E. Celik, "Improving Parkinson's Disease Diagnosis with Machine Learning Methods," 2019 Sci. Meet. Electr. Biomed. Eng. Comput. Sci., pp. 1–4, 2019.

47. A. Anand, J. Sahaya, and R. Alex, "Evaluation of Machine Learning and Deep Learning Algorithms Combined with Dimentionality Reduction Techniques for Classification of Parkinson's Disease," 2018 IEEE Int. Symp. Signal Process. Inf. Technol., pp. 342–347, 2018.

48. D. Swain, V. Hiwarkar, N. Motwani, and A. Awari, "Predicting the Occurrence of Parkinson's Disease Using Various Classification Models," 2018.

49. C. Caramia et al., "IMU-based Classification of Parkinson's Disease from Gait: A Sensitivity Analysis on Sensor Location and Feature Selection," vol. 2194, no. c, 2018, doi: 10.1109/JBHI.2018.2865218.

50. A. Dinesh and J. He, "Using Machine Learning to Diagnose Parkinson's Disease from Voice Recordings." IEEE MIT Undergraduate Research Technology Conference (URTC), pp. 1–4, 2017.

51. R. Prashanth, S. Dutta Roy, P. K. Mandal, and S. Ghosh, "High-Accuracy Detection of Early Parkinson's Disease through Multimodal Features and Machine Learning," Int. J. Med. Inform., vol. 90, pp. 13–21, 2016, doi: 10.1016/j.ijmedinf.2016.03.001.

52. D. J. Cook, M. Schmitter-Edgecombe, and P. Dawadi, "Analyzing Activity Behavior and Movement in a Naturalistic Environment Using Smart Home Techniques," IEEE J. Biomed. Heal. Informatics, vol. 19, no. 6, pp. 1882–1892, 2015, doi: 10.1109/JBHI.2015.2461659.

53. Y. Wang, A. Wang, Q. Ai, and H. Sun, "Biomedical Signal Processing and Control: An Adaptive Kernel-Based Weighted Extreme Learning Machine Approach for Effective Detection of Parkinson's Disease," Biomed. Signal Process. Control, vol. 38, pp. 400–410, 2017, doi: 10.1016/j.bspc.2017.06.015.

54. R. E. Burke, "Evaluation of the Braak Staging Scheme for Parkinson's Disease: Introduction to a Panel Presentation," Mov. Disord., vol. 25, no. SUPPL. 1, 2010, doi: 10.1002/mds.22783.

55. W. Hazzard, J. Blass, J. Halter, J Ouslander, and M. Tinetti, Principles of Geriatric Medicine and Gerontology. New York City, McGraw-Hill Education, Jul. 2003.

56. N. Kour, Sunanda, and S. Arora, "Computer-Vision Based Diagnosis of Parkinson's Disease via Gait: A Survey," IEEE Access, vol. 7, pp. 156620–156645, 2019, doi: 10.1109/ACCESS.2019.2949744.

57. O. Cigdem, A. Yilmaz, I. Beheshti, and H. Demirel, "Comparing the performances of PDF and PCA on Parkinson's Disease Classification Using Structural MRI Images," 26th IEEE Signal Process. Commun. Appl. Conf. SIU 2018, pp. 1–4, 2018, doi: 10.1109/SIU.2018.8404697.

58. D. Impedovo and G. Pirlo, "Dynamic Handwriting Analysis for the Assessment of Neurodegenerative Diseases" IEEE Rev. Biomed. Eng., vol. 12, pp. 209–220, 2019, doi: 10.1109/RBME.2018.2840679.

59. T. Khan, D. Nyholm, J. Westin, and M. Dougherty, "A Computer Vision Framework for Finger-Tapping Evaluation in Parkinson's Disease," Artif. Intell. Med., vol. 60, no. 1, pp. 27–40, 2014, doi: 10.1016/j.artmed.2013.11.004.

60. A. Ul Haq, J. Li, Md. H. Memon, J. Khan, S. Ud Din, I. Ahad, R. Sun, Zhilong Lai, "Comparative Analysis of the Classification Performance of Machine Learning Classifiers and Deep Neural Network Classifier for Prediction of Parkinson Disease," 2018 15th Int. Comput. Conf. Wavelet Act. Media Technol. Inf. Process. ICCWAMTIP 2018, vol. l, pp. 101–106, 2019, doi: 10.1109/ICCWAMTIP.2018.8632613.

61. L. Zahid et al., "A Spectrogram-Based Deep Feature Assisted Computer-Aided Diagnostic System for Parkinson's Disease," IEEE Access, vol. 8, pp. 35482–35495, 2020, doi: 10.1109/ACCESS.2020.2974008.

62. Y. Dai, Z. Tang, Y. Wang, and Z. Xu, "Data Driven Intelligent Diagnostics for Parkinson's Disease," IEEE Access, vol. 7, pp. 106941–106950, 2019, doi: 10.1109/ACCESS.2019.2931744.

## CHAPTER 10

Breast Cancer Detection: A Survey

Esraa Hassan, Fatma M. Talaat, Zeinab Hassan, and Nora El-Rashidy

Faculty of Artificial Intelligence, Kafrelsheikh University, Egypt

CONTENTS

B REAST CANCER is a disease that develops when the cells in the breast grow out of control as a result of abnormalities in the DNA. These aberrant cells build up and eventually form a tumor or a tumor that can be felt. Breast cancer is caused primarily by advancing age and a family history of the disease. As a result, faster detection of breast cancer is required to reduce the number of fatalities linked to this disease. As a result of recent discoveries in artificial intelligence, deep learning models have made significant gains in computer vision, e-commerce, cyber-security, and healthcare because of deep learning. Numerous programs have developed effective solutions to aid radiologists in medical image processing. Automatic lesion identification and categorization in mammograms, for example, are still regarded as critical issues that necessitate a more exact diagnosis and study of problematic lesions. This survey looks at several commonly used breast cancer datasets, whether they are text or images, as well as related works demonstrating previous efforts for breast cancer diagnosis.

DOI: 10.1201/9781003251903-10

10.1 INTRODUCTION

Breast cancer (BC) is the most frequent cancer that affects women all over the world. In 2020, over 279,000 instances were reported in the United States, with a 15% fatality rate compared to other cancers. It is an uncontrollable cell division that can spread to other parts of the body. It usually begins in the epididymis that leads to the nipple. Dimpling on the breast skin and a lump in one of the armpits is followed by a change in the size of one or both breasts. There are two types of breast tumors: Benign and malignant. Benign tumors are non-cancerous since they do not spread into the breast tissue around the duct and do not pose a threat to one's life. The traditional technique for cancer diagnosis relies heavily on medical specialists' experience and visual examinations. Consequently, this form of assessment takes a while and is prone to human error. As a result, there is a need to build an automatic diagnosis method for cancer early detection. Several studies have used machine learning approaches to create pattern recognition and classification models for clinical diagnosis. Machine Learning (ML) is a subset of Artificial Intelligence (AI) that permits a system to learn by feeding it a set of facts and learning through experience rather than significant programming. Most researchers used three types of data to aid in the early detection of breast cancer: Image data, text data, and biomarker data. Firstly, for text data, Shiny et al. [1] proposed a Support Vector Machine (SVM) discriminate ratio. They compared three of the most popular machine learning (ML) algorithms. The methods used in the original Wisconsin BC dataset were Random Forest (RF), SVM, and Naïve Bayes (NB). The results show that SVM has the best performance and accuracy, with 97.2% accuracy. In addition to KNN, NB, and DT, Jiande et al. [2]used the SVM algorithm. Using an ML approach, they used gene expression data to classify (triple-non-triple) negative BC patients. The implementation of various algorithms revealed that SVM had the best performance, with an accuracy of 90%. Mücahid et al. [3] proposed a model for the detection of BC based on age, biomarkers, glucose, and body mass index using Artificial Neural Networks (ANNs) and (NB). Feature selection (FS) techniques are also used to improve the model's performance. The goal of FS was to reduce data dimensionality and achieve high BC classification accuracy. The proposed system was a two-stage mechanism. First, the 22 features from the WDBC dataset were chosen using FS techniques. The ANN classifier with 15 neurons (15-neuron ANN) achieved 96.4% accuracy for BC classification, with 99.9% sensitivity, 98.4% precision, 1.6% false-positive rate, and 0.42-second processing time. In the initial trial conducted without FS, the suggested 15-neuron ANN classifier attained a classification accuracy of 98.8%. The accuracy of categorization improved by 0.6% on average after using FS. Secondly, image datasets, two types of images can be used to detect breast cancer: (i) Mammogram images and (ii) histology images. Mammography screening has long been considered as the most successful method of detecting abnormalities in the breast tissue, with the most common results being breast masses and calcifications, both of which can indicate the presence of cancer [4]. We may use deep learning algorithms on both sorts of photos, but the latter, i.e. histology images, has a greater accuracy because the images contain considerably more cell information and are high resolution [5]. As a result, in this scenario, we'll use histology images. Our task is to detect the mitosis cell in the image. Mitosis count is a critical indicator for the diagnosis of breast cancer. Many machine learning and deep learning techniques like SVM, Decision Tree, CNN, etc. have been applied [6,7]. But very few deep learning models which have a larger number of layers have been applied due to various computation and data-related problems. In this case study we are going to apply U-Net for image segmentation and then Faster R-CNN for object detection. Early identification is critical for BC because it improves survival rates, reduces treatment risks, and lowers mortality. This chapter is organized as follows: Section 10.2 describes the several types of breast cancer datasets. Section 10.3 describes related works and concludes in Section 10.4.

10.2 DATASETS

Collecting datasets for breast cancer cases is a critical task for assisting researchers in their diagnostic tasks. The Breast Carcinoma Subtyping dataset, the PatchCamelyon dataset, the Breast Cancer Semantic Segmentation dataset, and the Breast Cancer Histopathological Dataset. The dataset is presented in this survey to explain the roles of each of them in several recent research studies on breast cancer diagnosis.

10.2.1 Wisconsin Breast Cancer Dataset (WBDC)

Wisconsin Breast Cancer Dataset (WBCD) downloaded from the UCI ML Repository in this part. This dataset is utilized to separate between cancerous and benign tumors. In Wisconsin, all features were derived from fine needle aspirate images that depict nucleus features. In Wisconsin hospitals, the WDBC contains the characteristics of 569 patients (357 benign and 212 malignant cases) and 32 qualities. The measurement result is represented by each property. The identifying number and patient diagnosis status are the first and second attributes, respectively. Standard error, mean, and the least of red nucleus traits are the remaining properties.

10.2.2 Breast Carcinoma Subtyping (BRACS)

The Breast Carcinoma Subtyping (BRACS) dataset is a vast collection of annotated Hematoxylin and Eosin (H&E)-stained pictures designed to aid in the diagnosis of breast lesions. There have been 547 Whole Slide Images (WSIs) and 4539 Regions of Interest (ROIs) retrieved from the WSIs in BRACS. Each WSI and its associated ROIs are categorized into several lesion types by a group of three board-certified pathologists. BRACS distinguishes between three types of lesions: Benign, malignant, and atypical, which are further subdivided into seven categories [8].

10.2.3 Breast Cancer Semantic Segmentation (BCSS) Dataset

The BCSS dataset comprises approximately 20,000 tissue region segmentation annotations from The Cancer Genome Atlas breast cancer pictures (TCGA). Using the Digital Slide Archive, pathologists, pathology residents, and medical students collaborated to annotate this large-scale dataset as shown in Figure 10.1. It allows for the creation of very accurate tissue segmentation machine-learning algorithms [9].

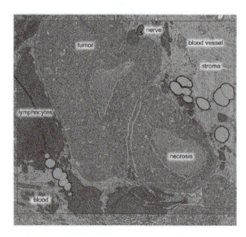

Figure 10.1 Samples of the BCSS dataset.

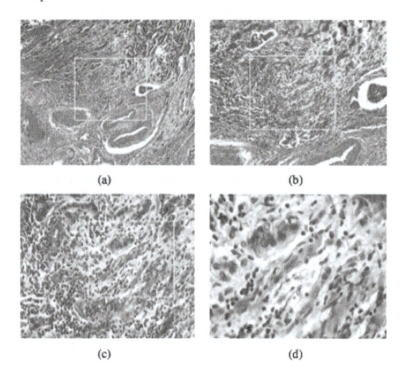

Figure 10.2 Samples of the BreakHis dataset.

10.2.4 Breast Cancer Histopathological Database (BreakHis)

The Breast Cancer Histopathological Image Classification (BreakHis) is made up of 9109 microscopic images of breast tumor tissue collected from 82 people at various magnification levels ($40\times$, $100\times$, $200\times$, and $400\times$) as shown in Figure 10.2. There are 2480 benign and 5429 malignant samples in this collection (700×460 pixels, 3-channel RGB, 8-bit depth in each channel, PNG format) [10].

Figure 10.3 Samples of the PCam dataset.

10.2.5 PatchCamelyon (PCam) Dataset

PatchCamelyon is a dataset for picture categorization. It is made up of 327.680 color pictures (96 × 96 px) taken from lymph node histopathologic scans. Each image has a binary label that indicates the existence of metastatic tissue as shown in Figure 10.3. PCam is a new machine learning benchmark that is larger than CIFAR10, smaller than ImageNet, and trainable on a single GPU [11].

10.2.6 Kumar Dataset

The Kumar dataset contains 30 1000 × 1000 picture tiles taken at 40 magnifications from The Cancer Genome Atlas (TCGA) database for seven organs (6 breast, 6 livers, 6 kidneys, 6 prostates, 2 bladders, 2 colons, and 2 stomachs) as shown in Figure 10.4. Each nucleus's boundary is completely marked within each image [12].

10.3 RELATED WORK

Graham et al. [6] propose Dense Steerable Filter CNNs (DSF-CNNs), which use group convolutions in a densely connected framework with multiple rotatable copies of each filter. In comparison to traditional filters, each filter is specified as a linear combination of steerable basis filters, allowing precise rotation, and reducing the number of trainable parameters. W. Lafarge et al. [7] focused on histopathology image analysis applications where it is desirable that the machine learning models do not collect the arbitrary global orientation information of the observed tissues. Three separate histopathology image analysis tasks are used to test the proposed system. Zhou et al. [8] developed a new smooth truncated loss that modifies losses to minimize outlier disturbance. As a result, the network may concentrate on learning from reliable and informative data, which improves generalization capability intrinsically. Experiments on the MICCAI Multi-Organ-Nuclei-Segmentation challenge in 2018 confirmed the efficacy of our proposed technique. On numerous independent multi-tissue histology picture datasets, Graham et al. [9] exhibit state-of-the-art performance relative to previous approaches. We present a new dataset of Hematoxylin and Eosin-stained colorectal adenocarcinoma picture tiles, which contain 24,319 exhaustively annotated nuclei with associated class labels, as part of this effort. Weiler et al. [10] extend He's weight initialization approach to filters specified as a linear

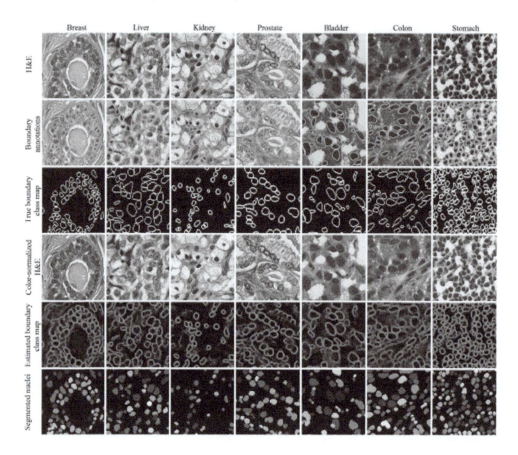

Figure 10.4 Samples of the Kumar dataset.

combination of atomic filters. Numerical experiments reveal that when the number of sampled filter orientations increases, the sample complexity increases significantly, indicating that the network generalizes learned patterns across orientations. S. Cohen et al. [11] propose Group-equivariant Convolutional Neural Networks (G-CNNs), a logical generalization of convolutional neural networks that exploits symmetries to reduce sample complexity. G-CNNs make use of G-convolutions, a novel form of layer with significantly more weight sharing than standard convolution layers. We describe a network and training technique that heavily depends on data augmentation to make better use of the given annotated samples. A contracting path is used to collect context, while a symmetrically expanding path is used to ensure exact localization.

10.4 CONCLUSION

Deep learning has made great progress in supervised learning and is continuing to improve radiologists' workflow and decision-making. This chapter included an overview of breast cancer datasets and the efficacy of several deep learning architectures for breast cancer screening for several related works. Furthermore, the goal of this work

TABLE 10.1 Breast Cancer Datasets and Specifications

Year	Authors	Model	Tasks	Dataset	Results (%)
2020	Graham et al. [5]	CIA-Net	Breast Tumour Classification Nuclear Segmentation Tumour Classification	PCam Kumar	F1-Score = 0.874 Dice = 0.891
2020	W. Lafarge et al. [13]	Roto-Translation Equivariant Convolutional Networks	Breast Tumor Classification Data Augmentation Multi-tissue Nucleus Segmentation Translation	PCam Kumar	Dice = 0.811
2019	Zhou et al. [14]	CIA-Net	Instance Segmentation Multi-Tissue Nucleus Semantic Segmentation	Kumar	Dice = 0.818
2018	Graham et al. [15]	HoVer-Net	Classification Multi-Tissue Nucleus Nuclear Segmentation	CoNSeP	Dice = 0.826
2018	Weiler et al. [16]	Steerable G-CNN	Breast Tumour Classification Segmentation Multi-Tissue Nucleus	PCam Kumar	Dice = 0.791
2016	S. Cohen et al. [17]	G-CNN	Breast Tumour Classification Segmentation Multi-tissue Nucleus	PCam Kumar	Dice = 0.856
2015	Ronneberger et al.	U-Net	Cell Segmentation Colorectal Microscopy Image Segmentation Lesion Segmentation Lung Nodule	CVC-ClinicDB Kumar	F1-Score = 0.7783 AUC = 0.9772

was to gain a clear grasp of current developments in deep learning for breast tumor identification.

Bibliography

1. D. Houfani, S. Slatnia, O. Kazar, N. Zerhouni, A. Merizig, and H. Saouli, "Machine learning techniques for breast cancer diagnosis: Literature review," Adv. Intell. Syst. Comput., vol. 1103 AISC, February, pp. 247–254, 2020, doi: 10.1007/978-3-030-36664-3_28.

2. F. Paquin, J. Rivnay, A. Salleo, N. Stingelin, and C. Silva, "Multi-phase semicrystalline microstructures drive exciton dissociation in neat plastic semiconductors," J. Mater. Chem. C, vol. 3, January 2019, pp. 10715–10722, doi: 10.1039/b000000x.

3. M. A. Rahman and R. C. Muniyandi, "An enhancement in cancer classification accuracy using a two-step feature selection method based on artificial neural networks with 15 neurons," Symmetry (Basel), vol. 12, no. 2, 2020, doi: 10.3390/sym12020271.

4. S. Dabeer, M. M. Khan, and S. Islam, "Cancer diagnosis in histopathological image: CNN-based approach," Informatics Med. Unlocked, vol. 16, no. May, p. 100231, 2019, doi: 10.1016/j.imu.2019.100231.

5. S. Graham, D. Epstein, and N. Rajpoot, "Dense steerable filter CNNs for exploiting rotational symmetry in histology images," IEEE Trans. Med. Imaging, vol. 39, no. 12, pp. 4124–4136, 2020, doi: 10.1109/TMI.2020.3013246.

6. S. Elmuogy, N. A. Hikal, and E. Hassan, "An efficient technique for CT scan images classification of COVID-19," vol. 40, pp. 5225–5238, 2021, doi: 10.3233/JIFS-201985.

7. E. Hassan, M. Shams, N. A. Hikal, and S. Elmougy, "Plant seedlings classification using transfer," no. July, pp. 3–4, 2021.

8. "BRACS," [Online]. Available: https://www.bracs.icar.cnr.it/.

9. BCSS, "No Title," [Online]. Available: https://github.com/PathologyData Science/BCSS.

10. "BreakHis," [Online]. Available: https://web.inf.ufpr.br/vri/databases/breast-cancer-histopathological-database-breakhis/.

11. "PCam," [Online]. Available: https://github.com/basveeling/pcam.

12. "Kumar," [Online]. Available: https://monuseg.grand-challenge.org/Data/.

13. M. W. Lafarge, E. J. Bekkers, J. P. W. Pluim, R. Duits, and M. Veta, "Roto-translation equivariant convolutional networks: Application to histopathology image analysis," Med. Image Anal., vol. 68, no. 2, 2021, doi: 10.1016/j.media.2020.101849.

14. C. I. Aggregation, Y. Zhou, O. F. Onder, Q. Dou, and E. Tsougenis, "CIA-Net: Robust nuclei instance segmentation," vol. 1, pp. 1–12.

15. S. Graham, Q. D. Vu, S. E. A. Raza, A. Azam, and Y. W. Tsang, "HoVer-Net: Simultaneous segmentation and classification of nuclei in multi-tissue histology images," pp. 1–18.

16. M. Weiler, F. A. Hamprecht, and M. Storath, "Learning steerable filters for rotation equivariant CNNs," Proc. IEEE Comput. Soc. Conf. Comput. Vis. Pattern Recognit., pp. 849–858, 2018, doi: 10.1109/CVPR.2018.00095.

17. M. Welling and M. W. U. V. A. Nl, "Group equivariant convolutional networks," vol. 48, 2016.

## CHAPTER 11

Review of Artifact Detection Methods for Automated Analysis and Diagnosis in Digital Pathology

Hossain Shakhawat

Department of Computer Science, American International University, Dhaka, Bangladesh

Sakir Hossain, Alamgir Kabir, S. M. Hasan Mahmud, and M. M. Manjurul Islam

Memorial Sloan Kettering Cancer Center, New York, New York

Faisal Tariq

James Watt School of Engineering, University of Glasgow, Glasgow, UK

CONTENTS

TRADITIONALLY, PATHOLOGICAL analysis and diagnosis is performed by observing biopsy specimen fixed on a glass slide using a microscope. In digital pathology, a scanner converts the glass slide specimen into a high-resolution digital image,

DOI: 10.1201/9781003251903-11

called whole slide image (WSI) which is observed on a computer screen. The WSI can be processed using image processing and artificial intelligence (AI) techniques for automatic analysis and diagnosis. The application of image processing and AI uncovers many useful information that remained unexplored in traditional pathology. This enabled more accurate analysis and precise selection treatment for better patient care. However, the accuracy of analysis could be affected if the quality of the WSI is compromised due to artifacts such as focus blur, tissue folds, or air bubbles. In traditional pathology, the specimen is checked for quality by manual review. A modern pathology requires an intelligent solution for reproducible automated analysis. This chapter introduces the whole slide imaging system and the major artifacts found in WSI, addresses the challenges of automated analysis caused by the artifacts, and reviews the existing artifact detection methods.

11.1 INTRODUCTION

Traditionally, the pathological analysis starts with a biopsy to remove tissue from the body, then the specimen is processed at different stages to place it on a transparent glass slide in a life-like state which is then examined under the microscope for analysis or to make a diagnostic decision. A pathologist compiles the entire specimen manually by adjusting the microscope to select suitable regions for examination during which artifact-affected areas are identified and ignored. Such manual selection and examination of regions are time-consuming, labor-intensive and the assessment may vary among pathologists. Figure 11.1 shows the approach of a traditional microscope-based pathological system. A biopsy specimen goes through a series of steps to prepare the glass slide which includes tissue fixation, fitting tissue into a cassette, tissue processing, sectioning, and staining. When the glass slide specimen is ready, it is observed using a microscope. Pathological analysis using a glass slide specimen and microscope involves checking the quality of the glass slide specimen, selecting the appropriate region, and compiling the observed information. In a microscope-based traditional pathology, all these tasks are performed manually by a pathologist.

In digital pathology, an advanced imaging device called whole slide imaging (WSI) scanner is utilized in place of the microscope. The WSI scanner converts the entire pathological specimen into a digital image which is observed on a computer screen for analysis and diagnosis using special software. The WSI scanner is considered the key to digital pathology. WSI system enables inch-perfect conversion of pathology specimen into a digital image, rapid transmission of the digital specimen for consultations, lossless preservation of the specimen, and user-friendly examination of the specimen on a digital display. Moreover, the incorporation of the digital specimen to pathology allows the application of pattern recognition and image analysis techniques for automated analysis and diagnosis. However, the quality of analysis using automated methods highly depends on the quality of the digital specimen which can be affected by different artifacts. The digital specimen includes the artifact contained in the glass slide, plus the artifacts generated while converting the glass slide to the digital image. Artifacts such as tissue fold and focus blur trouble the automated tools and thus mislead the analysis. In traditional pathology, artifacts are detected and ignored

Slide preparation

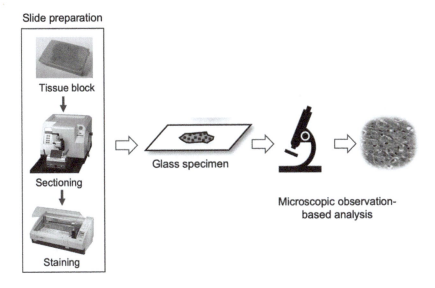

Figure 11.1 Traditional pathology for microscope-based subjective analysis.

manually by the pathologist which is time prohibitive and vulnerable to fatigue. For a reproducible automated analysis and diagnosis system, a more practical approach is required. Therefore, the need for an automated quality evaluation system that can estimate the usability of a WSI and locate the artifacts to eliminate them during analysis is indisputable.

This chapter portrays the WSI system as the basis of automated analysis and diagnosis for smart healthcare. Then, the chapter presents the major artifacts which trouble the automatic analysis using WSI. Finally, existing artifact detection methods are reviewed and their application for the automatic analysis is assessed to design a practical system.

11.2 AUTOMATED ANALYSIS AND DIAGNOSIS IN DIGITAL PATHOLOGY

Digital pathology is an image-based environment where pathological operations such as sharing, managing, analysis, and diagnosis are performed using information technology and a digital version of pathological specimen called WSI. WSI scanner transforms the entire pathological specimen into a digital image. The concept of the digital pathological specimen was first introduced for remote pathological works, also called telepathology. The digital specimen was shared over the network for analysis or consultation by a pathologist of a remote location. The technology of producing digital specimens significantly improved over the years with the technology of telepathology which includes store-forward, robotic microscopy, and WSI. A robotic microscope produces a digital image for a selected area of the glass slide specimen adaptively where the WSI scanner converts the entire specimen into a digital image. The WSI system has two major components: A scanner that produces the digital image and a viewer that is used for observing the image on a computer screen. For many years, the WSI system was considered impractical and discouraged for clinical-pathological works. Major issues were the scanning time, image quality, image size, image format, color variation, interactive handling, and data sharing [2,3]. At present time, many advanced scanners are available that are capable of converting the entire specimen into a digital image in less than a minute. The quality of the image has improved significantly in recent time and the pixel size is reduced from 1 μm to 0.23 μm. The viewing software is improved which allows seamless zooming, panning, and handling of the WSI. In 2017, Food and Drug Administrator in the United States approved Phillips's scanner for primary diagnosis which promoted the adaptation of the WSI system in the hospitals. A laboratory in Sweden has implemented a WSI system for large-scale routine diagnosis [4]. However, there are some issues that remained unsolved and worthy to report, mainly related to image quality. The quality of the WSI depends on the quality of the glass slide specimen and the ability of the scanner to produce a high-quality image. Faulty specimen preparation introduces tissue artifacts such as tissue fold and air bubbles in the glass slide which eventually compromises the quality of WSI. The glass slide specimen preparation involves manual operation at different stages, though automatic sectioning machines and staining machines are available in some laboratories. The surface of the specimen on the glass slide is not entirely flat which troubles the focusing system of the scanner to find the optimal focus by adjusting the objective lens. As a result, the focusing system fails sometimes which introduces out-of-focus artifacts in the image. Noise is another crucial artifact caused by the scanning system. The image format is another important issue for the practical implementation of the WSI scanner. The WSI scanner converts the entire specimen into a digital image which is very large in dimensions and requires a large space to store. The size of an uncompressed WSI file ranges between 1GB and 30GB and there is no standard file format. Thus, it is not difficult to understand why a WSI is tricky to handle. The WSI system utilizes a pyramid model to allow the user to access the image at different magnifications as illustrated in Figure 11.2.

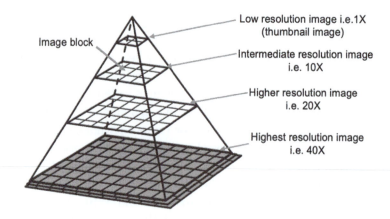

Figure 11.2 Pyramid model for a multi-resolution whole slide image.

The top layer is the thumbnail image which provides the lowest resolution such as 1× magnification and the bottom layer contains the highest-resolution image such as

40× or 60× magnification. Each image layer is divided into fixed-size image blocks and the coordinates of the blocks are synchronized so that a particular image block can be accessed at different magnification. This multi-resolution image structure allows access to the specimen at different magnification or resolutions depending on the necessity of image analysis application. A higher magnification image contains finer details of the specimen, and the access time increases with the magnification as well. Usually, a high-resolution image is more efficient for an image analysis method, but it takes a longer time to process. There is a tradeoff between accuracy and processing time. For some tasks, a low-resolution image is sufficient such as tissue area segmentation and large object detection on the slide. However, analyzing the tissue structure or cell morphology and detecting small biomarker signals require a high-resolution image which could be time-consuming. Figure 11.3 shows the structure of a digital pathology pipeline for automated image analysis and diagnosis systems for breast cancer patients. In a digital pathology pipeline, automated sectioning and staining systems are utilized for sectioning and staining the specimen. This allows the massive production of glass slides in a short time from a tissue block. After that, the glass slide is scanned using the WSI scanner. Modern scanners support a batch scanning system, allowing the operator to load multiple slides at a time which ranges from 10 to 400 slides depending on the scanner model. Then the profile for scanning the batch is set by the operator which includes the focus point selection, tissue area detection threshold, white balancing, magnification, image compression, and other scanner settings.

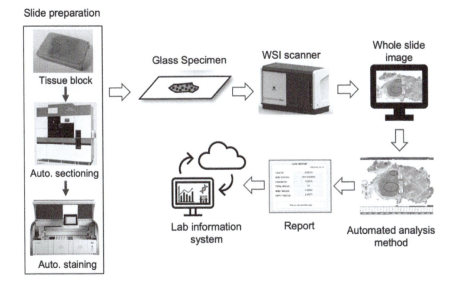

Figure 11.3 Digital pathology workflow for automated analysis.

After that, the glass slides are scanned and stored in the given directory without demanding any input from the operator. This system reduced the operator's work significantly compared to the previous scanner which requires loading one slide at a time. Once the scanning is completed, the operator manually checks the quality of the WSI. The quality is evaluated subjectively by the operator to select a WSI for analysis. If the quality is judged satisfactory the WSI is forwarded for analysis, otherwise, the glass slide is rescanned. The operator mainly judges the quality of WSI in terms of the amount of out-of-focus regions, tissue area detection and color. This manual evaluation is time-consuming, vulnerable to fatigue and not practical for routine clinical applications where a large number of slides are scanned daily. Moreover, if there is a useless artifact in the image such as tissue fold, pen mark, or dust the pathologist must locate and exclude them manually during the analysis. Therefore, it is necessary to evaluate the quality of WSI automatically. After that, the WSI is observed on a computer screen using a special viewer. In digital pathology, automated image analysis tools are utilized which automated analysis which can be broadly categorized into two types: Morphological image analysis and molecular image analysis.

11.2.1 Morphological Image Analysis

The application of digital image analysis, pattern recognition and machine learning techniques enable detailed analysis of tissue architecture and morphology which lead to better diagnosis. Features such as nuclei shape, size, and distortion are effective in determining the degree of malignancy for a cancer patient. Hematoxylin and Eosin (H&E) staining is considered the standard staining method in histopathology. Hematoxylin dye stains the acidic components of the tissue such as nuclei in blue color and the eosin stains all other components of the tissue in pink which includes the cytoplasm, cell walls and extracellular fibers. This interpretation is often sufficient for many diagnoses which are based on the structure of the cells. The H&E WSI is used routinely in digital pathology for the morphological assessment using automated algorithms to understand cell and tissue structure [5–10]. In the breast cancer analysis, malignant and invasive cancer regions are identified on the H&E WSI. Figure 11.4 shows an H&E WSI of breast tissue. Although H&E is useful for detecting the changes in the cells, one major limitation of the H&E slide is the lack of molecular data associated with a cell.

11.2.2 Molecular Image Analysis

In order to visualize more detailed information of the molecules present in the specimen, special stains are used such as immunohistochemistry (IHC), chromogenic in situ hybridization, fluorescence in situ hybridization and others. IHC expresses antibody-antigen interactions and is commonly used for subtype cancer classification. In IHC-DAB (diaminobenzidine)-stained images, positively stained regions are brown and negative are blue. The number of positive and negative nuclei is counted to estimate the percentage of positive nuclei to analyze the cancer tissue. IHC is also used for assessing the grade of breast cancer where the staining strength in the cell membrane is scored to identify HER2-positive cancer. The presence of estrogen receptor (ER), progesterone receptor (PR), HER2, and proliferation marker protein Ki67 in IHC is used for the diagnosis of breast cancer. There are several automated and semiautomated image analysis methods available for IHC-based image analysis [11–13].

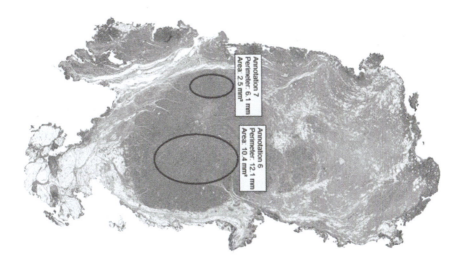

Figure 11.4 H&E WSI of a breast tissue.

The CISH and FISH are in situ hybridization techniques for detailed molecular assessment which allows counting the presence of the HER2 gene inside the nucleus. Figure 11.5 shows the example of molecular image analysis using FISH and CISH WSI. An example of an automated image analysis system is the breast cancer decision support system proposed by Yagi et al. to assist pathologists to select a patient for giving anti-HER2 therapy [14]. This system utilized H&E WSI in the first step to detect invasive region automatically and then CISH WSI is used for quantifying the HER2 amplification status. Saito et al. proposed a fully automated image analysis system to detect malignant cancer regions from gastric, colorectal and breast cancer specimens [15]. Another example of an automated image analysis system is the hepatocellular carcinoma assessment system developed by Yamashita et al. [16].

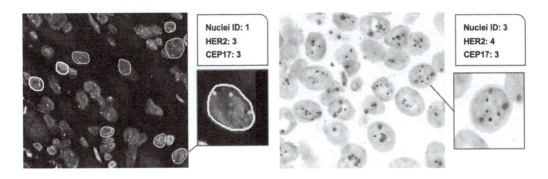

Figure 11.5 HER2 and CEP17 biomarker detection from FISH (left) and CISH (right) WSI.

11.3   MAJOR ARTIFACTS

Artifacts are introduced at different stages of specimen preparation. In microscope-based pathology, artifacts are mainly caused by faults that happened while preparing the glass slide specimen. In digital pathology, the WSI scanner scans the glass slide specimens to convert them into digital images, and artifacts can be introduced during this transformation. One way to categorize the artifacts is based on when they are produced as tissue artifacts and scanning artifacts. Artifacts that are produced at different stages of glass slide preparation are considered tissue artifacts and the artifacts produced while scanning the specimen are scanning artifacts. Both types of artifacts require trouble analysis. However, scanning artifacts can be fixed by rescanning the specimen but rescanning could not help fix the tissue artifacts. Therefore, an intelligent system should rescan the specimen if the image is poor due to scanning artifacts and detect tissue artifacts to eliminate them from analysis if they are affected by tissue artifacts. Sometimes, when the entire specimen is affected by tissue artifacts, the specimen is prepared again from a different section of the tissue block which is time consuming and requires a lot of work.

11.3.1   Scanning Artifacts

Scanning artifacts are generated mainly due to the hardware failure of the WSI scanner. A study in Dutch digital pathology documented that 5% of the scanned slides suffer scanning artifacts [1]. The out of focus and noise are the major scanning artifacts found in whole slide images. Figures 11.6 and 11.7 show an example of out-of-focus and noise artifacts, accordingly.

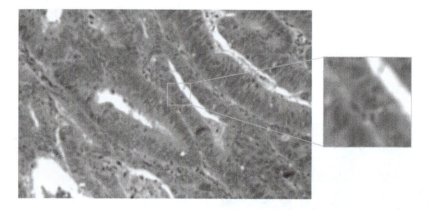

Figure 11.6   Out-of-focus artifact.

A WSI scanner selects a limited number of focus points automatically or allows the operator to select them manually for scanning a specimen. Then the scanner scans the specimen either following the line scanning method or the tile scanning method. In the case of the line scanning method, the slide is divided into a number of specific lines, wherein the tile scanning method the slide is divided into tiles. The limited focus points are assigned to the lines or tiles. Figure 11.8 depicts the tile and

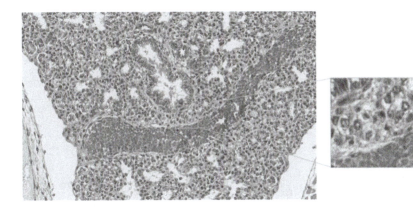

Figure 11.7 Noise artifact.

line-based scanning approach of the WSI scanner. If the scanner assigns a focus point to each patch, the focus of the image will be more accurate, and the resultant image will be high in quality. But a higher number of focus points will increase the scanning time significantly which is not practically feasible. Therefore, it is necessary to select an optimal number of points for which the quality is satisfactory and scanning time is minimal. Selecting the optimal focus points for sufficient quality in a minimal scanning time is the main challenge while scanning a specimen. As a result, sometimes parts of the image become out of focus. After assigning the focus points, the scanner adjusts the focal planes for the selected patches either by adjusting the objective lens position and the slide stage. Now, if the scanner selects focus points from some regions that are not properly aligned with the height, then those areas will be blurry in the WSI. The difference in distance between the erroneously selected focus point and the ideal focus points is directly proportional to the amount of blurriness. Appropriate stage and optics alignment are necessary to avoid focus problems. Figure 11.9 shows how autofocusing works in the WSI scanner.

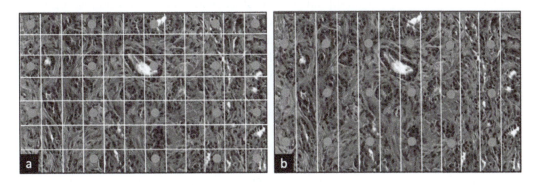

Figure 11.8 Tile-based (a) and line-based (b) scanning methods of a WSI scanner. Solid circles indicate focus points.

Noise is the other common scanning artifact found in WSI. The noise of a digital image can be classified broadly into two types. Firstly, the noise which is independent

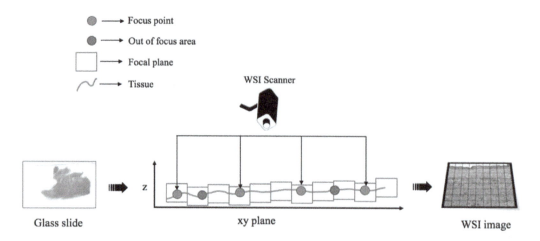

Figure 11.9 Focusing mechanism of a WSI scanner.

of pixel values and the noise which is not independent of pixel values. Noises caused by the imaging sensor are dependent on the pixel values. An imaging sensor converts the photons into the electrons, which are then converted into pixels values. Examples of sensor-based noises are fixed pattern noise, dark current noise, shot noise, amplifier noise, and quantization noise. Detecting sensor-based noises is difficult as their values are similar to other pixels. In the case of WSI scanners, it is difficult to identify the source of the noise. Therefore, a noise detection approach that considers noise as a random variable, independent of the pixel's values, is suitable for estimating noise from a WSI.

11.3.2 Tissue Artifacts

The glass slide specimen preparation is a lengthy process that starts with surgical biopsy followed by fixation, tissue processing, embedding, sectioning, and staining and ends with placing a coverslip on the top of the slide. Tissue fixation is done soon after removing the tissue after surgery to preserve the tissue in its original state. Usually, 10% buffered formalin is used as a fixative to prevent decay due to autolysis and putrefaction. Tissue fixation is crucial as the investigation of molecules or tissue structure depends on the accurate preservation of the specimen. After fixation, the specimen is cut using a scalpel to fit it into a cassette where it is stored until processed. Tissue processing is the next step, which is a combination of three tasks dehydration, cleaning, and embedding. Dehydration is performed to remove water and formalin from the tissue by immersing the specimen in alcohol. After that xylene is used to remove the alcohol used earlier. This allows the infiltration with paraffin wax and the tissue is enclosed by the paraffin wax. Then the tissue block is embedded into a supporting medium. After embedding the tissue is cut into sections and placed on slides. Before sectioning, the wax is removed to expose the tissue. Then the tissue is sectioned into thin slices of 4–5 μm using a microtome or an automatic sectioning machine and placed on a glass slide. After that, the slide is stained and protected with a coverslip to prevent it from drying out. Faults in the preparation cause artifacts in the glass slide specimen. Tissue artifacts alter tissue morphology and cytology features which lead to misinterpretation of analysis. These artifacts may cover only a small area of the specimen or could cover the entire specimen. Tissue artifacts are considered useless and ignored for analysis by practitioners. Some artifacts are easy to identify in a slide while some are difficult. One common way to categorize tissue artifacts is based on the stages of glass slide preparation. Tissue-fold, air bubble, tissue tear, dust, pen mark, and curling artifacts are the major artifacts found in histopathological specimens while tissue fold and air bubbles are the most common ones. In the case of molecular analysis using CISH slides, HER2 dye noise is a major problem that affects the HER2 assessment of breast cancer patients.

The tissue fold is the multiple layers of the thin tissue slice. It happens at the tissue processing stage during the lifting of the tissue section. Due to the poor blade or fatty tissue sometimes, the tissue adheres to the surface of the blade and overlaps on the layer of tissue. Tissue folds alter the features of the specimen and negatively affect the diagnosis [17]. They have higher thickness compared to the normal tissue areas and soak a higher amount of dye during the staining. As a result, tissue folds have higher saturation and lower brightness compared to their neighboring areas which could affect the automated analysis tools. Due to the higher thickness, tissue folds attract the automated focusing algorithm. If focus points are selected from a fold, it makes the neighboring regions out of focus. Figure 11.10 shows the example of tissue folds. Tissue fold is considered the most common artifact in histopathology, and it is very difficult to avoid even with greater care.

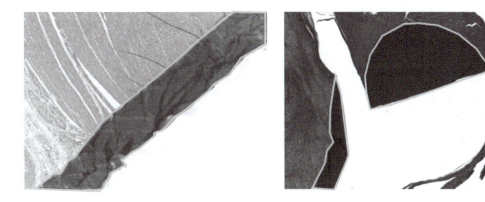

Figure 11.10 Shows the tissue folds (dark regions marked by the white stripes) for two different stains.

Air bubbles are the air trapped under the coverslip of the glass slide. Usually, this happens at the mounting stage when the coverslip is placed on the stained section caused by the thin mounting medium. If the bubbles are generated due to the very thin mounting medium, it can be avoided by using a mounting medium of adequate thickness. Bubbles can be generated and trapped under the section due to poor floatation techniques as well. The use of boiled water in a floatation bath may help fix this issue. Sometimes the trapped air is absorbed by the tissue which makes the tissue dry and it fails to adhere to the glass slide properly. Air bubbles hide the details of the tissue structure making it difficult to extract significant features for the analysis. Therefore, air bubbles-covered areas are considered useless and avoided by pathologists. However, it grabs the attention of the automatic focusing system of the scanner. If the focus points are selected for such areas, it makes the neighboring normal tissue regions out of focus. Air bubble artifacts usually have dark boundaries as the air has a different refraction index compared to its surrounding areas. The tissue areas inside an air bubble look washed out as shown in Figure 11.11.

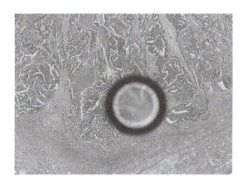

Figure 11.11   Air bubbles are shown.

Ink marks are used in histopathology for several purposes to assists practitioner in their work. The tissue edges are marked to identify the orientation and the face of the tissue during embedding to ensure the correct placement of the specimen. It can be used to locate small lesions in a larger tissue section, tabular structures, and other components in a specimen. Sometimes the ink marker invades the tissue structure which is ignored by the pathologists during analysis. Therefore, it is necessary to detect ink marks automatically and eliminate such areas from the analysis.

Curling artifact is the crimping and bending of the tissue, usually seen in incisional biopsies. They are similar to the tissue folds but considered less of a problem due to their smaller size. Curling artifacts are generated due to the shrinking by formalin fixation [18]. A crush artifact is the distortion or mutilation of tissue due to the compression caused by the surgical instrument such as forceps or others [19]. It changes the tissue morphology which makes it difficult to recognize the structure. The nuclei become distorted and could mislead the analysis. Another artifact caused by the compression of tissue is hemorrhage which is often misinterpreted as the morphological change [19].

An HER2 noise artifact is a common artifact for CISH slides. HER2 noise looks similar to a HER2 signal. As a result, it becomes difficult to differentiate HER2 signals from a HER2 noise. HER2 noise is caused due to the prolonged hybridization time. In molecular analysis, it is crucial to eliminate false-positive signals and quantify only efficient signals to ensure the reliability of the analysis. Therefore, it is necessary to detect HER2 noise and eliminate them from quantification otherwise the result could be misleading. Figure 11.12 shows a CISH slide with HER2 noise.

Examples of other artifacts that could affect the automated analysis include tissue dust, formalin pigment, incomplete staining, fulgeration artifact, injection artifacts,

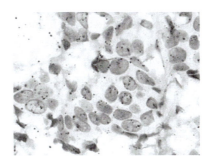

Figure 11.12   HER2 (tiny dots) noise in the background which looks similar to a HER2 signal.

over dehydration, under dehydration, and others. Tissue specimens could be contaminated by dust, dirt, or fingerprints during the preparation. Tissue tear is the splitting of tissue specimen into parts which is caused by the faulty blade. Pigment artifact is the deposit of formalin pigments on the tissue. Long exposure of tissue to formalin solution produces formaldehyde pigments as a result of the reaction between formalin and hemoglobin.

11.4   CHALLENGES FOR AUTOMATED ANALYSIS DUE TO ARTIFACTS

In digital pathology, analysis is performed by examining the WSI using automated methods. The accuracy of the analysis depends on how well the method extracts morphological features or detects the biomarkers which require an artifact-free quality WSI. The analysis could be misleading and unproductive if the WSI contains artifacts, and such areas are included in the analysis. Markiefka et al. performed the diagnosis of deep learning-based classification method. In the experiment, they have intentionally added artifacts in the test set to show that the performance of the deep learning model degrades significantly in the presence of artifacts in WSI. [20] In a similar investigation, we have generated digitally degraded images by adding focus blur and noise. The focus blur artifact was produced by using a Gaussian blur filter where the standard deviation was varied from 2 to 14 with a step size of 2. The noise artifact was generated by varying the standard deviation of the Gaussian noise filter from 2 to 14 with a step size of 2. We also produced excessive sharp images by using unsharp masking, the characteristic of the Gaussian filter was the same. After that these artifact-affected images were utilized for molecular analysis and morphological analysis. The morphological analysis identified a significant difference in the texture features between the original image and the digitally blurred image. In the molecular analysis, the HER2 (human epidermal growth factor receptor 2) detection and nuclei detection methods failed to detect the HER2 gene from blurry, noisy, and excessively sharp images where it successfully detected the gene from the original images, illustrated in Figure 11.13.

In order to investigate the effect of tissue artifacts, Haralick's gray level co-occurrence matrix (GLCM) features were derived from artifact-affected areas and compared with its neighboring regions which are not out of focus. A significant difference was noticed in the feature values which indicates the change caused by artifacts. This experiment was conducted for tissue-fold and air bubble artifacts where Haralick texture features were utilized. The texture is a useful indicator for morphological image analysis such as identifying objects or classifying structures. Haralick features are one of the earliest methods for analyzing images using texture information. Other popular methods for texture-based image analysis include the scale-invariant feature transform (SIFT), speed-up robust feature (SURF), and local binary patterns (LBPs).

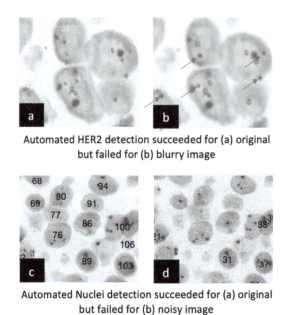

Automated HER2 detection succeeded for (a) original
but failed for (b) blurry image

Automated Nuclei detection succeeded for (a) original
but failed for (b) noisy image

Figure 11.13 Automated image analysis tools fail in the presence of artifacts.

11.5 REVIEW OF EXISTING ARTIFACT DETECTION METHODS

Several artifact detection methods have been developed for digital pathology. Most of the methods detect either tissue artifacts or scanning artifacts while a few methods incorporate both types of artifact detection. Again, most of the tissue artifact detection methods only detected the tissue-fold artifacts considering it as the only major tissue artifact. Methods for detecting air bubble, dust, ink mark, hemorrhage, and crush artifacts are very few. On the other hand, most of the papers that identified the scanning issues for quality evaluation considered focus blur as the major artifact and detected them. In this chapter, the artifact-detection methods are reviewed by categorizing them into three groups based on the used technology as image enhancement-based approach, machine learning approach, and others. Methods that modify or enhance the image information for detecting the artifacts are considered image enhancement-based approaches. Machine learning-based artifact detection methods utilize some representative features extracted from the artifact and use them to teach a model to differentiate artifacts from normal tissue areas. The machine learning methods can be further divided into two sections, the traditional machine learning approach and the deep learning approach. In this chapter, the traditional machine learning approach means the methods that utilize SVM, Naïve Bayes, K-means clustering, linear regression, logistic regression, or similar models where the selection of features to feed the model is made by the user or by a separate method. Traditional machine learning models are efficient for solving classification, regression, clustering, and dimension reduction problems especially when the dataset is limited. The deep learning method means the methods where the selection of features to train the model is made by the method itself. These models can learn features from the given data such as images or videos and identify significant features for the given task without experts explicitly identifying them. Recent research has shown that the deep learning-based method particularly CNN achieves better performance than traditional machine learning-based methods while it requires a significantly larger dataset, higher computation power and time for training. Examples of some popular CNN models include VGG16, ResNet, InceptionNet, GoogleNet, and MobileNet. There are some other methods for detecting artifacts that rely on different techniques such as the Tanengrad operator, Laplacian operator, the magnitude of Sobel gradient or the behavior of point spread function (PSF) of the scanner. Table 11.1 shows the list of the reviewed artifact detection methods in this chapter. Some commercial artifact detection applications are also available, but the details of these works are not sufficient for reproducing the experiment. Therefore, they are not included in this table.

11.5.1 Image Enhancement-Based and Other Methods

The tissue folds contain multiple layers of tissue, have a higher thickness and soak more dye compared to the normal tissue areas. As a result, the tissue fold appears dark in color having higher saturation and lower brightness compared to the normal tissue areas. Most of the image enhancement-based methods are motivated by this fact and utilize the saturation difference between the tissue fold and normal tissue as the major property to identify tissue-fold pixels. The saturation difference property of the tissue fold motivated the development of a color-shifting method to enhance the color metric difference in folded and non-folded areas. One effective way to enhance the color metric differences without changing its hue is to transform the RGB image to HSV (hue, saturation, and value) image and then modify the saturation or luminance component. Increasing the saturation decreases the luminance of the image and vice-versa.

Pinky A Bautista et al. proposed a similar color enhancement-based tissue-fold detection method [21]. In their method, they have shown that the shifting and scaling can be directly applied to the RGB values of the pixel to enhance the luminance and saturation accordingly without transforming the RGB image to the HSV space. This method modified the saturation by adding some constant to the RGB values of the pixel which is considered as color enhancement by shifting the saturation. However, adding a constant to the RGB values decreases the saturation and increases the luminance of the pixel. On the contrary, adding a constant to the saturation value of the pixel in HSV space increases the saturation and decreases the luminance.

TABLE 11.1   List of Reviewed Papers

SI	Paper	Detected Artifacts	Approach	Magnification	Evaluation
1	Palokangas et al. 2007	TA: Tissue fold SA: N/A	TML	20×	Mis segment rate 14.54%
2	Bautista et al. 2009	TA: Tissue fold SA: N/A	IME	Thumbnail	N/A
3	Zerbe et al. 2011	TA: N/A SA: Focus blur	IME	High resolution	N/A
4	Hashimoto et al. 2012	TA: N/A SA: Focus blur & noise	IME & TML	20× & 40×	Pearson corr. coef. 0.94
5	Kothari S el al. 2013	TA: Tissue fold SA: N/A	IME	Thumbnail	Average true rate 0.77
6	Lopez et al. 2013	TA: N/A SA: Focus blur	TML	20×	Accuracy 0.88
7	Shrestha P et al. 2016	TA: N/A SA: Focus blur, brightness & contrast	Other	40×	N/A
8	Senaras C et al. 2018	TA: N/A SA: Focus blur	DL	40×	Accuracy 93.2%
9	M. S. Hosseini et al. 2018	TA: N/A SA: Focus blur	Other	40×	Pearson corr. coef. 0.8556
10	A. Foucart et al. 2018	TA: Tissue fold SA: N/A	DL	1.25× & 0.625×	Accuracy 91%
11	Babaie et al. 2019	TA: Tissue fold SA: N/A	DL	20×	Accuracy 81%
12	Janowczyk et al. 2019	TA: Tissue fold, ink mark & air bubble SA: Focus blur	DL	40×	Average agreement 95%
13	Kohlberger T et al. 2019	TA: N/A SA: Focus blur	DL	40×	Spearman rank coef. 0.94
14	Hossain et Al. 2020	TA: Tissue fold & air bubble SA: Focus blur & noise	IME & TML	1× & 20×	Average accuracy 97.7%
15	Moyes et al. 2020	TA: Stain artifacts SA: N/A	DL	20×	N/A
16	Smit et al. 2021	TA: Tissue fold, air-bubble, ink mark & dust SA: Focus blur	DL	Low-resolution	Dice score 0.91

Abbreviations: SI: Serial; TA: Tissue artifact; SA: Scanning artifact; TML: Traditional machine learning; DL: Deep learning; IME: Image enhancement; N/A: Not available.

Both operations enhance the color metric features by changing the saturation and intensity while it keeps the hue unchanged. In their method, the constant was derived adaptively based on the saturation-luminance difference of a pixel to enhance a tissue-fold pixel more than a normal pixel. After that, a fixed threshold was applied on the luminance change to identify a tissue-fold pixel that has higher change than a normal pixel. The fixed threshold may not work well for all slides due to the technical and biological variation in slides. The color of the WSI depends on the staining protocol of the laboratory which varies among the laboratories. A tissue fold in a lightly stained slide can be similar to a normal region in a highly stained slide. The color also varies for different WSI scanners. Therefore, it is necessary to demonstrate a color enhancement-based detection method for multiple scanners and multiple slides prepared by different laboratories. This method was demonstrated for multiple scanners, but only H&E slides were used. This method utilized a low-resolution thumbnail image which is efficient for practical, but only tissue-fold artifacts were detected. Another method for detecting the tissue fold based on the saturation is proposed by Kothari et al. where two adaptive thresholds are applied to differentiate the folded regions from the normal tissue [22]. They have shown that a saturation and intensity difference is a major indicator to identify tissue fold by investigating connectivity of pixels in a slide based on their saturation-intensity difference for H&E specimens. They created a connectivity map of pixels where two pixels were considered connected if they had the same saturation-intensity difference and were in the 8×8 neighborhood. The connected pixels' count was highest for the nuclei regions that have an intermediate saturation-intensity difference while for the folded regions the difference was higher and the connected pixels' count was lower. Two threshold values were derived from the map to locate the tissue-fold pixels from nuclei and other parts of the specimen. This saturation-intensity-based feature was selected from a list of 461 features derived from 105 H&E-stained WSIs using a statistical method that ensures the efficacy of saturation-based features for tissue-fold detection. This method achieved better detection accuracy compared to the Pinky et al. method when applied on different slides. This method was demonstrated on a cancer specimen, and it was found that the cancer prediction models performed better after eliminating the tissue folds. However, a single colorimetric feature-based method may not be robust for a reproducible image analysis system. Moreover, it is necessary to demonstrate the method for other stains.

Both image enhancement-based methods demonstrated the effectiveness of the saturation-based features for tissue-fold detection. Both utilized a low-resolution thumbnail image which ensured fast detection, important for practical implementation. However, these methods only considered tissue artifacts and detected only tissue folds in their work, and it is necessary to detect other popular artifacts such as air bubbles, ink marks, and dust for automated analysis.

In another study, Zerbe et al. proposed a method that utilizes the Tanenbaum gradient to estimate the sharpness of image patches to assess the presence of focus blur and classified them as excellent, okay, review, or defective. However, this method did not consider the tissue artifacts [23]. Shrestha et al. proposed a reference-based image quality evaluation method that compares the focus quality, brightness, and contrast of an image with the reference to judge the usability [24]. In pathological image analysis, it is not possible to find an ideal reference image. Therefore, it is not practicable to evaluate the focus quality of an image by comparing it with a standard reference image. A detection or evaluation method that doesn't rely on reference is more suitable for practical use. MS Hosseini et al. proposed a no-reference focus quality evolution method that boosts the high-frequency information of the image to estimate the focus problem [25].

11.5.2 Machine Learning-Based Methods

One of the earliest methods for detecting tissue folds was proposed by Palokangas et al. [26]. This method also utilized the saturation-intensity difference image but relied on the k-means clustering method other than the threshold-based techniques to identify the folded regions. The k-means clustering is an unsupervised method meaning there is no need of labeled data. One of the major challenges of applying machine learning techniques on medical image-based data for object detection or segmentation is annotating a significant number of images to prepare label data for training as it requires a significant amount of time and effort. However, an unsupervised method such as K-means can categorize the unlabeled input data such as the pixels into a specified number of clusters and is useful for detecting objects from unannotated images. Pixels with similar properties are kept in the same cluster and each cluster is considered as a candidate for a possible class. Nevertheless, the performance of K-means clustering depends on the selection of the number of clusters which is the value of K and the initialization of centroids. A pathological specimen may contain different types of components in a slide such as nuclei, cytoplasm, background, blurred regions, tissue folds, or other irregularities. An optimal number of clusters should be selected to separate the tissue-fold regions from other components of the slide. Palokangas et al. determined the cluster number based on the variance of the cluster containing folds. The second issue when applying the K-means is its random initialization of initial centroids. A poor selection of initial centroids may lead to inefficient clustering. The computation time could increase significantly as well. Another major limitation of the k-means method is that it cannot handle well data with varying sizes and densities, unlike the Gaussian mixture model. Palokangas et al. utilized a post-processing step to compensate for the misclassifications in their method. This method detects only the prominent tissue folds and results in high false positives in the absence of folds in the slide. Moreover, this method relied on a 20× magnification image and was demonstrated for only H&E slides. Although the unsupervised methods provide the advantage of using an unlabeled image, it could be confused by other irregularities of the specimen. Therefore, a supervised method such as template matching, SVM or CNN is more suitable for detecting artifacts from histopathology slides.

Hashimoto et al. proposed a quality evaluation method that derives the quality of the image block using a linear regression model based on the amount of blurriness and noise. This method was compared with the pathologist's rank scores and mean square errors for the selected image blocks. In both cases, the proposed method had high concordance. This method was designed for evaluating only selected regions of interest, training and test images were captured from the same slide, was demonstrated on a single H&E slide, and resulted in false positives for the tissue artifact-affected image blocks. Therefore, the practical usability and the generalizability of the method require more investigation [27]. Lopez et al. proposed a blur detection method in which they identified a set of features which include Haralick features and a Tenengrad operator [28]. After that, a decision tree classifier was utilized to detect an image block with a focus blur problem. This method included H&E and IHC slides scanned by a single scanner. This method achieves good accuracy if applied only to H&E or IHC slides, but the generalized accuracy is lower.

Hossain et al. proposed another quality evaluation method that incorporated tissue artifact detection with the quality evaluation based on blurriness and noise [29,30]. They proposed a supervised artifact detection method in which they have utilized SVM with optimally selected features. The features were selected from a features bank which included saturation-based colorimetric features, texture features, shape features and other physical property-based features. The rotation-invariant classification performance was also investigated for the feature selection. They have shown in their experiment that traditional machine learning techniques such as SVM with optimally selected handcrafted features can achieve better classification performance than deep learning models such as CNN when the dataset is significantly small. In their work, they have considered scanning artifacts and tissue artifacts both types of artifacts for evaluating the quality of a WSI for automated analysis. Tissue folds and air bubbles were detected as tissue artifacts using two binary SVM classifiers. While the focus blur and noise artifacts were detected to identify the scanning failure. Focus blur was identified based on the width of the edges. The noise was identified based on the assumption that it is a random value that is different from its neighbors. One important aspect of this work is the investigation of issues for practical implementation of an automated image analysis system in hospitals and the incorporation of pathologists' feedback to design a user-friendly system. They have analyzed the time requirement of the artifact detection and quality evaluation system. The system was demonstrated for 52 different slides by two different scanners which includes multiple organs, and different stains to ensure its robustness and practical use. They have experimented with different resolution images for tissue artifact and scanning failure detection and finally selected the resolution for which the accuracy is optimal, and time is minimal. Their proposed system relies on a thumbnail image for tissue artifact detection and a 20× image for scanning failure detection. However, it is necessary to consider other tissue artifacts while designing a fully automated system. The focus blur was detected based on the width of the edges which requires a lot of time. Recently, CNN-based models were found to be effective for detecting focus blur in a comparatively short time even using lower magnification such as 10×. Moreover, some CNN-based models are found productive when the dataset is limited. Transfer learning can also be utilized by using the bottleneck features and training only the final layers.

11.5.3 Deep Learning-Based Methods

Two major tasks while designing a machine learning solution are identifying the discriminative features and training the machine with those features. Traditional machine learning techniques rely on human intuition or a separate algorithm for feature selection while deep learning methods such as CNN can extract and identify features on their own. This autonomous approach of incorporating the feature selection process with parameter tuning while training the model enables deep learning methods to solve many complex problems. CNN-based models were found to be very effective for processing visual information to convert them into meaningful features to produce the desired outcome. In recent times, many CNN-based models have been demonstrated for several medical image analysis tasks such as biomarker detection, anomaly detection, cancer classification, nuclei segmentation, grade assessment and others.

Senaras et al. proposed a CNN-based method to detect focus blur problems of WSIs and demonstrated it on a diverse dataset [1]. The image dataset used for training the model included 8 different labels of focus blurs, generated by varying the focus plane and extracting from 16 differently stained slides at 40× magnification. This method outperforms Lopez's method in terms of accuracy and speed. However, in the comparative study, they used a 20× image downsampled from a 40× image while Lopez's result was generated for a 20× image. The image quality of the WSI scanner can be explained using the sensor resolution such as the sampling rate and optical resolution such as numerical aperture. WSI scanned at 40× have higher numerical aperture and lower pixel size such as 0.23 micron/pixel than 20×. When a 40× image is downsampled to a 20× image, the sample rate or the pixel size of the image is converted to a 20× equivalent one such as 0.46 micron/pixel. However, the numerical aperture means the optical resolution remains unchanged. As a result, 20× down-sampled images usually provide finer details than actual 20× images. This method takes more than 10 minutes to evaluate a WSI while the modern WSI scanners can scan a slide within a minute. Time is an important consideration for the practical implementation, and it will be efficient if the quality of the WSI is evaluated in less than a minute without adding any extra time to the scanning process. Another CNN-based focus blur-detection method is proposed by Kohlberger et al. where they incorporated the clinical relevancy of the focus problem and categorized the focus blurs accordingly [31]. This method digitally generated the focus blur-affected images to increase the diversity of the training data. This synthetic data generation saves time and money for training a machine learning model whose performance highly depends on using a significantly large and diverse dataset. However, the impact of such data generation on the performance of the trained model requires more investigation. This method was demonstrated on two different scanners (spearman rank coefficients were 0.81 and 0.94 for two scanners) and the result differs significantly for them. Therefore, it is necessary to consider more scanners to ensure the reproducibility of the method.

Foucart et al. proposed a residual CNN-based method to identify tissue-fold artifacts that can utilize an imprecisely and inaccurately annotated dataset [32]. In medical image analysis, annotating data for training machine learning models is a time-consuming and arduous task. Moreover, it requires proper training or experience. Weekly supervised methods are very useful for this type of application where data annotation is difficult. They investigated the performance of the residual network for both detection and segmentation on the low-resolution image such as 1.25× image by varying the residual units. The model tends to provide higher accuracy for segmentation than the detection which may be caused by the erroneous annotation. The segmentation network took less than two minutes to locate the tissue ffold artifacts from a WSI when implemented using a GPU-enabled computer.

Babaie et al. investigated the performance of different machine learning methods for detecting tissue-fold artifacts when the annotated dataset is limited [33]. This method utilized the transfer learning technique for 5 different CNN models which include VGG16, Google Net, Inception V3, ResNet 101 and Dense Net 201. After that the performance of traditional machine learning methods such as decision trees, SVMs, and KNNs were compared with the traditional method. Further, the feature extraction capability of CNN was combined with the SVM classifier which depends on a limited datapoint to classify an image. The Dense Net 201 and SVM-based classifiers outperformed others providing an accuracy higher than 96%. However, the generalized performance was 81% which is significantly lower. In the experiment, multiple tissue organs were considered. Including images from different stains and scanners is important to ensure the robustness of the method.

Janowczyk et al. proposed an open-source tool for checking the quality of WSI images for automated analysis [34]. This tool relies on the supervised method using a set of image features such as brightness, contrast, edge and others to identify image regions that are suitable for analysis. The method was demonstrated on 450 images at 40× magnification to identify artifact-free regions in the presence of tissue-fold, air-bubble, ink mark, and focus blur artifacts. This experiment was performed using a machine with four hyperthreaded core processors and it took 130 minutes for evaluating 450 images. The high computation time is expected as the method relies on a high-resolution image. The method established 95% concordance with the pathologist's manual evaluation. However, the detail of the method is not available for reproducing the results, to the best of our knowledge. The performance of the method degrades in the presence of compression artifacts, stromal regions, or faintly stained regions in the WSI. Moyes A et al. presented another unsupervised method to separate the stains in a histological slide and detect stain-dependent artifacts such as color pigments [35]. Smit et al. proposed a multi-class pixel-wise semantic segmentation method that identified the source of the quality degradation in the WSI so that appropriate actions can be taken to fix the issues. This work utilized a diverse dataset which included images from multiple tissues, stains, and scanners and segmented the major artifacts such as tissue fold, air bubble, ink mark, dust, and focus blur. However, 85% data was used for training and only 15% for the evaluation of where the highest dice coefficient score was 0.91.

11.6 DISCUSSION

In digital pathology, the WSI serves as the basis for analysis and diagnosis. For a reproducible automated whole slide imaging system, it is necessary to eliminate unnecessary parts and useless artifacts of the WSI thereby analyzing only those areas that are sufficient in quality and relevant clinically. The first article regarding artifact detection was published less than two decades ago. Since then, not many articles published on this subject. A few methods utilizing different image analysis techniques are proposed for detecting artifacts from digital pathology specimens, these methods range from classifying image patches with an artifact to locating the precise boundary of artifact for segmentation. Most of these methods either detect tissue artifacts or scanning artifacts but for a complete-reproducible automated image analysis system, both artifacts should be considered for quality assurance of WSI. All these papers have shown promising results for detecting specific artifacts, but these results are not sufficient to warrant the method's generalized performance to unseen data. Therefore, it is necessary to undertake a comprehensive investigation of how well a method performs for detecting all the common tissue and scanning artifacts from unseen images of different organs, different stains slide prepared at different labs, and scanners by different scanners using the different scan profiles. Moreover, only a few papers were intended for developing a practical system by integrating the quality evaluation with the analysis system. Time is an important consideration for making a system practical. However, detecting some artifacts such as focus blur and noise from a low-resolution image is challenging and using a high-resolution image will increase the detection time significantly, making the system less practical. On the other hand, artifacts such as tissue folds, air bubbles, and ink marks can be detected from a low-resolution image at high precision. Therefore, an efficient approach could be utilizing the multi-resolution imaging system of WSI where the different resolutions will be used for detecting different artifacts depending on the nature of the artifact. Lastly, the design of the quality assurance system should be practical and user-friendly for its successful integration with the automated image analysis and diagnosis system in the pathology pipeline. Furthermore, the results of the quality evaluation system can be utilized for efficient focus point selection by integrating the quality assurance system with the WSI scanning system. This type of pipeline of integrated systems is useful for smart healthcare where an automated system evaluates the competency of the specimen, selects suitable regions from the specimen and examines the selected regions automatically without requiring the user to intervene.

11.7 CONCLUSION

In this chapter, the major issues related to the quality of WSI and their impact on analysis using automated methods were discussed. The limitations of existing artifact detection methods were investigated to use as a terminal quality assurance system. Other important issues for developing a complete reproducible analysis and diagnosis system were also addressed.

Bibliography

1. Senaras C, Niazi MKK, Lozanski G, Gurcan MN (2018) DeepFocus: Detection of out-of-focus regions in whole slide digital images using deep learning. *PLoS ONE* 13(10): e0205387. https://doi.org/10.1371/journal.pone.0205387

2. Ghaznavi F, Evans A, Madabhushi A, Feldman M. Digital imaging in pathology: Whole-slide imaging and beyond. *Annu Rev Pathol.* 2013 Jan 24;8:331-59. doi: 10.1146/annurev-pathol-011811-120902. Epub 2012 Nov 15. PMID: 23157334

3. Farahani N, Parwani A, Pantanowitz L. Whole slide imaging in pathology: Advantages, limitations, and emerging perspectives. *Pathology and Laboratory Medicine International.* 2015;7:23-33 https://doi.org/10.2147/PLMI.S59826

4. Thorstenson, S., Molin, J., & Lundström, C.F. (2014). Implementation of large-scale routine diagnostics using whole slide imaging in Sweden: Digital pathology experiences 2006–2013. *Journal of Pathology Informatics*, 5.

5. Le H, Gupta R, Hou L, Abousamra S, Fassler D, Torre-Healy L, Moffitt RA, Kurc T, Samaras D, Batiste R, Zhao T, Rao A, Van Dyke AL, Sharma A, Bremer E, Almeida JS, Saltz J. Utilizing automated breast cancer detection to identify spatial distributions of tumor-infiltrating lymphocytes in invasive breast cancer. *Am J Pathol*. 2020 Jul;190(7):1491–1504. doi: 10.1016/j.ajpath.2020.03.012. Epub 2020 Apr 8. PMID: 32277893; PMCID: PMC7369575.

6. Veta M, van Diest PJ, Kornegoor R, Huisman A, Viergever MA, Pluim JPW (2013) automatic nuclei segmentation in H&E stained breast cancer histopathology images. *PLoS ONE* 8(7): e70221. https://doi.org/10.1371/journal.pone.0070221

7. Wienert, S., Heim, D., Saeger, K. *et al.* Detection and segmentation of cell nuclei in virtual microscopy images: A minimum-model approach. *Sci Rep* **2**, 503 (2012). https://doi.org/10.1038/srep00503

8. Cruz-Roa, A., Gilmore, H., Basavanhally, A. *et al.* Accurate and reproducible invasive breast cancer detection in whole-slide images: A deep learning approach for quantifying tumor extent. *Sci Rep* 7, 46450 (2017). https://doi.org/10.1038/srep46450

9. Cruz-Roa A, Gilmore H, Basavanhally A, Feldman M, Ganesan S, Shih N, et al. (2018) High-throughput adaptive sampling for whole-slide histopathology image analysis (HASHI) via convolutional neural networks: Application to invasive breast cancer detection. *PLoS ONE* 13(5): e0196828. https://doi.org/10.1371/journal. pone.0196828

10. S. Kwok "Multiclass classification of breast cancer in whole-slide images" *International Conference Image Analysis and Recognition*, Springer, Cham, Switzerland (2018), pp. 931–940

11. Van Eycke, YR., Allard, J., Salmon, I. *et al.* Image processing in digital pathology: an opportunity to solve inter-batch variability of immunohistochemical staining. *Sci Rep* 7, 42964 (2017). https://doi.org/10.1038/srep42964

12. Kaplan, K. Quantifying IHC data from whole slide images is paving the way toward personalized medicine. *MLO Med Lab Obs*. 47, 20–21 (2015)

13. Helps, S. C., Thornton, E., Kleinig, T. J., Manavis, J. & Vink, R. Automatic nonsubjective estimation of antigen content visualized by immunohistochemistry using color deconvolution. *Applied Immunohistochemistry & Molecular Morphology* 20, 82–90 (2012)

14. Md Shakhawat Hossain, Matthew G. Hanna, Naohiro Uraoka, Tomoya Nakamura, Marcia Edelweiss, Edi Brogi, Meera R. Hameed, Masahiro Yamaguchi,

Dara S. Ross, and Yukako Yagi "Automatic quantification of *HER2* gene amplification in invasive breast cancer from chromogenic *in situ* hybridization whole slide images," *Journal of Medical Imaging* 6(4), 047501.https://doi.org/10.1117/1.JMI.6.4.047501

15. Akira Saito, Eric Cosatto, Tomoharu Kiyuna, Michiie Sakamoto, "Dawn of the digital diagnosis assisting system, can it open a new age for pathology?" *Proc. SPIE 8676, Medical Imaging 2013: Digital Pathology*, 867602 (29 March 2013); doi: 10.1117/12.2008967

16. Yoshiko Yamashita, Tomoharu Kiyuna, Michiie Sakamoto, Akinori Hashiguchi, Masahiro Ishikawa, Yuri Murakami, Masahiro Yamaguchi, "Development of a prototype for hepatocellular carcinoma classification based on morphological features automatically measured in whole slide images", *Analytical Cellular Pathology*, vol. 2014, ArticleID 817192, 2 pages, 2014. https://doi.org/10.1155/2014/817192

17. Bindhu, P., Krishnapillai, R., Thomas, P., Jayanthi, P.: Facts in artifacts. *Journal of Oral and Maxillofacial Pathology*: JOMFP 17(3), 397 (2013)

18. Rastogi V, Puri N, Arora S, Kaur G, Yadav L, Sharma R. Artefacts: A diagnostic dilemma—a review. *J Clin Diagn Res.* 2013 Oct;7(10):2408–13. doi: 10.7860/JCDR/2013/6170.3541. Epub 2013 Oct 5. PMID: 24298546; PMCID: PMC3843421

19. Taqi SA, Sami SA, Sami LB, Zaki SA. A review of artifacts in histopathology. *J Oral Maxillofac Pathol.* 2018 May–Aug;22(2):279. doi: 10.4103/jomfp.JOMFP_125_15. PMID: 30158787; PMCID: PMC6097380.

20. Schömig-Markiefka, B., Pryalukhin, A., Hulla, W. *et al.* Quality control stress test for deep learning-based diagnostic model in digital pathology. *Mod Pathol* 34, 2098–2108 (2021). https://doi.org/10.1038/s41379-021-00859-x

21. Bautista PA, Yagi Y. Improving the visualization and detection of tissue folds in whole slide images through color enhancement. *J Pathol Inform* 2010:1:25

22. Kothari S, Phan JH, Wang MD. Eliminating tissue-fold artifacts in histopathological whole-slide images for improved image-based prediction of cancer grade. *J Pathol Inform* 2013;4:22

23. Zerbe, N., Hufnagl, P., & Schlüns, K. (2011). Distributed computing in image analysis using open source frameworks and application to image sharpness assessment of histological whole slide images. *Diagnostic Pathology*, 6 (Suppl 1), S16. https://doi.org/10.1186/1746-1596-6-S1-S16

24. Shrestha P, Kneepkens R, Vrijnsen J, Vossen D, Abels E, Hulsken B. , A quantitative approach to evaluate image quality of whole slide imaging scanners. *J Pathol Inform* 2016;7:56.

25. Hosseini, M. S., Brawley-Hayes, J. A., Zhang, Y., Chan, L., Plataniotis, K., & Damaskinos, S. (2020). Focus quality assessment of high-throughput whole slide imaging in digital pathology. *IEEE Transactions on Medical Imaging, 39*(1), 62–74. https://doi.org/10.1109/tmi.2019.2919722

26. Palokangas, S., Selinummi, J., & Yli-Harja, O., Segmentation of folds in tissue section images. *2007 29th Annual International Conference of the IEEE Engineering in Medicine and Biology Society* (2007).

27. Hashimoto, N., Bautista, P.A., Yamaguchi, M., Ohyama, N., Yagi, Y. (2012). Referenceless image quality evaluation for whole slide imaging. *Journal of Pathology Informatics*, 3:9.

28. Moles Lopez X, D'Andrea E, Barbot P, Bridoux A-S, Rorive S, et al. (2013). An automated blur detection method for histological whole slide imaging. *PLoS ONE* 8(12): e82710. doi:10.1371/journal.pone.0082710

29. Hossain M. Shakhawat, Tomoya Nakamura, Fumikazu Kimura, Yukako Yagi, Masahiro Yamaguchi. Automatic quality evaluation of whole slide images for the practical use of whole slide imaging scanner, *ITE Trans. on MTA*, Vol. 8, No. 4, pp. 252–268 (2020).

30. Hossain M. Shakhawat, Tomoya Nakamura, Fumikazu Kimura, Yukako Yagi, Masahiro Yamaguchi. Practical image quality evaluation for whole slide imaging scanner, *4th Biomedical Imaging and Sensing Conference* 2018 (BISC2018), *Proceedings of SPIE*, 10711, 107111S, Apr. 2018.

31. Kohlberger T, Liu Y, Moran M, Chen PC, Brown T,Hipp JD, et al. Whole-slide image focus quality: Automatic assessment and impact on AI cancer detection. *J Pathol Inform* 2019;10:39.

32. A. Foucart, O. Debeir and C. Decaestecker, "Artifact identification in digital pathology from weak and noisy supervision with deep residual networks," *2018 4th International Conference on Cloud Computing Technologies and Applications (Cloudtech)*, 2018, pp. 1–6, doi: 10.1109/ CloudTech.2018.8713350.

33. Babaie, M., & Tizhoosh, H.R. (2019). Deep features for tissue-fold detection in histopathology images. *ECDP*.

34. Janowczyk A., Zuo R., Gilmore H., Feldman M., Madabhushi A. (2019). HistoQC: An open-source quality control tool for digital pathology slides, *JCO Clinical Cancer Informatics*, 2019.

35. Moyes A., Zhang K., Ji M., Zhou H., Crookes D. (2020) Unsupervised deep learning for stain separation and artifact detection in histopathology images. In: Papież B., Namburete A., Yaqub M., Noble J. (eds) *Medical Image Understanding and Analysis*. MIUA 2020. Communications in Computer and Information Science, vol 1248. Springer, Cham. https://doi.org/10.1007/978-3-030-52791-4_18

## CHAPTER 12

Machine Learning-Enabled Detection and Management of Diabetes Mellitus

Shafiqul Islam

School of Computing Science, University of Glasgow, UK

Faisal Tariq

James Watt School of Engineering, University of Glasgow, UK

CONTENTS

D IABETES MELLITUS (DM) is a metabolic disorder which affects millions of people around the world. DM occurs when the glucose levels get elevated in blood stream. If the DM remains undetected and uncontrolled for prolonged period, many severe complications can arise in various body organs leading to life threatening situations. Therefore, prevention or delayed onset of diabetes can provide life changing experience for many people. However, it can only be achieved if a rigorous screening process can be developed which can identify individuals who are at risk of developing diabetes in near future. Machine learning techniques can be used to extract various diabetes-related features and then exploiting them to predict the long-term possibility of diabetes development. In this chapter, machine and deep learning techniques for advanced prediction of hemoglobin A1c (HbA1c), a fundamental index used for diabetes management will be explained in detail along with its rigorous performance evaluations.

DOI: 10.1201/9781003251903-12

12.1   INTRODUCTION TO DIABETES

Diabetes is one of the fastest-growing diseases that develops from rising blood glucose (BG) levels. The human body organ pancreas generates insulin which maintains BG levels. The insufficient insulin in the blood or the incapability of the body to utilize insulin helps to develop diabetes [1]. The consequences of diabetes result in an increase of healthcare expenditure. The International Diabetes Federation (IDF) estimated that about 425 million people lived with diabetes in 2017 worldwide. The global summary report of diabetes is adapted from [2]. IDF has also forecasted 629 million people are going to develop diabetes by the year 2045.

The diabetes scenario for the Middle East and North Africa (MENA) region is alarming. It is the most vulnerable area for diabetes, with 39 million in 2017, which will reach 67 million by 2045 with an increment of 78%. Qatar, with 14.1% diabetics of the total population, is one of the countries most affected by diabetes. This disease will worsen in the next few decades due to the increase in the elderly population and the high level of obesity. According to the Qatar diabetes survey conducted in 2015, the prevalence of diabetes in Qatar was 17%, and 11–23% had pre-diabetes. One-third of diabetic people in Qatar were unaware of their disease. Approximately 23% of women living in Qatar developed diabetes during their pregnancy. A Qatari population-based clinical study [3] conducted to forecast diabetes prevalence warned that 24% Qatari population would develop diabetes by the year 2050. Qatar's annual healthcare cost was 1.5 billion in 2015, and it is estimated that by 2055 cost will reach 8.4 billion. The Qatar national vision 2030 envisions a healthy population. Early-onset detection of diabetes and a proper diabetes management plan might help fulfill their vision of 2030.

About $727 billion is spent yearly by diabetes patients, which is equivalent to one in every eight dollars spent on healthcare. The people of the working-age group (20–64 years) are highly affected by diabetes, with 326.5 million people currently living with diabetes. Gender-wise, 221 million men and 204 million women developed diabetes. The study of IDF also found that there are 280 million diabetic people in an urban setting, while the figure for rural counterparts is 145 million. The number of people living with diabetes in the urban setup will reach 472 million by 2045. The most alarming part of the study by IDF is that about half of the diabetic people are unaware of their disease. Hence, priority should be given to facilitating screening, diagnosing, and providing necessary healthcare to diabetes patients. People who go undiagnosed with diabetes incur a larger expenditure on healthcare than those diagnosed with diabetes.

Hyperglycemia and hypoglycemia are two significant hallmarks of diabetes. If these two are not appropriately monitored, there is a high risk of developing health complexities such as cardiovascular diseases. If taking appropriate measures would have been possible beforehand, these health complications can be tackled. One study [4] highlighted that advanced detection of diabetes can lower the risk of health complexities by taking necessary intervention in advance. Another critical biomarker used for diabetes management is glycated hemoglobin (HbA1c). Another study [5] found that a low HbA1c test value is correlated with the reduced microvascular difficulties developed from diabetes. However, a higher HbA1c test value is associated with the progression of diabetes-related comorbidities. The proper management of diabetes is vital as it can facilitate the patients to take practical steps early and can stop health complications development. The development of diabetes can be stopped or slowed down its onset if a person maintains a healthy lifestyle. Certain people, who are overweight, age over 45 years, have a family history of diabetes, and are physically less active, are at high risk of developing diabetes in their lifetime than others.

12.2   TYPES OF DIABETES

Classification and diagnosis of diabetes are two complex tasks. There are mainly three types of diabetes such as type 1 diabetes (T1D), type 2 diabetes (T2D), and gestational diabetes mellitus (GDM). T1D occurs when the immune system destroys beta cells in the pancreas [6]. These beta cells are responsible for insulin production. Due to beta cells' damage, the body generates insulin in a lower amount which results in insulin deficiency. Insulin is responsible for carrying glucose to the body cells. Cells use this glucose as a fuel to generate energy. The energy production process is shut off due to the damage of beta cells as glucose is not transferred to the body cells in the absence of insulin. Consequently, glucose accumulates in the bloodstream and causes high blood sugar.

Children and adolescents are most susceptible to developing T1D. Although the exact reason for T1D is unknown to researchers, gene plays a role in this harmful development [7]. Environmental factors such as viral infection, toxin encounters, or food habits are also associated with T1D. A family history of diabetes increases the risk of developing the condition among the progeny. If T1D goes untreated, serious health complications, sometimes life-threatening problems could arise. T1D patients are at risk of developing eye disease, kidney damage, nerve cell damage, and cardiovascular disease.

T2D is the most common category of diabetes that develops due to insulin resistance [8]. The root causes responsible for T2D diabetes are not fully understood, but researchers have observed a correlation between T2D and obesity, age, ethnicity, and a family history of diabetes [9]. A study [10] showed that eating calorie-rich refined foods and beverages and too few whole fruits, vegetables, and whole-grain can significantly elevate the risk of developing T2D. Although T2D is commonly seen among older people, children, adolescents, and younger adults are also at significant risk of T2D development due to physical inactivity, poor dieting, and rising level of obesity. Symptoms for T2D are similar to that of T1D, except the progression of the symptoms are slowly appears, and onset time is challenging to determine in the case of T2D. Some patients are first diagnosed with T2D only when they develop health complications, such as foot ulcers, a change in renal failure vision.

According to the American Diabetes Association (ADA), diabetes is detected between 24 and 28 weeks of pregnancy is known as GDM [11]. GDM elevates BG levels which can affect the newborn baby's health. In the case of GDM, the BG level goes to normal soon after delivery. However, there is a risk of developing T2D if someone develops GDM during pregnancy. There is a high chance of developing

GDM at a later birth for women with hyperglycemia during their first pregnancy. About 50% of pregnant women with hyperglycemia develop T2D after 5-10 years of their delivery [12]. During pregnancy, the placenta produces other hormones that impair insulin action in the body cells. As a result, the BG level rises.

12.3   HOW IS DIABETES DIAGNOSED

The World Health Organization (WHO) provides guidelines to diagnose diabetes as summarized in Figure 12.1. The BG value of <100 mg/dL is considered normal, BG values between 100 and 126 mg/dL are considered pre-diabetes, and BG values ≥126 mg/dL are confirmed as the development of diabetes. Alternately, ADA considers the level of HbA1c <5.7% as normal, 5.7% to 6.4% as pre-diabetes, and ≥6.5% as diabetes.

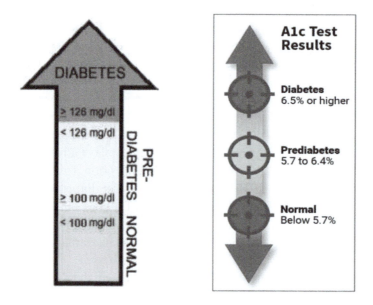

Figure 12.1   Illustration of diabetes diagnosis and HbA1c test ranges.

12.4   MANAGEMENT OF DIABETES

The proper management of diabetes refers to maintaining BG levels near normal by adjusting food intake, medication, or physical activity. In recent years, there has been much advancement in electronic health technology, which has proved to be a cost-effective method for monitoring health-related data. The advancement of e-health technology allows for a cohort of health-related data to be collected. For diabetes diagnosis and management, BG measurement plays a vital role in reducing diabetes-related health complications. Different laboratory tests such as oral glucose tolerance test (OGTT) and HbA1c measures are used for diabetes management.

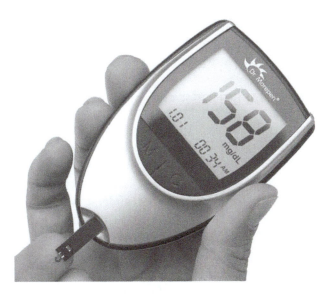

Figure 12.2 Glucometer.

Advancement in sensor technology facilitates the daily monitoring of BG. These sensor devices used to measure BG are three types: Invasive, semi-invasive, and non-invasive. A brief overview of each category is presented in the subsequent sections.

For self-management, a BG measurement device was not available before the 1960s. The first device easy to use at home was developed in 1981, known as a Glucometer [13]. This meter was based on the same basic principle used for today's measurement. At first, the blood sample is exposed to an enzyme, which oxidizes the blood (releases electrons). Then released electron passes through an electrode, which reads the current value proportional to the oxidized glucose. The more glucose present in the blood, the higher the oxidation, leading to higher glucose levels. Over the years, the device's ability to measure BG accurately has been improved, and nowadays, advanced glucose meters are available for use, as shown in Figure 12.2.

One of the significant advancements in the BG monitoring system is developing the continuous glucose monitoring (CGM) system. CGM is an approach to measuring BG continuously throughout the day and night, which can provide a complete picture of a user's BG level necessary for better treatment decisions and glucose control [14]. A tiny electrode known as a glucose sensor is inserted under the skin to measure the interstitial fluid's glucose level. The most well-known CGM devices available are Dexcom and Freestyle Libre. However, CGM measures BG differently from the traditional gold standard method. The CGM system measures the glucose from the interstitial fluid, not directly from the blood. The main drawback is that this CGM reading lags behind the actual BG value by 10–15 minutes [15].

A needle-free alternative to the finger prick test is a long-lasting dream device for many patients with diabetes. In the early 2000s, Cygnus incorporated Glucowatch [16]. It used a weak electrode to draw glucose from the interstitial fluid into an auto sensor. Approved by the FDA in 2002, the device was supposed to provide regular and painless measurement, but soon it gained a bad reputation due to its painful rashes in the skin's contact point, inaccurate reading, and long warm-up time. Despite promises from many companies, no product has taken its place in reality. Non-invasive glucose monitoring system remains a dream device among diabetes patients.

12.5 CURRENT CHALLENGES AND LIMITATIONS

In recent years, there have been praiseworthy advancements in sensor technology, which proved cost-effective and secured their utilization in support of health and health-related fields. These advancements facilitate generating a significant amount of health-related data. A key challenge is a lack of any readily available insight into those health data. Therefore, for a deeper understanding or, more specifically, for knowledge discovery from raw healthcare data, machine learning (ML) models are frequently implemented [17]. We have witnessed healthcare professionals' interest in utilizing available data from diabetes patients for diabetes management by applying ML and data mining techniques. However, today's healthcare practitioners are already overwhelmed with various professional and administrative responsibilities. Therefore, to extract any useful information from such data, we require to build models to facilitate the practitioner by providing a more in-depth insight and a "second opinion" based on the data.

One of the fundamental challenges in diabetes diagnosis and management is that little work is done in the early-onset detection of diabetes progression. Another crucial challenge in diabetes management is whether the approach should be reactive or proactive. Reactive responses are often recommended after the development of diabetes, while proactive activity is carried out before the development of the disease. Early intervention through proactive responses can reduce healthcare burdens and complications through early-onset detection of diabetes and proper diabetes management. Although ML techniques had been applied for disease diagnosis, few works addressed the early onset detection of T2D. Furthermore, the current approach of HbA1c estimations is used to take reactive actions for diabetes management. The estimations of HbA1c are sometimes way off from the actual levels. Moreover, these estimations are derived using current BG measurements. The advanced CGM technology has not been utilized for HbA1c prediction. There is no previous work on the early prediction of HbA1c that may facilitate proactive responses for diabetes management.

12.6 MACHINE LEARNING FOR DIABETES MANAGEMENT

This section aims to investigate previous ML-based works related to diabetes prediction and management. We start with a brief introduction to the ML. Then we summarized ML techniques implemented for diabetes detection and prediction. Finally, we present a summary of findings in the literature with their limitation and our future work scope.

Arthur Samuel first coined the term ML in 1959 [18]. ML enables a machine or computer to learn from data without being programmed explicitly. The ML approach trains a computer to behave like humans and animals: Learn from experience. ML algorithms use statistical or computational methods to develop a relationship between data and outcomes without predetermined equations or models. The algorithm tries to find a pattern in data and make better decisions and predictions. Accurate prediction of a specific event occurring in the future, based on past data, has seen a surge in interest due to advances in data mining and ML techniques. Lately, we have seen such approaches applied in many fields, including business management, industrial automation, sports analytics, and healthcare. In the present era of big data, ML techniques are applied for solving problems in areas such as image processing for face and object detection, drug discovery, disease detection, DNA sequencing, predictive maintenance, voice recognition, and machine translation [19]. ML tasks are divided into two main categories: (a) Supervised learning, where an algorithm learns from available training data, also known as ground truth, and (b) Unsupervised learning, in which an algorithm tries to discover insight about unlabeled data. By taking into consideration the severe impacts of diabetes, ML and data mining-based approaches have recently attracted enough attention among the diabetes research community.

12.6.1 Machine Learning Algorithm Used for Diabetes Management

A support vector machine (SVM) is one of the most popular supervised ML approaches used for classification and regression. SVM finds a hyperplane to maximize the margin between the groups by utilizing the Lagrangian optimization technique [20]. SVM uses a kernel technique that transforms the data to distinguish between different classes of data. Yu et al. [21] implemented SVM to test its potential for diabetes detection. Their study used a total of six years of health records from the National Health and Nutrition Examination Survey (NHANES) conducted on the US population. An AUC score of 83.47% and 73.18% was reported for 10-fold cross-validation on diabetes and pre-diabetes detection. Barakat et al. [22] proposed an intelligible SVM for the diagnostic of diabetes. The data was sourced from the National Survey of Diabetes, 1991, in the Sultanate of Oman. Optimized SVM with RBF kernel and gamma, regularization parameters were chosen based on the training data's best performance.

An ensemble learning method that generates multiple classifiers and provides a final result by combining results from all classifiers based on voting. There are two methods of ensembling, namely boosting and bagging. In boosting approach, misclassified instances are given more weight during later iterations. However, trees are built upon a bootstrap sample during the implementation bagging method. One particular example of the bagging method is a random forest (RF) algorithm proposed by Breiman [23]. Researchers also [24] utilized the RF model to select a single-nucleotide polymorphism (SNPs) responsible for the development of diabetes. The experiment was carried out with a dataset of 677 subjects, 429 healthy and 248 people with diabetes. An AUC score of 85.3% was achieved for the selected SNP feature.

Logistic regression (LR) is a statistical method where log-odds of the probability of an event are linear combinations of independent variables [25]. Although the model outputs the probability of an event, it is used in the classification task by applying a threshold. The logistic regression approach's outcome is binary, such as positive or 1 (diabetes) and negative or 0 (non-diabetes). Logistic regression tries to develop a relationship between feature and outcome by finding the best descriptive fitting model [26]. There are two ways of learning this function. A discriminating model learns the function directly to compute class posterior, while a generative model learns the conditional class probability, class prior, followed by applying the Bayes rule. A modified alternative to discriminative and generative models is to merge probability altogether to learn the discriminative function that directly maps input to output. Devi et al. [27] proposed a modified LR model for detection and finding the most relevant predictor of T2D. The study has used a dataset consisting of 739 patients with 31 detailed features such as age, gender, random blood sugar, fasting blood sugar, cholesterol, lipoprotein, HbA1c. An accuracy of 90.4% was reported for the sigmoid activation function.

A Naïve Bayes (NB) classifier is a candid and compelling model used for classifying data based on the Bayes theorem [28]. NB classifiers assume that all attributes are conditionally independent of the given class label. The goal of this classifier is to learn a representative function from a given training labeled dataset. Nongyao et al. [29] compared primary ML classifiers such as ANN, decision trees (DT), NB, LR, and RF. The diabetes data came from 26 primary care units in Thailand, consisting of 12 features such as smoking behavior, BMI, family history of hypertension, and diabetes. They have built unbiased models using a cross-validation method and achieved the highest accuracy of 85.56% for random forest classifiers.

An artificial neural network (ANN) is one of the dominant tools used in ML. As the name 'neuron' suggests, ANN is a brain-inspired system that tries to replicate the human brain [30]. ANN consists of an input and output layer and a hidden layer (in most cases) to transform input into some form that the next layer can use. ANN is handy in finding a pattern or feature extraction from data considered complicated or laborious for a human. The success of ANN is due to a technique known as "backpropagation," which allows changing the weight of the hidden layer if there is any misclassification. The fundamental advantage of ANN is that it does not require in-depth knowledge about the relationship between variables. Instead, it tries to recognize a pattern in the dataset and store those patterns as a weight for later use for the test case. One study [31] developed a deep neural network-based model to monitor BG levels. Their model was evaluated using Diabetes Research in Children Network (DirecNet) data consisting of 25 patients with type 1 diabetes for 30 minutes prediction. They reported accuracies of 64.88% for diabetes detection.

12.6.2 Type 2 Diabetes Progression Prediction

One of our recent studies presented an ML-based framework for early-onset detection of T2D and implemented workflow is summarized in Figure 12.3. The OGTT data gathered from a clinical trial known as the San Antonio Heart Study (SAHS) were pre-processed to remove anomalies and noise. Two novel feature extraction techniques were proposed, and the best features were selected through feature selection techniques. Finally, the ML framework was optimized for advanced T2D progression prediction. The performance of the developed model is discussed and benchmarked

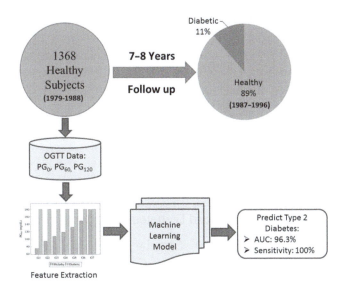

Figure 12.3 A ML-based framework for advanced prediction of type 2 diabetes.

with the literature. The work done in this study resulted in the scholarly output at *IEEE Access journal* [32].

The SAHS dataset was collected during 1978 (baseline) and 1988 (follow-up), where people from the Mexican American (MA) and non-Hispanic White (NHW) groups participated [33]. Different biomarkers such as age, ethnicity, fasting BG values, BG values at half an hour, one hour, and two hours of OGTT were collected in baseline and follow-up study. We studied a total of 1368 participants, where 171 developed T2D during the follow-up. We pre-processed raw OGTT to fill missing data using the mean values of respective biomarkers, followed by feature extraction. The body's glucose digestion capability features have been extracted using a fractional derivative approach [34]. We attempted to find salient features related to the future progression of T2D and implemented filter, wrapper, and embedded feature selection methods. The feature selection was followed by proposing and optimizing ML models such as SVM, NB, and RF for separating diabetic from non-diabetic.

TABLE 12.1 A 10-Fold CV Performance Comparison of T2D Prediction Models

Model	Accuracy	Sensitivity	Specificity	AUC
SVM	92%	51.9%	99.5%	74.8%
Random Forest	94.07%	58.5%	99.2%	90.7%
Bagging	94.15%	57.9%	99.3%	88.3%
Boosting	94.3%	59.1%	99.3%	87.1%
NB	81.65%	81.9%	81.6%	88.8%
A1DE	84.46%	84.8%	84.1%	89.3%
A2DE	92.94%	93.4%	92.5%	89.5%
Ensembling	**95.94%**	**100%**	**91.5%**	**96.3%**

The classification results of the implemented approaches are highlighted in Table 12.1. The ensembling approach achieved the highest accuracy of 95.94%. We aimed to improve sensitivity compared to specificity as misclassifying a T2D case is considered more severe than misclassifying a non-diabetic case. Our model detected all the T2D progressors (sensitivity, 100%) but misclassified 8.5% (specificity, 91.5%) healthy cases.

Our findings from this study are significant to identify whether individuals have a chance to be T2D progressors in the future. This advanced prediction will facilitate the prevention of T2D, or at least delay the onset of T2D by taking effective steps.

12.6.3 HbA1c Prediction for Enhanced Diabetes Management

Periodic monitoring of HbA1c is essential for the proper management of diabetes. An above-normal range of HbA1c levels increases the risk of developing diabetes-related complexities such as cardiovascular disease. To further facilitate diabetes management, we developed an HbA1c prediction model that can predict HbA1c levels two–three months in advance through applying ML techniques on time-series CGM data as summarized in Figure 12.4. The results presented in this study appear in the *IEEE Sensor Journal* [35].

We initiated a retrospective data collection with Sidra Medicine, Qatar, to collect BG data from 200 participants using CGM sensors. The Institutional Review Board (IRB) of Sidra Medicine approved our study (IRB Number, 1536761-1). We used CGM sensors known as Freestyle Libre for 90–120 days for each participant. The CGM sensor provides 96 measurements of BG levels per day. The HbA1c levels were collected separately at the laboratory and correlated with BG levels by extracting novel features such as time in range (TIR) based on:

$$\text{TIR}_{70-180} = \frac{\sum_{i=1}^{N}(C(x_i) \geq 70 \wedge \leq 180)}{N} \tag{12.1}$$

$$\text{TIR}_{250-300} = \frac{\sum_{i=1}^{N}(C(x_i) \geq 250 \wedge \leq 300)}{N} \tag{12.2}$$

$$\text{TIR}_{300-350} = \frac{\sum_{i=1}^{N}(C(x_i) \geq 300 \wedge \leq 350)}{N} \tag{12.3}$$

where C represents overall counts, x stands for individual BG values, and N represents the sample size. TIR_{70-180} represents how much time a participant spends in the BG ranges of 70–180 mg/dL. A previous study shows that TIR and HbA1c are correlated significantly [36].

The proposed three-staged MSMC framework is highlighted in Figure 12.5. The framework predicts HbA1c levels two–three months in advance time frame using BG data. We split participants based on HbA1c ranges into four distinct patients groups. These patient groups belong to control group (class C1, $\leq 7.5\%$), medium control group (class C3, $7.5\% < \text{HbA1c} \leq 9\%$), low control group (class C5, $9\% < \text{HbA1c} \leq 12.5\%$), and uncontrolled group (class C6, $> 12.5\%$), respectively.

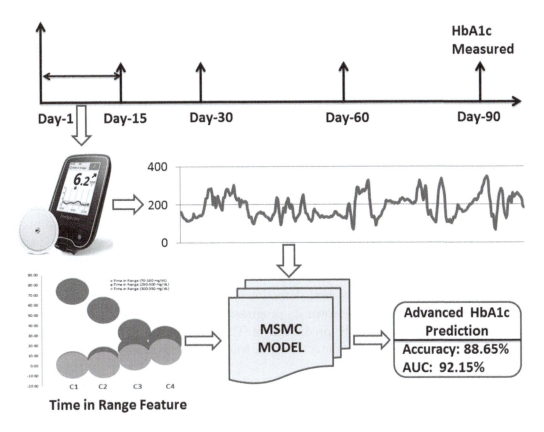

Figure 12.4 The proposed multi-stage CGM data analysis framework.

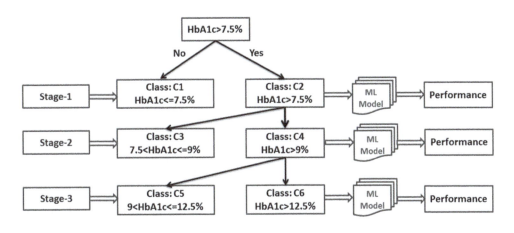

Figure 12.5 Split of participants based on HbA1c levels for HbA1c prediction.

We performed 10-fold cross-validation, and an accuracy of 90.67% was observed using the ensemble model in stage-1. The SVM displayed accuracy of 83.49% while separating groups in stage-2. In the final stage-3, an accuracy of 91.80% was observed for A1DE, which provided an overall accuracy of 88.65%. To the best of our knowledge, this is the first time in literature that HbA1c prediction has been attempted by applying an MSMC classification framework. The studies [37, 38] attempted to estimate HbA1c levels only. On the other hand, our proposed framework has potential applications for both doctors and patients to arrange preemptive measures for better management of diabetes.

12.7   CONCLUSION

The current strategy for managing diabetes is the reactive approach which only intervenes after the disease occurs. A proactive approach can bring substantial health benefits to the patients by starting early interventions long before the disease progresses. The OGTT and HbA1c tests are performed widely to diagnose and manage diabetes. The ML model can detect individuals at an increased risk of developing the disease in the future and thus facilitate the prevention or delay of its progression. Developing a framework for early-onset detection of T2D has a huge benefit for starting early intervention and slowing down its progression. To this end, we have developed a novel approach for the early prediction of T2D progression. The developed ML pipeline has the capability to detect people who are going to develop T2D in the future.

The HbA1c test measures the average amount of glucose for the previous 3–4 months. A test value of HbA1c $\geq 6.5\%$ is used as a threshold to diagnose diabetes, and healthcare professionals often recommend maintaining HbA1c levels $\leq 7.5\%$ to remain in the good controlled group. The HbA1c test levels are extensively utilized to identify patients with uncontrolled or poorly controlled diabetes. Uncontrolled or poorly controlled diabetes causes health difficulties in the future. These complications can be avoided with the proactive intervention and treatment plan by accurately predicting HbA1c levels in advance. Therefore, advanced prediction of HbA1c facilitates the management of diabetic patients. In the literature, several attempts were made to only *estimate* instantaneous HbA1c using BG measurement data. We devised a novel approach for the advanced prediction of HbA1c levels. The framework achieved an accuracy of 88.65%. This research work also compared and discussed the existing works related to the current HbA1c estimations. This advanced HbA1c prediction is the pioneer in facilitating futuristic knowledge about the patient's glycemic profile so that necessary intervention can be taken. Thus health complexities that arise from diabetes can be reduced.

Several areas came across in this chapter that were beyond this research and could be interesting to explore in the future. First, one limitation of this research work was the lack of evaluation of the model on other similar OGTT datasets. Second, as the risk of cardiovascular disease progression is significant among T2D patients, an extension of this research could be to predict T2D leading to cardiovascular disease using deep learning approach. Third, predicting a smaller range of HbA1c is crucial to avoid misclassification of patients. In our MSMC model, the range was 1.5%. The further split of the patients into smaller ranges such as 0.5%, 0.25%, and 0.1% have the potential to make the model generalized. The prediction of a narrower HbA1c range could be a potential research direction that was beyond our research due to the smaller data size.

Bibliography

1. K. G. M. M. Alberti and P. F. Zimmet, "Definition, diagnosis and classification of diabetes mellitus and its complications. Part 1: diagnosis and classification of diabetes mellitus. Provisional Report of a WHO Consultation," *Diabetic medicine*, vol. 15, no. 7, pp. 539–553, 1998.

2. K. Ogurtsova, J. da Rocha Fernandes, Y. Huang, U. Linnenkamp, L. Guariguata, N. Cho, D. Cavan, J. Shaw, and L. Makaroff, "IDF diabetes atlas: global estimates for the prevalence of diabetes for 2017 and 2045," *Diabetes research and clinical practice*, vol. 128, pp. 40–50, 2017.

3. S. F. Awad, M. O'Flaherty, J. Critchley, and L. J. Abu-Raddad, "Forecasting the burden of type 2 diabetes mellitus in qatar to 2050: a novel modeling approach," *Diabetes research and clinical practice*, vol. 137, pp. 100–108, 2018.

4. S. J. Griffin, K. Borch-Johnsen, M. J. Davies, K. Khunti, G. E. Rutten, A. Sandbæk, S. J. Sharp, R. K. Simmons, M. Van den Donk, N. J. Wareham, et al., "Effect of early intensive multifactorial therapy on 5-year cardiovascular outcomes in individuals with type 2 diabetes detected by screening (addition-Europe): a cluster-randomised trial," *Lancet*, vol. 378, no. 9786, pp. 156–167, 2011.

5. L. Agrawal, N. Azad, G. D. Bahn, L. Ge, P. D. Reaven, R. A. Hayward, D. J. Reda, N. V. Emanuele, V. S. Group, et al., "Long-term follow-up of intensive glycaemic control on renal outcomes in the Veterans Affairs Diabetes Trial (VADT)," *Diabetologia*, vol. 61, no. 2, pp. 295–299, 2018.

6. M. A. Atkinson, G. S. Eisenbarth, and A. W. Michels, "Type 1 diabetes," *Lancet*, vol. 383, no. 9911, pp. 69–82, 2014.

7. H. Hakonarson, S. F. Grant, J. P. Bradfield, L. Marchand, C. E. Kim, J. T. Glessner, R. Grabs, T. Casalunovo, S. P. Taback, E. C. Frackelton, *et al.*, "A genome-wide association study identifies kiaa0350 as a type 1 diabetes gene," *Nature*, vol. 448, no. 7153, pp. 591–594, 2007.

8. S. Chatterjee, K. Khunti, and M. J. Davies, "Type 2 diabetes," *Lancet*, vol. 389, no. 10085, pp. 2239–2251, 2017.

9. I. Shai, R. Jiang, J. E. Manson, M. J. Stampfer, W. C. Willett, G. A. Colditz, and F. B. Hu, "Ethnicity, obesity, and risk of type 2 diabetes in women: a 20-year follow-up study," *Diabetes care*, vol. 29, no. 7, pp. 1585–1590, 2006.

10. R. L. Duyff, *American Dietetic Association complete food and nutrition guide.* Houghton Mifflin Harcourt, 2012.

11. A. D. Association et al., "Gestational diabetes mellitus," *Diabetes care*, vol. 27, p. S88, 2004.

12. H. Herath, R. Herath, and R. Wickremasinghe, "Gestational diabetes mellitus and risk of type 2 diabetes 10 years after the index pregnancy in sri lankan women—a community based retrospective cohort study," *PloS one*, vol. 12, no. 6, p. e0179647, 2017.

13. S. Clarke and J. Foster, "A history of blood glucose meters and their role in self-monitoring of diabetes mellitus," *British journal of biomedical science*, vol. 69, no. 2, pp. 83–93, 2012.

14. D. C. Klonoff, "Continuous glucose monitoring: roadmap for 21st century diabetes therapy," *Diabetes care*, vol. 28, no. 5, pp. 1231–1239, 2005.

15. C. D. Block, J. Vertommen, B. Manuel-y Keenoy, and L. V. Gaal, "Minimally-invasive and non-invasive continuous glucose monitoring systems: indications, advantages, limitations and clinical aspects," *Current diabetes reviews*, vol. 4, no. 3, pp. 159–168, 2008.

16. M. J. Tierney, J. A. Tamada, R. O. Potts, L. Jovanovic, S. Garg, C. R. Team, et al., "Clinical evaluation of the Glucowatch® biographer: a continual, non-invasive glucose monitor for patients with diabetes," *Biosensors and bioelectronics*, vol. 16, no. 9-12, pp. 621–629, 2001.

17. J. Wu, J. Roy, and W. F. Stewart, "Prediction modeling using EHR data: challenges, strategies, and a comparison of machine learning approaches," *Medical care*, pp. S106–S113, 2010.

18. A. L. Samuel, "Some studies in machine learning using the game of checkers," *IBM Journal of research and development*, vol. 3, no. 3, pp. 210–229, 1959.

19. P. Domingos, "A few useful things to know about machine learning," *Communications of the ACM*, vol. 55, no. 10, pp. 78–87, 2012.

20. I. Steinwart and A. Christmann, *Support vector machines*. Springer Science & Business Media, 2008.

21. W. Yu, T. Liu, R. Valdez, M. Gwinn, and M. J. Khoury, "Application of support vector machine modeling for prediction of common diseases: the case of diabetes and pre-diabetes," *BMC medical informatics and decision making*, vol. 10, no. 1, p. 16, 2010.

22. N. Barakat, A. P. Bradley, and M. N. H. Barakat, "Intelligible support vector machines for diagnosis of diabetes mellitus," *IEEE transactions on information technology in biomedicine*, vol. 14, no. 4, pp. 1114–1120, 2010.

23. A. Liaw, M. Wiener, et al., "Classification and regression by random forest," *R news*, vol. 2, no. 3, pp. 18–22, 2002.

24. B. López, F. Torrent-Fontbona, R. Viñas, and J. M. Fernández-Real, "Single nucleotide polymorphism relevance learning with random forests for type 2 diabetes risk prediction," *Artificial intelligence in medicine*, vol. 85, pp. 43–49, 2018.

25. D. R. Cox, "The regression analysis of binary sequences," *Journal of the Royal Statistical Society. Series B (Methodological)*, pp. 215–242, 1958.

26. F. E. Harrell, "Ordinal logistic regression," in *Regression modeling strategies*, pp. 311–325, Springer, 2015.

27. M. N. Devi, A. alias Balamurugan, and M. R. Kris, "Developing a modified logistic regression model for diabetes mellitus and identifying the important factors of type ii DM," *Indian journal of science and technology*, vol. 9, no. 4, 2016.

28. K. P. Murphy et al., "Naïve Bayes classifiers," *University of British Columbia*, vol. 18, 2006.

29. N. Nai-arun and R. Moungmai, "Comparison of classifiers for the risk of diabetes prediction," *Procedia computer science*, vol. 69, pp. 132–142, 2015.

30. T. M. Mitchell, "Artificial neural networks," *Machine learning*, vol. 45, pp. 81–127, 1997.

31. H. N. Mhaskar, S. V. Pereverzyev, and M. D. van der Walt, "A deep learning approach to diabetic blood glucose prediction," *Frontiers in applied mathematics and statistics*, vol. 3, p. 14, 2017.

32. M. S. Islam, M. K. Qaraqe, S. B. Belhaouari, and M. A. Abdul-Ghani, "Advanced techniques for predicting the future progression of type 2 diabetes," *IEEE access*, vol. 8, pp. 120537–120547, 2020.

33. J. P. Burke, K. Williams, S. P. Gaskill, H. P. Hazuda, S. M. Haffner, and M. P. Stern, "Rapid rise in the incidence of type 2 diabetes from 1987 to 1996: results from the San Antonio Heart Study," *Archives of internal medicine*, vol. 159, no. 13, pp. 1450–1456, 1999.

34. A. Atangana, H. Jafari, S. B. Belhaouari, and M. Bayram, "Partial fractional equations and their applications," *Mathematical problems in engineering*, vol. 2015, 2015.

35. M. S. Islam, M. K. Qaraqe, S. Belhaouari, and G. Petrovski, "Long-term HbA1c prediction using multi-stage CGM data analysis," *IEEE sensors journal*, vol. 21, no. 13, pp. 15237–15247, 2021.

36. R. W. Beck, R. M. Bergenstal, P. Cheng, C. Kollman, A. L. Carlson, M. L. Johnson, and D. Rodbard, "The relationships between time in range, hyperglycemia metrics, and HbA1c," *Journal of diabetes science and technology*, vol. 13, no. 4, pp. 614–626, 2019.

37. C. L. Rohlfing, H.-M. Wiedmeyer, R. R. Little, J. D. England, A. Tennill, and D. E. Goldstein, "Defining the relationship between plasma glucose and hba1c: analysis of glucose profiles and HbA1c in the Diabetes Control and Complications Trial," *Diabetes care*, vol. 25, no. 2, pp. 275–278, 2002.

38. D. M. Nathan, J. Kuenen, R. Borg, H. Zheng, D. Schoenfeld, and R. J. Heine, "Translating the A1c assay into estimated average glucose values," *Diabetes care*, vol. 31, no. 8, pp. 1473–1478, 2008.

## CHAPTER 13

IoT and Deep Learning-Based Smart Healthcare with an Integrated Security System to Detect Various Skin Lesions

Khairul Islam and Zahidul Islam
Department of Information & Communication Technology, Islamic University, Kushtia, Bangladesh

Al Amin
Department of Computer Science & Engineering, Prime University, Dhaka, Bangladesh

Mojibur Rahman Redoy Akanda
Department of Computer Science & Engineering, Texas A&M University, Texas, USA

Shabuj Hossen
Department of Electrical & Electronic Engineering, Prime University, Dhaka Bangladesh

Feroza Naznin
Department of Computer Science & Engineering, Green University of Bangladesh, Dhaka, Bangladesh

Mohammad Ali Moni
School of Health and Rehabilitation Sciences, Faculty of Health and Behavioural Sciences, University of Queensland, Queensland, Australia

CONTENTS

DOI: 10.1201/9781003251903-13

THE IMPLEMENTATION of Artificial Intelligence (AI) and Internet of Things (IoT) are extensively increasing in different field of expertise for automation of the processes. And the use of AI and IoT in the field of healthcare contributed automatic devices which greatly serve modern civilization in reducing time and cost. However, there is a security concern when employing autonomous devices that can disclose one's personal information. Therefore, in this study we propose a state-of-the-art pipeline to build smart healthcare systems with integrated security. In our study, we have proposed an app, which will collect skin lesion images of patients with proper authorization of the patients from any place. Collected data will be sent to the cloud where a Computer-Assisted Diagnosis (CAD) system is deployed to classified the skin lesion condition. Our proposed CAD established using deep learning-based technology to identify a dermoscopic image such as skin lesions. In this study, our focus was mainly on classifying seven various types of skin lesions, particularly early-stage melanoma, in order to suggest to patients a recommendation system for taking the necessary steps. Our deployed CAD was trained on the "HAM10000" dataset using a Convolutional Neural Network (CNN) [1]. We have used fine-tuning of deep learning models such as VGG16, ResNet152, InceptionV3 and DenseNet201. DenseNet201 produced the best accuracy with 85% in classifying among seven different skin lesions. The result will be sent to the patient's app profile and provide necessary steps for access. As a result, our smart healthcare systems will be capable of diagnosing patients with a high degree of security.

13.1 INTRODUCTION

Cyber-security is the most critical sort of security that individuals must be concerned about in the digital age. Electronic information is stored and organized in banks, schools, hospitals, corporations, governments, and practically every other modern institution one can think of. This implies that anyone with a computer, an internet connection, and profound computer knowledge may view, steal, or manipulate all of your most sensitive information, including credit card numbers and checking accounts, medical records, and phone bills. Hacking is the term used to describe the acts aimed at compromising digital devices like computers, cellphones, tablets, and even entire networks; and the more current concept of hacking revolves around the difficulty of understanding and manipulating technology [2]. Computer infections generated by hackers cost losses of about $55 billion at various organisations, nearly doubling the harm they caused in 2002 (https://www.trustwave.com). In modern times, less human interference and more interaction are allowed by utilizing underlying technologies such as pervasive computing, communication capabilities, internet protocols, and apps. Therefore, a new technology combining all the above facilities called IoT turned our way of life from traditional to smart. Various protocols are required to identify desired tasks by the IoT devices, such as spoken language instead of the structure exchanges of messages. Moreover, it is essential to determine the appropriate limitations that comply with each device while performing multiple parallel tasks [3]. As a result, confirming the secrecy of an IoT device is critical for user data security.

Skin lesions are serious conditions that affect a large portion of people all over the world. Early diagnosis of melanoma using noninvasive dermatoscopy methods boosts the survival percentage substantially [4]. And it is one of the most common cancers in the white population, hence computer-assisted approaches for classifying skin lesions based on dermoscopic images are of great interest. Skin cancer has the highest chance of occurrence with a substantial malignancy risk among all cancer forms. Deep learning has been recently introduced in the fields of skin lesion segmentation and classification.

Melanoma has a greater number of fatalities than non-melanoma skin cancer. Therefore, discriminating between cancerous and non-cancerous melanoma skin images has received a lot of attention nowadays [5]. In 2008, a total of 8694 patients' skin excisions were studied. Basal cell cancer (BCC) had 72.7% PPV (Positive Predictive Value), squamous cell carcinoma (SCC) had 49.4% PPV, and cutaneous melanoma (CM) had 33.3% PPV. The sensitivity score was about 63.9%, 41.1%, and 33.8%, respectively, for BCC, SCC and CM in typical clinical diagnosis. Although only 6.3%

of clinically diagnosed common naevi (skin discoloration) turned out to be melanoma, around 21% of all melanomas were categorized as usual naevi [6]. The annual number of cases of skin cancer has increased to 53%, and melanoma affects one in every 52 women and one in every 32 men in the United States. It is estimated that about 10,000 individuals died from it [7]. In the United States, a total of 178,560 new cases of melanoma have been reported since 2018. Furthermore, 9320 people have died due to melanoma, in which 3330 were women and 5990 were men [8].

CNNs have shown great promise, and they're frequently employed in various contests such as the ImageNet challenge, in which researchers attempt to categorize lots of different natural objects [9, 10]. In this study, we have fine-tuned and trained various neural network architectures such as VGG16, DenseNet201, InceptionV3, NasNetLarge, and ResNet to identify multiple skin lesions successfully. However, an existing CNN architecture such as VGGNet with a very deep neural Network and layers have the strength to classify complex classification like skin lesions [11]. In [12], a group of researchers achieves the best accuracy with 84.00% accuracy on the "HAM10000" dataset. "HAM10000" is a well-known dataset that researchers widely use for classifying skin lesions. Various machine learning and image processing concepts and applications were deployed on the dataset to develop a robust computerized skin lesion classifier. There are 10,015 dermatoscopic images in the dataset, which is released as a training set for academic machine learning purposes. More than half of the lesions have been confirmed by pathology, while the remaining cases have either been followed up or confirmed by in vivo confocal microscopy. The dataset is publicly available through the ISIC archives on the Harvard Dataverse [13].

Although in our earlier study, we have identified the risk of cancer in skin related patient by evaluating genetic information of the patient [14], this particular study follows non-invasive strategy of identifying the risk of skin lesions. Therefore, in this study, our main aim is to develop a framework for pre-screening skin lesions at an early stage so that the patient can take proper steps before the condition gets worse. In our framework, we have hypothesized that the patient would be able to use the device at home for free as well as the data, which is well secured from unauthorized access.

13.1.1 Related Work

There are numerous works on building Computer-Assisted Diagnosis (CAD) for skin lesions. However, very few studies proposed an IoT-based skin lesion detection, more specifically with enough safety measures of the patient's data. As a result, we proposed a framework to determine a combination of an IoT-based CAD with proper safety focus of the acquired data and results.

In 2019, Kiran Pai et al. trained a Convolutional Neural Networks for classifying skin lesions on the "HAM10000" dataset. They have utilized VGG16 architecture, and they were able to achieve an accuracy of 79% [15]. In another research study, Ardan Adi Nugroho and his team presented an identification system for skin cancer using the "HAM10000" skin cancer dataset, with training and validation accuracy of 80% and 78%, respectively [13]. Research published by Imen Ben Ida in 2016 provides a survey on IoT safety in the context of eHealth and clouds. In our previous work, we established an IoT-based data collection system for skin lesions. However, we didn't propose any security system for the data collector. We have also developed a CAD for skin classification using ensemble models [16]. Moreover, in another study, we have hypothesized a remote monitoring system for smart healthcare with numerous facilities for the patient such as pre-screening of cervical cancer [17]. Although the above study only considers the image-based pre-screening of various diseases, we have also created smart diagnosis assistant by using medical history of a patient in a separate study for early-stage diabetes prediction [18].

Security requirements are discussed in the context of healthcare, focusing on topics such as data interchange vulnerability, unauthorised access, multiple vulnerabilities, cryptography, eHealth Cloud Security damage by viruses [19]. In [20], IoT's Security, current Status, challenges, and prospective measures was discussed. They have also discussed Perception Layer, Network Layer, Application Layer, and the Security Features of IoT and demonstrated how to provide security. These two studies are separated as: IoT-based skin lesion detection and IoT security. In our research, we've integrated both the two.

13.2 METHODOLOGY

The working procedure for IoT-based skin cancer detection is listed in this section. In Figure 13.1, we have graphically depicted the technique of our proposal for this work.

To use the app, the patient must first create an account by entering their name, age, phone number, NID (or birth registration ID), and email address. The smart phone would take a picture of the diseased skin and a picture of the face with the planned software to detect the existence of cancer and for security purposes, respectively. The protection components are included in the program. After you register for this app, it will create a random code for security purposes. All data will be sent to a central repository via apps on the patient's smartphone that are linked to the internet and have already been collected. Even then, the image of the skin disease will be combined with generated random code by the app using a special algorithm, making it impossible for an unknown to retrieve the target image after obtaining this encoded data. This random code will be gotten from the central repository where a random code generator algorithm will be worked. The data will be sent to the central repository via the "WiFi Module ESP32," also known as the Antenna. After decoding the combination data with the same algorithm as the app, the "ARM Cortex-M CPU" will retrieve the target image and randomly generated code. The image will be sent to the "CAD" device, and via the "GSM SIM800M" antenna, the randomly generated code will be sent to the patient's phone. This cloud computing's "SSD Storage Device" can store all data, including an image for face protection and a copy of a randomly generated code for data distribution with these two security confirmations. The picture will be prepossessed in the CAD system to delete unnecessary details and improve the accuracy of the required information. The picture will then be segmented to extract important features and determine the area of interest. The classifier can

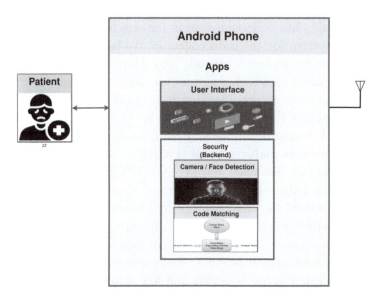

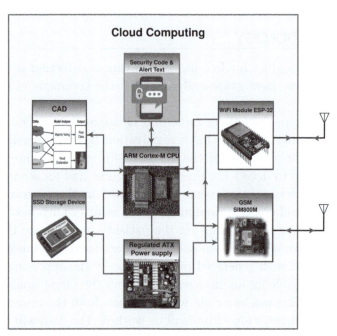

Figure 13.1   The proposed methodology is represented graphically.

decide whether or not cancer has been found as a result of this procedure. The "SSD Storage Device" will then be used to save the data. In cloud computing, the "Security Code & Alerting Text" block will confirm the protocol stack and notify the patient that the patient's result is ready. Following receipt of the alerting text, the patient will enter the code that received and provide his face image to the apps. This data will be sent to cloud storage once further by the apps. And the Security Code &

alerting Text block will compare the security code to the face detection algorithm. In that block, a face recognition algorithm will indeed be generated to validate is not whether the patient's request has been accomplished. If these two protection components match, the result would be sent to the apps via the "WiFi Module ESP-32" antenna, and the patient will receive this information from the app.

13.2.1 Patient

A patient is the one that is undergoing medical care at a doctor's office or maybe even a clinic. And this patient who is probably affected with a skin disease will provide data to the application like an image of the skin, phone number, email, and fingerprint in to register the app and get the result.

13.2.2 Sign-Up Information and Sign-In for Skin Cancer Detection

To register for this app, you will need to provide your name, phone number, age, email, and NID (or birth registration ID). The patient will receive a code after enrolling for this app, which will be used as the user's log-in code. As a result, the user will be able to log in to the app using their user name and log-in code.

13.2.3 Smartphone

A smartphone is a small electronic computer that can be used for more than just making phone calls, sending text messages, and taking pictures. A consumer would be able to connect to the network and use a variety of applications and services as a result of this. Currently, there are about 3 billion smartphone users, and these numbers are expected to continue to increase by a few hundred million annually over the next few years [21]. According to our proposed framework, a designed app will be installed and run by the patient for capturing the skin images. The smartphone's built in camera will be used to take an image of the diseased skin which will later be analysed by the CAD to detect the existence of cancer. However, initially the patient is required to scan their face for ensuring that they are an authorized user of the app.

13.2.4 Application

Application software is a concept that refers to software that serves a particular function. It's usually a system and perhaps a series of programs that produce desired results according to the users' needs. Application is playing an important role in society nowadays. A user can use the application easily from anywhere [22].

There are two parts to the application, one is the user interface, and another is security.

13.2.4.1 User Interface

Your application's user interface is all the user needs to view and engage with, and that this is constructed like a hierarchical structure of layouts and widgets for an

Android application. The layouts are group views, containers that control the position of your child's views on the screen.

13.2.4.2   Backend

Mobile App Development: involves the development on the server side, and in the backend, all processes are processed like random code generation, photography, etc. It is necessary to store, secure, and process the data in a mobile app.

Camera: By this, camera images of skin diseases and patients faces are taken to send to cloud computing to detect skin lesions and secure the process.

Encryption/Combined Face Images and Random Code: Random code and face images will be combined for security purposes as a hacker won't be able to retrieve real data, and won't differentiate the target data. The random code will be gotten from the central repository. After encrypted, this data will be sent to the central repository to detect cancer. It's one of the important parts.

13.2.5   Cloud Computing

The distribution of various services through the Internet is known as cloud computing. These services include data storage, servers, databases, networking, and software, and it's called that because the data is accessed remotely in the cloud or a virtual space. One of the most valuable advantages of cloud computing is the ability to store data in the cloud. The cloud will store any related business data, making it more available and usable. Data and programs are being snatched from desktop computers and corporate server rooms and transferred to the compute cloud [23]. The maximum process will be done in this cloud computing because all data will be stored here, and the patient will be able to access his profile after crossing the security process.

13.2.5.1   Antena for Transmitting and Receiving Information

WiFi Module ESP-32:   The ESP32 is a low-cost, low-power module on a chip microcontroller with implemented Wi-Fi and dual-mode Bluetooth that can function as a stand-alone system or as a slave computer to a host MCU, minimizing communication system latency on the main application processor. It can bind to WiFi networks and support Arduino libraries. It's simple to connect to a WiFi network, and the Tensilica Xtensa LX6 microprocessor powers this highly integrated structure [24].

GSM SIM800M: Since GSM carriers store customer details on a disposable SIM card, it's far simpler to switch phones on GSM networks and it serves as a link between both the user and the device network layer. In this project, GSM SIM800M is used. SIM800 is a quad-band GSM/GPRS module that operates on frequencies of 850 MHz GSM, 900 MHz EGSM, 1800 MHz DCS, and 1900 MHz PCS. It was used in this project. It also has GPRS multi-slot class 12/class 10 (optional) and supports GPRS coding schemes CS-1, CS-2, CS-3, and CS-4 [25].

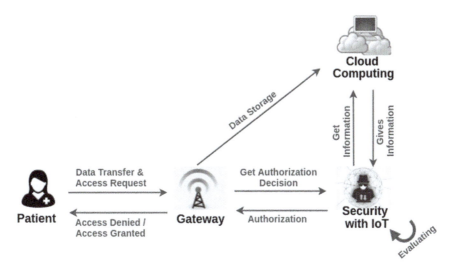

Figure 13.2 Security of the data.

13.2.5.2 ARM Cortex-M CPU

The ARM Cortex-M processor cores are a set of 32-bit RISC ARM processor cores. Cortex-M cores are commonly used as dedicated microcontroller chips, but they are often "hidden" within SoC chips as power management controllers, I/O controllers, system controllers, touch screen controllers, smart battery controllers, and sensors con. In the industry, the Cortex-M architecture is the most common and widely used processor family. Variation in the number of microcontrollers is now approaching 3000. CMSIS was created by ARM to be the fundamental library for communicating with ARM Cortex-based microcontrollers [26].

13.2.5.3 Data Storage and Data Security for Secure Transmission

Data Security: Data security is the protection system used to protect from unauthorized access and it is about keeping your data safe [27].

We've shown in Figure 13.2 how the data of the patients will be stored in cloud computing or through a secure gateway.

If an unauthorized individual can retrieve data, it poses a serious risk to the user. Of course, no one wants data to fall into the wrong hands. When everyone in the world is dependent on the internet nowadays, the most important aspect is to secure data for users on every platform.

In this chapter, two security systems are presented: Face Recognition security system and Matching security system by code generation.

Face Matching: If a user's face photo matches, they can access their profile; otherwise, they won't. This will detect intruders in inaccessible or high-security environments, as well as minimize error rates. The stability of today's institutions is a major concern [28].

Face recognition algorithms include Local Binary Pattern, HoG, Fisherface, as well as deep neural network models such as FaceNet, Probabilistic Face Embedding, ArcFace, and SpherFace. For image recognition, however, the "FaceNet" model is used in this article. FaceNet is a one-shot model that learns a mapping from face images to a compact Euclidean space where distances are directly proportional to face similarity. Rather than using an intermediate bottleneck layer as in previous deep learning methods, FaceNet uses a deep convolutional network to directly optimize the embedding itself [29]. Faces are normally aligned with simple affine transformations before being fed into convolutional neural networks to extract identity-preserving features in a traditional deep face recognition system [30]. Face recognition is becoming increasingly common for use in a variety of applications, including security systems and from the dataset Labeled Faces in the Wild (LFW), the AceNet can provide accuracy of up to 99.63%, and 95.12% on the YouTube Faces DB [31].

Figure 13.3 depicts the security stages that a patient must go through to obtain their results or access to their profile in the data store.

Code Generation For security purposes, a random code will be created in this application. All processes will be handled by this application. So this randomly generated code will be sent to the patient to encrypt the face images data. At first, the application will be installed and registered by the patient. And then images will be taken by performing continuous command of the application so that it make sure the security of the data and also send data in the word cloud by using a secure channel. It's the most important part of this work. To generate random code in this project, we used the Pseudorandom Code Generator (PRCG). Any software that uses math to simulate randomness is referred to as a pseudorandom number generator (PRNG). A deterministic random bit generator is another name for it (DRBG). It is produced by a computer and is required by many computer applications, including games and protection. These generators employ a numerical algorithm to generate a series of numbers that exhibit many of the characteristics of genuinely random numbers. Despite the fact that the sequence of pseudorandom numbers is not completely random, a good generator can generate numbers that are nearly identical to truly random numbers. The algorithm will produce the pseudorandom pattern after selecting an initial number or seed [32]. When a user crosses their security system with their code, it is the security system [33]. The user may receive a code from the central repository, which they must match with the code in their application to obtain the desired result. This is the most critical protection system for both the device and cloud.

```
FROM random IMPORT seed
FROM random IMPORT randint
FOR_IN range(6):
    SET value TO randint(0, 9)
    OUTPUT(value)
```

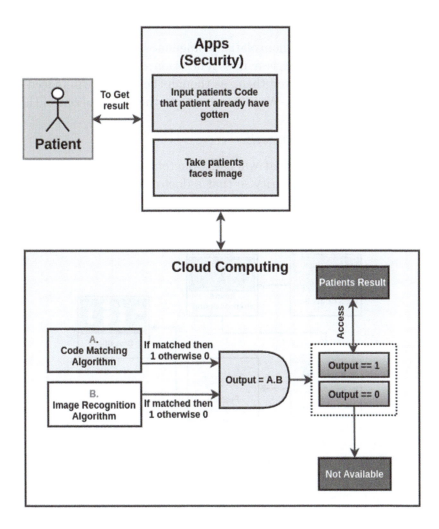

Figure 13.3 Patients must go through security to access their results.

Code Matching If a patient wants their results then they must, of course, enter the code through the alert message. If the correct code is entered or matched, then the patient can pass this security. We have suggested using the "Karp–Rabin Algorithm" in this project to match passwords or strings. The Karp–Rabin Algorithm is one of the algorithms used to detect the similarity levels of two strings and is a string searching algorithm made by Richard M. Karp and Michael O. Rabin [34].

13.2.5.4 Regulated ATX Power Supplies

There's a signal from the motherboard that controls all ATX power supplies. Also, they warn the motherboard when the DC voltages are correct so the system will be able to boot.

13.2.5.5   SSD Storage Device

A solid-state drive is a form of nonvolatile storage media that stores permanent data on solid-state flash memory. It is a new generation of storage devices. And all of the patient's records will be saved there, as though the patient will indeed be able to retrieve both of his past results. We may assume it's the profile of one of the patients.

13.2.5.6   Computer-Assisted Diagnosis

We want to work on a project that will assist doctors in detecting skin lesions on an IoT platform with sufficient security so that a doctor or pathologist can understand or diagnose the skin condition. This process explains how to use deep learning to detect skin lesions.

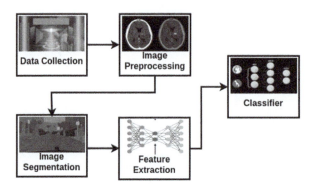

Figure 13.4   CAD for the classification model.

Figure 13.4 presents the steps that should be followed to classify skin lesions for this model.

Data Collection   Data collection is described as the method for accurate insights to be collected, measured, and analyzed by standard validated methods for research, and data collected from patients.

Image Pre-Possessing   Pre-possessing is a operation on images to develop images. Pre-processing is generally used to optimize images by reducing undesirable distortions or optimizing image features that are critical for image acquisition. There are many types of image pre-processing techniques like Pixel brightness transformations, Geometric Transformations, Image Filtering, Segmentation, Fourier transforms, and Image restoration [35].

Image Segmentation   Image segmentation is the method of separating an information into various areas based on pixel properties to define artifacts or borders in simplifying and explore it more accurately. This method allows us to get a much more detailed understanding of the objects in the picture. And with this technique, we can easily split our target images from other parts and we can easily remove extra parts of images [36]. *Example*: Affected skin can be isolated easily by image segmentation. There are two types of image segmentation techniques: Region-Based Segmentation and Edge Detection Segmentation.

Feature Extraction Feature extraction is a part of the dimensionality reduction process [37]. Features describe an image's actions, indicating its position in terms of space taken up, classification performance, and, of course, time consumption.

Classifier The classifier can be any algorithm that can categorize a set of data into classes. The classification algorithm is an assumption as well as separable method for assigning class labels for particular data points [38]. There are many types of classifiers in deep learning like MobileNet, DensNet121, ResNet101, VGG16, Xception, etc.

13.2.5.7 Alert Texts

Alerts seem to be a straightforward, one-button format that works well for brief, insightful messages. When the classifier detects a patient's final result, this portion will send an alert message to the patient's phone via wifi module, informing him that his result is ready, along with a random code that was produced and transmitted to the application before encrypting the patient's face image for security purposes.

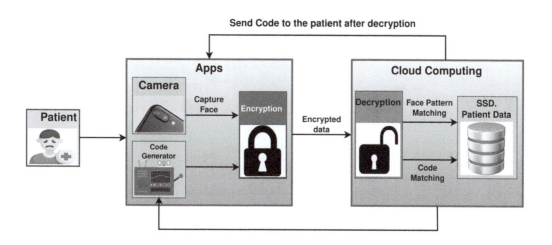

Figure 13.5 Representation of where encryption and decryption are done.

Figure 13.5 interprets the SMS alerts when a patient receives after undergoing a certain operation. This file will be decrypted after receiving the combined files of facial images and randomly generated code. After decryption, the model will generate a result, and then the alarm message will be transmitted to the patient via cloud computing.

This will ensure that the patient knows that his result is finalized when the message box is opened, and that he can do so by gathering his outcome data with two layers of protection.

13.2.6 Deep Learning-Based Techniques

The use of deep learning-based classifiers is commonplace because they can perform classification tasks for objects, in general, better than humans can. When dealing with unstructured data, deep learning's capacity to process vast amounts of features makes it incredibly strong. Deep learning techniques, on the other hand, may be overkill for simpler tasks since they need access to a large amount of data to be productive. Deep learning enables computational models with several processing layers to learn multiple degrees of abstraction for data representations [39]. VGG, AlexNet, ResNet, and Xception are some of the most recent deep learning models developed.

13.2.7 CNN with Fine-Tuning

A CNN is a type of deep neural network used to evaluate visual imagery in deep learning, and it has been demonstrated by CNN that it has a number of success stories in the fields of computer vision and machine learning [40].

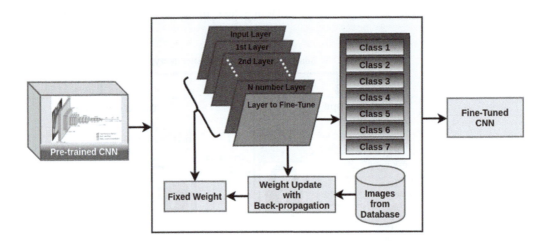

Figure 13.6 The architecture of the model after fine-tuning.

One of the most popular deep neural networks is CNN. It is named after the mathematical operation that applies matrices and performs a convolution when applied to two matrices, called convolution [41].

The architecture of fine-tuning of the CNN model is shown in Figure 13.6.

Applying advanced tuning techniques to improve the accuracy of a new neural network model will involve incorporating the results of an existing neural network. We can utilize what the model has already learned, as opposed to starting from scratch, by using an artificial neural network that has already been designed and trained. Pre-trained CNNs, which have been fine-tuned for a specific task, is a simple yet highly effective approach to solving an assigned work [42].

The equation for the CNN model is defined as:

$$h_{w,b}(x) = f(W^t x) = f(\sum_{i-1}^{n} W_i x_i + b) \tag{13.1}$$

13.3 EXPERIMENTS AND RESULTS

What was the experiment and outcome by our various models for this study in this section? There are descriptions of accuracy, loss from a different perspective, parameters, and hyperparameters. The outcome, along with a planned project, is discussed here.

13.3.1 Datasets

There are 10,015 skin lesions in the HAM10000 dataset, which is a significant collection of multi-source dermatoscopic images of common pigmented skin lesions.

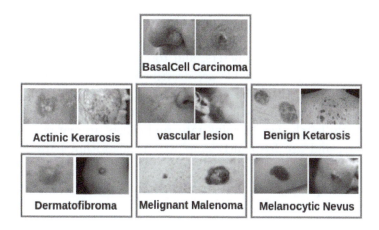

Figure 13.7 Images of different skin lesions.

Malignant melanoma (MEL), melanocytic nevus (NV), basal cell carcinoma (BCC), actinic keratosis (AKIEC), benign keratosis (BKL), dermatofibroma (DF), and vascular lesion are among the seven skin lesion types represented in this dataset (VASC) [1]. In Figure 13.7, images of seven types of skin lesions are shown.

13.3.2 Accuracy and Loss

The corresponding accuracy score and loss value for all the selected models are shown in Table 13.1.

13.3.3 Parameters and Hyperparameters

The parameters of each layer are called weights and biases in a CNN. When we consider the total weight and bias values, the total number of parameters is just the

TABLE 13.1   Four Deep Learning Models Are Compared to Reveal Which One Is the Most Accurate

Classification Model	Train Accuracy	Test Accuracy	Train Loss	Test Loss	Precision	Recall	F1-Score
VGG16	0.961	0.844	0.108	0.749	0.841	0.846	0.842
ResNet152	0.997	0.841	0.010	1.428	0.833	0.841	0.834
InceptionV3	0.961	0.844	0.108	0.749	0.841	0.846	0.842
DenseNet201	0.998	0.850	0.004	1.120	0.846	0.850	0.844

TABLE 13.2   These Parameters and Hyperparameters Are Utilized to Categorize Skin Lesions in These Models

Classification Models	Total Layers	Trainable Layers	None-Trainable Layers	Batch Size	Epochs	Optimizer	Learning Rate	Total-Trainable Params
VGG16	16	13	3(first)	64	50	Adam	0.842	17,039,367
ResNet152	152	122	30(first)	64	50	Adam	0.0001	58,049,216
InceptionV3	48	45	3(first)	64	50	Adam	0.0001	14,675,968
DenseNet201	201	171	30(first)	64	50	Adam	0.0001	17,936,064

sum of all weights and biases. Every layer in the convolutional layer has a weight, except for pooling and nonlinearity layers which don't have weights [41]. Table 13.2 show the parameters, hyperparameters, and total layers of models with which layers were trainable and non-trainable are described. Network structure and training procedures are determined by the network's hyperparameters. As hyperparameters perform better, the better ones will require a human setup of various hyperparameter values [43]. Epoch, batch size, learning rate, and others are parameters that control hyperparameters, and are frequently employed in algorithm parameter estimation processes.

13.3.4 Evaluation Metrics

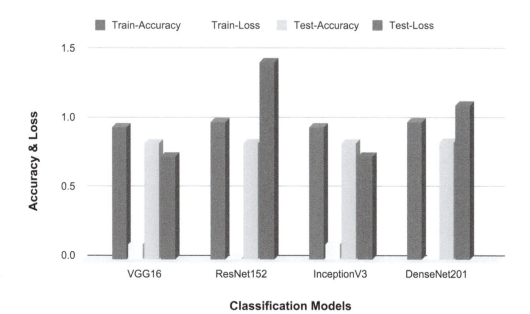

Evaluation metrics are utilized as a vital component of any databases science project for measurement of the quality for the statistical model or machine learning model and to assess the accuracy of the model generalized to future information.

13.3.4.1 *Training Loss and Validation Loss*

Training loss is the mistake on the training dataset, and validation loss is the mistake of the validation dataset running on the trained network, and both are perfect if the loss is 0.

13.3.4.2   Recall

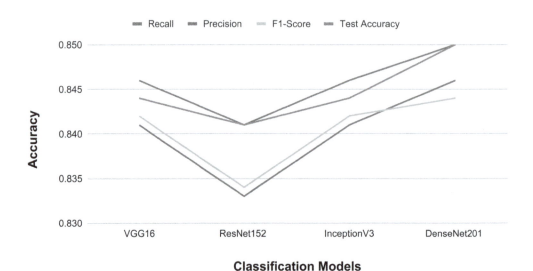

Classification Models

Measurement metrics entail the use of various distinct assessment metrics to eval-uate a model or algorithm and it is essential to ensure that your model is working properly and optimally, which is essential to get the optimal classification throughout the classification training [44].

The remainder is the measure of our method which correctly identifies True Pos-itives, whereby TP is the number of true positives and FN is the number of false negatives, and it is also the proportion of relevant instances found. We should re-member, the best recall value is 1 and the worst recall value is 0.

$$Recall = \frac{TP}{TP + FN} \qquad (13.2)$$

13.3.4.3   Precision

Precision is the proportion of positive observations correctly predicted with the total positive observations predicted. The prediction accuracy is used to understand when the cost of false positives is high. Let's assume that rare disease detection is the problem. If we were to use a model with a low accuracy level, many patients are told they have a disease that might lead to misdiagnosis.

$$Precision = \frac{TP}{TP + FP} \qquad (13.3)$$

13.3.4.4   F1-Score

F1-Score is the weighted mean of accuracy and recall and its score, often known as an F1-Score, measures the accuracy of the model on a dataset. This is used for assessing categorical classification systems, that categorize instances as correct or incorrect, and is frequently more helpful than precision, in particular, if your class distribution is unequal.

$$F1 - score = 2 \times \frac{Precision \times Recall}{Precision + Recall} \qquad (13.4)$$

13.3.4.5 Accuracy

The accuracy of classification is essentially the rate of proper classification, for either an independent test set or with a change in the idea of cross-validation, and has been used to evaluate which model identifies connections and correlations between variables in a dataset based on data input as well as training.

$$Accuracy = \frac{TP + TN}{TP + FP + TN + FN} \qquad (13.5)$$

13.3.5 Results

All models perform well with 84% accuracy after fine-tuning in evaluation matrices, however, DenseNet201 is our best model for this job with 85% test accuracy, 1.12 test loss, 84% precision accuracy, 84% recall accuracy, and 84% F1-Score accuracy. The DenseNet201 model has a training accuracy of 99%. There isn't much of a distinction between train and test accuracy. As a result, we may conclude that our model is neither underfitting nor overfitting. DenseNet201 has a total of 201 layers, but the first 30 layers were fine-tuned in this study, leaving only 171 layers to be trained. This model uses the Adam optimizer, which has a batch size of 64, 50 epochs, a learning rate of 0.0001, and a total trainable parameter count of 17,936,064.

13.3.6 Discussion

On the HAM10000 dataset, we trained deep learning models to identify skin lesions on IoT based on security. Four classification models were trained, and the accuracy of these models was good. The main focus of this research is on ensuring sufficient security for IoT-based skin lesion detection. Models were tested, however, we recommend using "IoT based on security" because the experiment was too costly. However, a method based on references is proposed. The main architecture of this project is depicted in Figure 13.1. Patients' data is shown in Figure 13.2 as it travels through a gateway, past security, and into cloud computing storage. Figure 13.3 depicts the security measures used to gain access to a patient's results. In this study, models were able to categorize skin lesions using a five-step process called CAD, as shown in Figure 13.4. The result will be ready in cloud computing after encryption and decryption, and then it will be transferred to the patient, as shown in the diagram in Figure 13.5. In our work, we've used CNN models with fine-tuning, and the architecture is depicted in Figure 13.6. Figure 13.7 depicts images of seven different forms of skin lesions. To understand the performance of all models, accuracy, loss, parameters, and hyperparameters are reported in Tables 13.1 and 13.2.

13.4 CONCLUSION

Skin lesion detectors based on the Internet of Things can assist doctors and pathologists in detecting skin lesions. If the patient stays connected to the internet while using this software, they will be able to detect their skin lesions effortlessly at any location. It helps you save both time and money. Patients may be protected from melanoma if they find it early. IoT-based skin lesion work is now being done in the research sector, but security is not incorporated in that work. Nowadays, security is paramount. When consumers utilize any program, they risk their data and the security of their results since it is dangerous if the user does not get the exact or accurate result. If an unauthorized person has access to the patients' records, he will be able to alter their outcomes. As a result, we must now provide safe IoT-based applications, particularly in the healthcare sector. It will be beneficial if patients or users feel confident in their results. And now, we've integrated these two key concepts: IoT-based skin detection and IoT-based security, and our work is focused on IoT-based skin lesion detection based on security.

ACKNOWLEDGEMENT

Muhammad Khairul Islam and Muhammad Al Amin contributed equally to this chapter.

Bibliography

1. Philipp Tschandl, Cliff Rosendahl, and Harald Kittler. The HAM10000 dataset, a large collection of multi-source dermatoscopic images of common pigmented skin lesions. *Scientific Data*, 5(1):1–9, 2018.

2. Robert G Morris. Computer hacking and the techniques of neutralization: An empirical assessment. In *Corporate hacking and technology-driven crime: Social dynamics and implications*, pages 1–17. IGI Global, 2011.

3. Mahmoud Ammar, Giovanni Russello, and Bruno Crispo. Internet of Things: A survey on the security of IoT frameworks. *Journal of Information Security and Applications*, 38:8–27, 2018.

4. Yuexiang Li and Linlin Shen. Skin lesion analysis towards melanoma detection using deep learning network. *Sensors*, 18(2):556, 2018.

5. Jeremy Kawahara, Aicha BenTaieb, and Ghassan Hamarneh. Deep features to classify skin lesions. In *2016 IEEE 13th International Symposium on Biomedical Imaging (ISBI)*, pages 1397–1400. IEEE, 2016.

6. Clare F Heal, Beverley A Raasch, PG Buettner, and David Weedon. Accuracy of clinical diagnosis of skin lesions. *British Journal of Dermatology*, 159(3):661–668, 2008.

7. Muhammad Attique Khan, Muhammad Younus Javed, Muhammad Sharif, Tanzila Saba, and Amjad Rehman. Multi-model deep neural network based fea-

tures extraction and optimal selection approach for skin lesion classification. In *2019 International Conference on Computer and Information Sciences (ICCIS)*, pages 1–7. IEEE, 2019.

8. Farhat Afza, Muhammad A Khan, Muhammad Sharif, and Amjad Rehman. Microscopic skin laceration segmentation and classification: A framework of statistical normal distribution and optimal feature selection. *Microscopy Research and Technique*, 82(9):1471–1488, 2019.

9. Olga Russakovsky, Jia Deng, Hao Su, Jonathan Krause, Sanjeev Satheesh, Sean Ma, Zhiheng Huang, Andrej Karpathy, Aditya Khosla, Michael Bernstein, et al. Imagenet large-scale visual recognition challenge. *International Journal of Computer Vision*, 115(3):211–252, 2015.

10. Md Khairul Islam, Most Nilufa Yeasmin, Chetna Kaushal, Md Al Amin, Md Rakibul Islam, and Md Imran Hossain Showrov. Comparative analysis of steering angle prediction for automated object using deep neural network. In *2021 9th International Conference on Reliability, Infocom Technologies and Optimization (Trends and Future Directions) (ICRITO)*, pages 1–7. IEEE, 2021.

11. Tallha Akram, Muhammad Attique Khan, Muhammad Sharif, and Mussarat Yasmin. Skin lesion segmentation and recognition using multichannel saliency estimation and m-SVM on selected serially fused features. *Journal of Ambient Intelligence and Humanized Computing*, pages 1–20, 2018.

12. Adria Romero Lopez, Xavier Giro-i Nieto, Jack Burdick, and Oge Marques. Skin lesion classification from dermoscopic images using deep learning techniques. In *2017 13th IASTED International Conference on Biomedical Engineering (BioMed)*, pages 49–54. IEEE, 2017.

13. Ardan Adi Nugroho, Isnandar Slamet, and Sugiyanto. Skins cancer identification system of HAM10000 skin cancer dataset using convolutional neural network. In *AIP Conference Proceedings*, volume 2202, page 020039. AIP Publishing LLC, 2019.

14. Md Khairul Islam, Md Habibur Rahman, Md Rakibul Islam, Md Zahidul Islam, Md Mainul Islam Mamun, AKM Azad, and Mohammad Ali Moni. Network-based systems biology approach to identify diseasome and comorbidity associations of systemic sclerosis with cancers. *Heliyon*, page e08892, 2022.

15. Kiran Pai and Anandi Giridharan. Convolutional neural networks for classifying skin lesions. In *TENCON 2019–2019 IEEE Region 10 Conference (TENCON)*, pages 1794–1796. IEEE, 2019.

16. Md Khairul Islam, Chetna Kaushal, and Md Al Amin. Smart home-healthcare for skin lesions classification with IoT-based data collection device. 2021.

17. Chetna Kaushal, Md Khairul Islam, Anshu Singla, and Md Al Amin. An iomt-based smart remote monitoring system for healthcare. *IoT-Enabled Smart Healthcare Systems, Services and Applications*, page 177, 2022.

18. Md Abu Rumman Refat, Md Al Amin, Chetna Kaushal, Mst Nilufa Yeasmin, and Md Khairul Islam. A comparative analysis of early stage diabetes prediction using machine learning and deep learning approach. In *2021 6th International Conference on Signal Processing, Computing and Control (ISPCC)*, pages 654–659. IEEE, 2021.

19. Imen Ben Ida, Abderrazak Jemai, and Adlen Loukil. A survey on security of IoT in the context of eHealth and clouds. In *2016 11th International Design & Test Symposium (IDT)*, pages 25–30. IEEE, 2016.

20. Rwan Mahmoud, Tasneem Yousuf, Fadi Aloul, and Imran Zualkernan. Internet of Things (IoT) security: Current status, challenges and prospective measures. In *2015 10th International Conference for Internet Technology and Secured Transactions (ICITST)*, pages 336–341. IEEE, 2015.

21. S O'Dea. Number of smartphone users worldwide from 2016 to 2021. *Statista Research Department*, 2020.

22. Rashedul Islam, Rofiqul Islam, and Tohidul Mazumder. Mobile application and its global impact. *International Journal of Engineering & Technology (IJEST)*, 10(6):72–78, 2010.

23. Brian Hayes. Cloud computing, 2008.

24. Alexander Maier, Andrew Sharp, and Yuriy Vagapov. Comparative analysis and practical implementation of the ESP32 microcontroller module for the internet of things. In *2017 Internet Technologies and Applications (ITA)*, pages 143–148. IEEE, 2017.

25. P Elechi, CO Ahiakwo, and ST Shir. Design and implementation of an automated security gate system using global system for mobile communication network.

26. Y Paunski and R Zahariev. Service robots control system, based on "arm cortex m" architecture microprocessor system". In *Pr. TU Sofia, XXVI International Conference ADP-2017*, pages 300–304, 2017.

27. Dorothy E Denning and Peter J Denning. Data security. *ACM Computing Surveys (CSUR)*, 11(3):227–249, 1979.

28. Michel Owayjan, Amer Dergham, Gerges Haber, Nidal Fakih, Ahmad Hamoush, and Elie Abdo. Face recognition security system. In *New Trends in Networking, Computing, E-learning, Systems Sciences, and Engineering*, pages 343–348. Springer, 2015.

29. Weiguo Wan and Hyo Jong Lee. FaceNet based face sketch recognition. In *2017 International Conference on Computational Science and Computational Intelligence (CSCI)*, pages 432–436. IEEE, 2017.

30. Mingjie He, Jie Zhang, Shiguang Shan, Meina Kan, and Xilin Chen. Deformable face net for pose invariant face recognition. *Pattern Recognition*, 100:107113, 2020.

31. Ivan William, Eko Hari Rachmawanto, Heru Agus Santoso, Christy Atika Sari, et al. Face recognition using FaceNet (survey, performance test, and comparison). In *2019 Fourth International Conference on Informatics and Computing (ICIC)*, pages 1–6. IEEE, 2019.

32. Majid Babaei and Mohsen Farhadi. Introduction to secure PRNGs. *International Journal of Communications, Network and System Sciences*, 4(10):616, 2011.

33. Nataliia Bielova, Nicola Dragoni, Fabio Massacci, Katsiaryna Naliuka, and Ida Siahaan. Matching in security-by-contract for mobile code. *Journal of Logic and Algebraic Programming*, 78(5):340–358, 2009.

34. Andysah Putera Utama Siahaan, Robbi Rahim, Mesran Mesran, and Dodi Siregar. K-gram as a determinant of plagiarism level in Rabin-Karp algorithm. 2017.

35. Milan Sonka, Vaclav Hlavac, and Roger Boyle. Image pre-processing. In *Image Processing, Analysis and Machine Vision*, pages 56–111. Springer, 1993.

36. Robert M Haralick and Linda G Shapiro. Image segmentation techniques. *Computer Vision, Graphics, and Image Processing*, 29(1):100–132, 1985.

37. Martin D Levine. Feature extraction: A survey. *Proceedings of the IEEE*, 57(8):1391–1407, 1969.

38. Ryo Asaoka, Hiroshi Murata, Aiko Iwase, and Makoto Araie. Detecting preperimetric glaucoma with standard automated perimetry using a deep learning classifier. *Ophthalmology*, 123(9):1974–1980, 2016.

39. Yann LeCun, Yoshua Bengio, and Geoffrey Hinton. Deep learning. *Nature*, 521(7553):436–444, 2015.

40. Jianxin Wu. Introduction to convolutional neural networks. *National Key Lab for Novel Software Technology. Nanjing University. China*, 5(23):495, 2017.

41. Saad Albawi, Tareq Abed Mohammed, and Saad Al-Zawi. Understanding of a convolutional neural network. In *2017 International Conference on Engineering and Technology (ICET)*, pages 1–6. IEEE, 2017.

42. Xiangnan Yin, Weihai Chen, Xingming Wu, and Haosong Yue. Fine-tuning and visualization of convolutional neural networks. In *2017 12th IEEE Conference on Industrial Electronics and Applications (ICIEA)*, pages 1310–1315. IEEE, 2017.

43. Nurshazlyn Mohd Aszemi and PDD Dominic. Hyperparameter optimization in convolutional neural network using genetic algorithms. *Int. J. Adv. Comput. Sci. Appl*, 10(6):269–278, 2019.

44. Mohammad Hossin and Md Nasir Sulaiman. A review on evaluation metrics for data classification evaluations. *International Journal of Data Mining & Knowledge Management Process*, 5(2):1, 2015.

## CHAPTER 14

Real-Time Facemask Detection Using Deep Convolutional Neural Network-Based Transfer Learning

Jakaria Islam Emon, M. M. Manjurul Islam, Syeda Amina Abedin and Shakhawat Hossain

Department of Computer Science, American International University-Bangladesh, Dhaka

Rupam Kumar Das

College of Science and Engineering, University of Glasgow, United Kingdom

CONTENTS

THE PANDEMIC of COVID-19 has caused severe harm and infected tens of millions of people across the globe. In response to the rapid outbreak of COVID-19, this chapter presents an automated facemask detection technology in a real-time video stream using a deep convolutional neural network (DCNN)-based transfer learning mechanism. The proposed framework has two blocks—the first module leverages the training phase which was built on visual geometry group network (VGG16), a variant of DCNN, to learn a model on a benchmark dataset. To build an accurate model,

DOI: 10.1201/9781003251903-14

VGG16 needs an abundance of data. Thus, we deploy a pre-trained model to be built on ImageNet, a large visual database, for fine-tuning the VGG16 model using knowledge transfer. In the second phase, we perform testing on offline facemask images and real-time video streaming using fine-tuned VGG16 model. The proposed methodology is successfully capable of identifying mask and nonmask people, yielding 41% (0.41 in ratio) and 20% (0.20 in ratio) of performance improvement in terms of accuracy from MobileNet V2 and RestNet50, respectively. The experimental results suggest that the proposed facemask detection algorithm can be deployed in a public place with satisfactory performance to contribute to controlling COVID-19 pandemics.

14.1 INTRODUCTION

In recent history, the greatest pandemic spread throughout the world is Coronavirus Disease 2019 (COVID-19), which occurred in late 2019. COVID-19 is a deadly infectious disease caused by the severe acute respiratory syndrome coronavirus 2 (SARS-CoV-2) [1]. As of October 2021, more than 246 million cases and 4.99 million deaths have been confirmed, making it one of the deadliest pandemics in history. Coronavirus infection is transmitted mainly by respiratory droplets formed when people breathe, talk, sneeze, or cough as per the Centers for Disease Control and Prevention (CDC) [2].

Therefore, many solutions namely lockdowns, shutdowns, and confinement are suggested by most of the world's governments to prevent the spread of COVID-19 infection rapidly. Beyond that, the activities in many parts of the world have slowed or stopped due to the human, economic, and social impacts of distancing and protection measures. In addition to various preventive measurements, the World Health Organization's (WHO) interim guidance suggested that the use of facemasks has clear links to reducing the spread of SARS-CoV-2 [3]. Specialists also endorse using masks in public places to reduce the transmission of infections [4]. As a preventive measure for the ongoing pandemic, this chapter seeks to develop a deep learning-based facemask detection system that is capable of identifying whether people in surveillance-type video streams are correctly wearing their masks. In the era of massive data, deep learning methods have been put to use to analyze these diverse data; these methods include representation learning for feature extraction and supervised learning for prediction. The accumulation of large volumes of data and the exploitation of parallel computing resources have accelerated the development of deep learning technology; over the past decade, these deep learning models have been used in various fields of academia and industry, and they are also being actively applied in computer vision [5].

Deep learning is a method of constructing and training a deep neural network with hidden layers to be mainly used for classification and prediction. The outstanding performance of the visual geometry group network (VGG16) [6]—an image classifier based on a deep convolutional neural network (DCNN)—won the ImageNet [7] challenge in 2014 for image classification with an overwhelming margin. The primary aim of this chapter is to build VGG16 classifier-based transfer learning process for automatic facemask detection in real-time video streaming.

The rest of the chapter is structured as follows. Section 14.2 deals with the review of related works in the past. Section 14.3 describes the materials and method of this research. In Section 14.4, we present the experimental results and findings of the proposed methodology. Finally, the conclusion of the proposed work is presented in Section 14.5.

14.2 RELATED WORKS

Image reconstruction and facial recognition for identity verification are one of the most common research topics [8]. The continuing epidemic has made research with similar intentions utterly popular. The main goal of our effort is to detect persons who are wearing masks or not in the real-time video stream. Researchers developed a method that can independently determine the existence of nose and mouth to detect facemask [9]. Their proposed system was efficient but it had a limitation—only complete front faces can be processed, and the sensors may be tricked simply by covering the mouth and nose by hand. Qin and Li developed a facemask recognition technique that can classify the pictures into three categories: Proper facemask wearing, wrong facemask wearing, and no facemask wearing using the SRCNet classification network [10]. Deb et al. proposed a different approach to the system that applies MobileNetV2 for facemask classification [11]. But this method is not representative of the different inter-ethnic, gender, and different face types.

Cabani et al. suggested a mask recognition system based on facial characteristic images with additional training data [9]. At the same time, the mobile application [12] was developed to educate individuals to use masks properly by detecting whether the masks of users covered their nose and their mouth simultaneously. For masked and non-masked face recognition, Ejaz et al. utilized the principal component analysis (PCA) technique [13]. While identifying faces without a mask, PCA's accuracy is 96.25%, but it drops to 68.75% when doing so while wearing a mask. Javed et al. created an interactive model, called MRGAN, which eliminates items like microphones from face pictures and recomposes the deleted area with an adversary network of the generating region [14]. A deep neural network architecture called DarkNet-19 was utilized by Li and colleagues for face identification and they used YOLOv3 for face detection [15]. An automated method for the identification of the presence or absence of the required surgical mask in operation rooms was proposed [16]. The goal of that system was to sound an alarm if a member of staff is not wearing a mask while on duty.

But relatively little research has been done on mask detection from the real-time video stream and these existing research works need further refinement for improved classification accuracy. Thus we propose a transfer learning strategy that leverages the pre-trained VGG16 model to enhance the facemask identification scheme from a live video stream to fight against COVID-19.

14.3 MATERIALS AND METHOD

Figure 14.1 illustrates the proposed facemask detection system, which comprises two main blocks. In the training phase, VGG16 classifier is fine-tuned using a limited medical facemask dataset with weights sharing from a large pre-trained ImageNet weight model. The fine-tuned model is further validated using a validation dataset for performance measurement. In the testing phase, models are tested with offline images and video streams for classification performance measurements. Detailed descriptions of the proposed methodology are as follows.

14.3.1 Dataset

We apply the benchmark facemask dataset from Kaggle to build and test the VGG16 model for classification accuracy measurement [17]. The dataset consists of 1500 mask and non-mask images. We randomly split this dataset into three sets such as training, validation, and testing sets. We kept 33% of the dataset (i.e., 500 images) for testing which is completely unknown to the model to ensure a higher generalization performance which means the testing set did not participate in model building. Note that 67% of the dataset (i.e., 1000 images) was used for model building in the training phase.

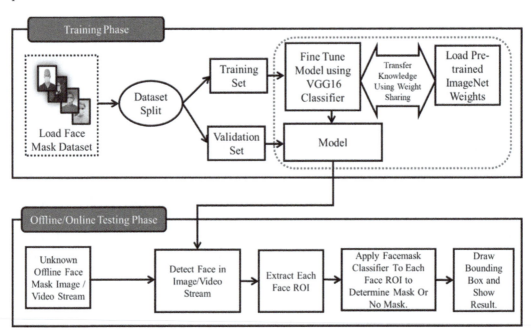

Figure 14.1 Schematic representation of the proposed work for identifying facemasks.

14.3.2 VGG16 Architecture

The proposed algorithm is based on the VGG16 deep learning model which is a variant of DCNN architecture [6]. VGG16 is regarded as one of the best computer vision model architectures to date. The main motivation of using VGG16 is that they focused on having convolution layers of 3 × 3 filter with a stride 1 and always used the same padding and max pool layer of 2 × 2 filter of stride 2 instead of having a large number of hyperparameters. It follows the arrangement of convolution and max pool layers consistently throughout the whole architecture. In the end, it has 3 fully connected layers (FC) followed by a SoftMax for the output layer. The 16 in VGG16 denotes its 16 layers that have weights. This network is a pretty large network with approximately 144 million parameters. The depiction of the proposed architecture of the VGG16 network is shown in Figure 14.2. Table 14.1 presents the dimensionality of each layer in the VGG16 architecture.

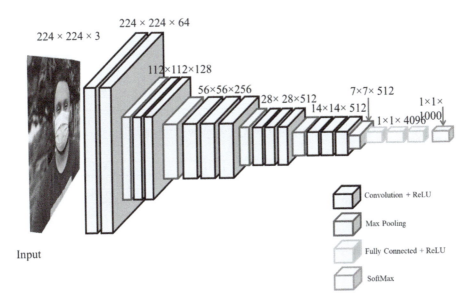

Figure 14.2 Proposed VGGNet16 network for facemask detection.

In the convolutional layer of VGG16, it performs mathematical operations on two functions in the form of integration of products among functions. The main motivation of convolution operation is to find salient features. If f is the input and w is the filter, the convolution operation can be defined as follows,

$$y = f * g = \int_{-\infty}^{\infty} f(t - \tau) w(\tau) d\tau, \tag{14.1}$$

where t defines an amount of shift. After the convolutional operation, we apply the activation function to introduce the nonlinearity of the model. Various activation functions namely rectified linear unit (ReLU), linear function, Sigmoid, or tanh mostly used, but we apply ReLU in this study to train the model since it ensures fast and near-global optimization of weights in comparison with other activation functions.

TABLE 14.1 The Dimensionality of the Proposed VGG16 Networks with Parameter Count

Layer (Type)	Output shape	Parameter
input_1 (Input Layer)	(None, 244, 244, 3)	0
block1_conv1 (Conv2D)	(None, 244, 244, 64)	1792
block1_conv2 (Conv2D)	(None, 244, 244, 64)	36928
block1_pool (Maxpooling2D)	(None, 112, 112, 64)	0
block2_conv1 (Conv2D)	(None, 112, 112, 128)	73856
block2_conv2 (Conv2D)	(None, 112, 112, 128)	147584
block2_pool (Maxpooling2D)	(None, 56, 56, 128)	0
block3_conv1 (Conv2D)	(None, 56, 56, 256)	295168
block3_conv2 (Conv2D)	(None, 56, 56, 256)	590080
block3_conv3 (Conv2D)	(None, 56, 56, 256)	590080
block3_pool (Maxpooling2D)	(None, 28, 28, 256)	0
block4_conv1 (Conv2D)	(None, 28, 28, 512)	1180160
block4_conv2 (Conv2D)	(None, 28, 28, 512)	2359808
block4_conv3 (Conv2D)	(None, 28, 28, 512)	2359808
block4_pool (Maxpooling2D)	(None, 14, 14, 512)	0
block5_conv1 (Conv2D)	(None, 14, 14, 512)	2359808
block5_conv2 (Conv2D)	(None, 14, 14, 512)	2359808
block5_conv3 (Conv2D)	(None, 14, 14, 512)	2359808
block5_pool (Maxpooling2D)	(None, 7, 7, 512)	0
flatten_1 (Flatten)	(None, 25088)	0
dense_1 (Dense)	(None, 2)	50178
Total Params: 144,764,866		
Trainable Params: 50,178		
Non-Trainable Prams: 144,414,688		

TABLE 14.2 The Comparison of Test Performance among Different Methodology

Models	PR	RR	AC
MobileNet V2	0.66	0.22	0.55
RestNet50	1.0	0.59	0.76
VGG16 (Proposed)	1.0	0.92	0.96

The ReLU is calculated as follows:

$$ReLU\ (y_i) = \max(0,\ y_i) \tag{14.2}$$

where y_i is the output of the convolutional layer.

The pooling layer comes next to convolutional layers that subside the features map by a factor of 2. The motivation of the pooling layer is to reduce overfitting and training time. A SoftMax function is used in the fully connected layer to classify images. An efficient optimization algorithm is required to learn such an extensive number of model parameters as reported in Table 14.2. We applied an adaptive moment estimation (Adam) algorithm to optimize model parameters that improve the root mean square propagation optimizer by calculating running averages of both the gradients and the second moments of the gradients [18].

14.3.3 Transfer Learning for Fine-Tuning of the VGG16 Model

As it is required to train more than 144 million parameters, the VGG16 network needs a massive amount of training dataset. It is always crucial to obtain quality data. Thus, we apply transfer learning which is a popular approach in deep learning, where a previously trained model is utilized to solve a new related problem. Transfer learning is useful when the training dataset images are small in numbers. To elevate the data scarcity issue, the ImageNet dataset was used to fine-tune models which contain 14 million images. First, a pre-trained model is built on the ImageNet dataset using VGG16 on the image. Then this pre-trained VGG16 model is used to fine-tune the model with the facemask dataset. Figure 14.3 illustrates the transfer learning mechanism using the pre-trained model.

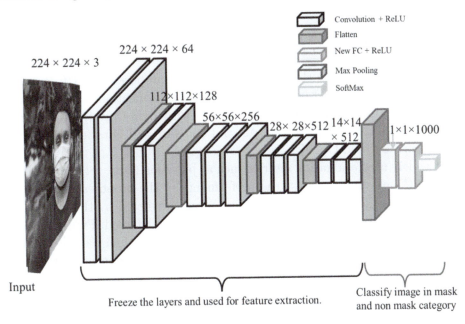

Figure 14.3 Transfer learning scheme of the proposed VGG16 model with a pre-trained model.

14.3.4 Testing Phase

To test the proposed facemask detection system, we import two files: One is the fine-tuned VGG16 model which is built using transfer learning and the XML classifiers. The proposed method is tested on both offline images and real-time video streams. To test the effectiveness of the proposed method on the real-time video stream, we set up the video acquisition using the webcam and pass each frame through the XML classifiers. It detects any face in the frame and a boundary box is drawn if the face is detected, and cropped face area is sent to the model for prediction. Region of interest (ROI) of face area is calculated, and the model returns its prediction after calculation. If the mask is predicted in the face, a green text "Mask" is set to appear at the top of the face. If not, "No Mask" will appear.

14.3.4.1 Video Acquisition Using OpenCV

Computer vision methods are included in an Open-Source Computer Vision Library (OpenCV). In this proposed solution, we capture a live stream video with a camera.

OpenCV provides a convenient interface to accomplish this task using its library. To capture a video, we construct the Video Capture object. The device index or the name of a video file can be used as a parameter. A device index is a number that identifies which camera is being used. A camera is usually attached. In this case, we just set the value to 0. The second camera is selected by passing 1, and so on. We set up an infinite while loop and use the read () method to read the frames. cv2.imshow() method is used to show the frames in the video. We can break the loop when we click a specific key.

14.4 RESULTS AND DISCUSSION

In this section, we evaluate the performance of the proposed VGG16 facemask detection system using a benchmark facemask dataset as mentioned in Section 14.3.1. Experimentation was carried out on a local PC with Intel Core i7 Processor, 32GB RAM, Nvidia GTX 980Ti GPU, and relevant software and libraries such as *Python*, Keras, TensorFlow, and OpenCV.

Most widely used evaluation metrics such as precision (PR), recall rate (RR), and accuracy (AC) are used to assess the performance of the proposed method [5]. The precision denotes the proportion of the real target predicted by the model, and the recall rate represents the proportion of all real targets detected. Predicted bounding boxes with the intersection over union (IOU) value greater than the threshold (0.5 by default) are defined as positive samples and vice versa. PR, RR, and AC are defined as follows:

$$PR = \frac{TP}{TP + FP} \tag{14.3}$$

where TP defines the number that correctly predicted positive class samples as positive, and FP defines the number that incorrectly predicts positive class but not positive.

$$RR = \frac{TP}{TP + FN} \tag{14.4}$$

where FN defines the number that incorrectly predicts negative class but positive.

$$AC = \frac{TP}{TP + FP + TN + FN} \tag{14.5}$$

Table 14.2 provides the comparison results among different algorithms in terms of PR, RR, and AC. According to the results in Table 14.2, it can be seen proposed method reports perfect accuracy in precision and 0.92 recall rate. Overall proposed VGG16 model significantly outperformed the state-of-the-art, yielding 41% (0.41 in ratio) and 20% (0.20 in ratio) of performance improvement in terms of accuracy from MobileNet V2 and RestNet50, respectively.

Figure 14.4 provides the results of the cross-validation approach during model building to ensure higher generalization performance in the testing phase. According to the results in the figure, it can be confirmed that the proposed VGG16 provides stable performance which justifies the higher test accuracy.

The results of the confusion matrices are provided in Figure 14.5. The confusion matrix visualizes the true vs predicted label deviation. In these results, it can be seen that the proposed method clearly outperformed the competing models with a few mis-classification rates.

The proposed method is also tested on the live video stream. Figure 14.6 presents the results of facemask detection. In the figure, a bounding box around a person's face was drawn with green or red text which denotes the existence of a mask and no mask, respectively.

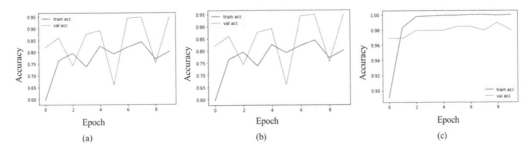

Figure 14.4 The trend of training and validation performance over each epoch. (a) MobileNet V2, (b) RestNet50, (c) VGG16 (Proposed).

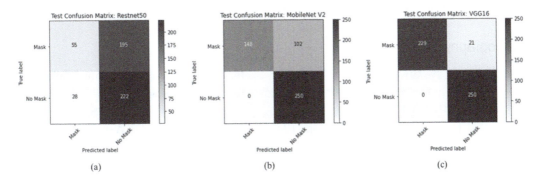

Figure 14.5 The confusion matrices of three different algorithms on the test dataset: (a) MobileNet V2, (b) RestNet50, (c) VGG16 (Proposed).

14.5 CONCLUSION AND FUTURE WORK

In this chapter, we developed a deep learning-based automated facemask detection system. The proposed methodology deployed a visual geometry group network (VGG16) to learn mask and non-mask benchmark datasets. But it is required an abundance of the dataset for accurate model building. Thus, we applied a transfer learning scheme that integrates a pre-trained VGG16 model on the ImageNet dataset to mitigate the data scanty situation. To verify the effectiveness of the proposed method, we performed rigorous experiments on benchmark offline datasets and real-time video streams. The experimental results suggest that the proposed facemask detection system is highly effective on both office and real-time images, yielding 41%

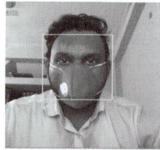

Figure 14.6 The test of the proposed VGG16 model on the real-time video stream. In the figure, the left and the middle images indicate mask, whereas the right one indicates no mask.

(0.41 in ratio) and 20% (0.20 in ratio) of performance improvement in terms of accuracy on MobileNet V2 and RestNet50, respectively. This significant performance improvement clearly indicates the potential application of the proposed facemask detection system in the public location to control the COVID-19 outbreak.

Train and test of the proposed system on large-scale images could be a potential future research scope. Further, identifying multiple faces from the stream of video frames is a promising research direction.

Bibliography

1. T. Acter, N. Uddin, J. Das, A. Akhter, T. R. Choudhury and S. Kim, "Evolution of severe acute respiratory syndrome coronavirus 2 (SARS-CoV-2) as coronavirus disease 2019 (COVID-19) pandemic: A global health emergency," Science of the Total Environment, vol. 730, pp. 138996–138996, 2020. W475W9026

2. K. A. Prather, C. C. Wang and R. T. Schooley, "Reducing transmission of SARS-CoV-2.," Science, vol. 368, no. 6498, pp. 1422–1424, 2020. W475W9026

3. R. u. o. p. p. e. (WHO, "World Health Organization" [Online]. Available: https://apps.who.int/iris/bitstream/handle/10665/331215/WHO-2019-nCov-IPCPPE_use-2020.1-eng.pdf. [Accessed 30 October 2021]. W475W9026

4. S. Feng, C. Shen, N. Xia, W. Song, M. Fan and B. J. Cowling, "Rational use of facemasks in the COVID-19 pandemic," Lancet Respiratory Medicine, vol. 8, no. 5, pp. 434–436, 2020. W475W9026

5. S. Sethi, M. Kathuria and T. Kaushik, "Facemask detection using deep learning: An approach to reduce risk of coronavirus spread," Journal of Biomedical Informatics, vol. 120, pp. 103848–103848, 2021. W475W9026

6. K. Simonyan and A. Zisserman, "Very deep convolutional networks for large-scale image recognition," in ICLR 2015: International Conference on Learning Representations 2015, 2015. W475W9026

7. J. Deng, W. Dong, R. Socher, L.-J. Li, K. Li and L. Fei-Fei, "ImageNet: A large-scale hierarchical image database," in 2009 IEEE Conference on Computer Vision and Pattern Recognition, 2009. W475W9026

8. A. Nestor, A. C. Lee, D. C. Plaut and M. Behrmann, "The face of image reconstruction: Progress, pitfalls, prospects," Trends in Cognitive Sciences, vol. 24, no. 9, pp. 747–759, 2020. W475W9026

9. A. Cabani, K. Hammoudi, H. Benhabiles and M. Melkemi, "MaskedFace-Net—A dataset of correctly/incorrectly masked face images in the context of COVID-19," Smart Health, vol. 19, pp. 100144–100144, 2021. W475W9026

10. B. Qin and D. Li, "Identifying facemask-wearing condition using image super-resolution with classification network to prevent COVID-19," Sensors, vol. 20, no. 18, p. 5236, 2020. W475W9026

11. B. N. Patel, S. Sipoliya and S. Mishra, Shreyansh and P. K. Sreelatha, "Face mask detection," International Journal of Advance Research, Ideas and Innovations in Technology, vol. 7, no. 4, pp. 548–551, 2021. W475W9026

12. K. Hammoudi, A. Cabani, H. Benhabiles and M. Melkemi, "Validating the correct wearing of protection mask by taking a selfie: Design of a mobile application "CheckYourMask" to limit the spread of COVID-19," CMES vol. 124, no. 3, pp. 1049–1059, 2020. W475W9026

13. M. S. Ejaz, M. R. Islam, Sifatullah and A. Sarker, "Implementation of principal component analysis on masked and non-masked face recognition," in 2019 1st International Conference on Advances in Science, Engineering and Robotics Technology (ICASERT), 2019. W475W9026

14. M. K. J. Khan, N. U. Din, S. Bae and J. Yi, "Interactive removal of microphone object in facial images," Electronics, vol. 8, no. 10, p. 1115, 2019. W475W9026

15. C. Li, R. Wang, J. Li and L. Fei, "Face detection based on YOLOv3," pp. 277–284, 2020. W475W9026

16. A. Nieto-Rodríguez, M. Mucientes and V. M. Brea, "System for medical mask detection in the operating room through facial attributes," in Iberian Conference on Pattern Recognition and Image Analysis, 2015. W475W9026

17. A. Jangra, "Face mask detection images dataset," 12 November 2021. [Online]. Available: https://www.kaggle.com/ashishjangra27/face-mask-12k-images-dataset.W475W9026

18. D. P. Kingma and J. L. Ba, "Adam: A method for stochastic optimization," in ICLR 2015: International Conference on Learning Representations 2015, 2015. W475W9026

19. S. A. Hussain and A. S. A. A. Balushi, "A real time face emotion classification and recognition using deep learning model," Journal of Physics: Conference Series, vol. 1432, no. 1, p. 12087, 2020. W475W9026

## CHAPTER 15

Security Challenges in Wireless Body Area Networks for Smart Healthcare

Muhammad Shadi Hajar and Harsha Kumara Kalutarage

School of Computing, Robert Gordon University, Aberdeen, UK

M. Omar Al-Kadri

School of Computing & Digital Technology, Birmingham City University, Birmingham, UK

CONTENTS

DOI: 10.1201/9781003251903-15

I N THE ERA of communication technologies, wireless healthcare networks enable innovative applications to enhance the quality of patients' lives, provide useful monitoring tools for caregivers, and allow timely intervention. However, security concerns are still holding back the widespread adoption of this promising technology. Insecure data communication violates the patients' privacy and may endanger their lives due to improper medical diagnosis or treatment. Although traditional security countermeasures, including authentication, encryption, and data integrity are essential to protect the network from external adversaries, more advanced AI-based security schemes are necessary to protect the network from internal threats.

This chapter starts with a concise introduction about wireless body area networks (WBANs) that comply with the IEEE 802.15.6 standard, which provides the reader with the necessary information to understand the rest of the contents in this chapter. Then, WBAN threats and countermeasures are comprehensively researched with a particular focus on AI-enabled methods. The potential attacks are widely investigated. Finally, traditional security countermeasures are discussed, followed by intrusion detection systems (IDS) and trust management system (TMS).

15.1   INTRODUCTION

A WBAN is a special kind of wireless sensor network (WSN) used mainly to monitor the body's physiological signs. It consists of tiny bio-medical sensor nodes that are distributed either on the human body or implanted inside it. The first and the only international standard for WBAN is defined in IEEE 802.15.6 [34], which was released in 2012. This standard defines reliable, low-power and short-range wireless communications with a vast range of data rates for a variety of healthcare applications. WBAN supports data rates starting from tens of Kbps (narrowband) up to 10 Mbps (ultra-wideband).

WBAN provides a promising technology to revolutionize future healthcare applications by providing real-time monitoring tools for caregivers and allowing timely medical interventions. Different kinds of sensor nodes, wearable, implantable medical devices (IDMs) and surrounding nodes are designed to sense the physiological signals of the human body and forward them to a remote medical server. These periodical medical readings include a vast range of bio-signals depending on the sensing unit of the sensor node (SN), such as blood pressure, glucose level, electrocardiogram (ECG), heart rate, electromyogram (EMG), body temperature, and oxygen saturation (SpO2). This continuous awareness of the patient's vital functions provides more flexibility and mobility to patients and enhances their life quality.

The widespread adoption of this revolutionized technology is driven by several factors. The rapid growth of the aging population across the globe is projected to reach around 1.5 billion in 2050, which is more than double the number in 2019 [84]. In the UK, for instance, the aging population over 85 is expected to double by the middle of 2041 [63]. Moreover, the overall expenditure of healthcare systems is increasing significantly, and the proportion of overloaded health professionals is also overgrowing. For instance, around 15% of the health budget is dedicated to diabetes, which will be one of the most causes of death by 2030 [89]. These reasons push firmly toward the adoption of this neoteric technology. However, security and privacy challenges are still holding back the wide adoption of this technology because any compromise could violate the patient's privacy and endanger their life. For instance, a sensor node with an insulin pump capability could receive a compromised order to inject an insulin overdose into the bloodstream. WBAN is vulnerable to vast kinds of security attacks and misbehavior activities. Although traditional security countermeasures, such as authentication, encryption, and integrity validation, are essential to protect the network from different kinds of threats, they are not enough to ensure a high level of security and privacy. Therefore, more advanced AI-based security countermeasures such as IDS and TDS are introduced in the literature to enhance the overall security and protect the network from potential innovative attacks which will be discussed further throughout this chapter [24].

This chapter sets off by providing an overview of AI in healthcare in Section 15.2, followed by a brief background about WBAN technology in Section 15.3. Then, Section 15.4 comprehensively presents the WBAN threats and vulnerabilities, while Section 15.5 explores different security countermeasures, including secure communication, intrusion detection, and trust management.

15.2 AI IN HEALTHCARE

The tremendous development in the field of artificial intelligence (AI) opens the door to think about adopting this revolutionized technology for healthcare applications. Body sensor nodes (BSNs) empowered by the advancements of AI are now able to collect the physiological signs of the human body and provide high-frequency and high-resolution remote monitoring tools. This technique helps physicians to diagnose, predict, and intervene when necessary. Moreover, the collected healthcare data is used to spot patterns, predict outcomes, verify hypotheses, and optimize operations. For instance, AI-based prediction models outperform doctors in predicting skin cancers by analyzing skin lesions using deep convolutional neural networks (CNNs) [16]. Another example, was a researcher [68] that built a prediction model to predict bacterial urinary tract infection (UTI) using a random forest algorithm. They also built a prescribing policy based on their prediction model to evaluate the physicians' prescriptions. The prediction model's performance had an AUC of 0.731, and the results showed a decrease of 7.42% of antibiotic use in Denmark, a one of the conservative countries of antibiotic use, which gives an indicator to better results for other countries.

Moreover, AI is widely used in protecting healthcare networks from security breaches [9, 30]. Supervised, unsupervised, and reinforcement learning are all introduced to enhance the overall security of healthcare networks. Supervised algorithms, such as support vector machine (SVM), random forest, and K-nearest neighbors (KNN) are introduced to detect network intrusions and spoofing attacks. The unsupervised learning algorithm, such as k-means clustering, has been used to detect denial of service (DoS) attacks. Furthermore, reinforcement learning is widely adopted for security and routing applications.

15.3 WIRELESS BODY AREA NETWORK

The standardization process of wireless sensor networks for healthcare applications is triggered by projecting its importance and critical role in the near future. Therefore, the Physical (PHY) and medium access control (MAC) layers are defined in the IEEE 802.15.6 [34] in order to ensure interoperability amongst devices from different vendors. The standard supports three kinds of physical layers namely narrowband (NB) PHY, ultra-wideband (UWB) PHY, and human body communication (HBC) PHY, which support different frequency bands and data rates in order to meet the requirements of different potential applications.

15.3.1 WBAN Topology

Each body area network (BAN) consists of one single sink node and a set of sensor nodes. The maximum number of nodes within one BAN is specified by 64 in the IEEE 802.15.6 standard. The star topology is adopted in the standard with two kinds of communications, simple and extended. In the simple one-hop star topology, all nodes should be within the direct communication range of the sink, while in the extended two-hop star topology, some nodes relay traffic to others as shown in Figure 15.1.

WBAN nodes could be classified according to their role, deploying location, and functionality [7]. Based on their roles, sensor nodes can be classified into:

- *Hub*: It could also be called sink or coordinator. It is the gateway of BAN's nodes to other BANs and networks. It usually has superior resources compared to other nodes as all BAN traffic goes through it.

- *Relay node*: In the extended star topology, some nodes have the relaying capability to help nodes that are not in the direct communication range.

- *End node*: These nodes are designed to sense the bio-signals and report them to the sink either directly if they are in the direct communication range, or via relay nodes.

15.3.2 WBAN Communication Architecture

There are different tiers of communication in the WBAN ecosystem where AI has been utilized for security in all these layers. Generally speaking, there are three tiers

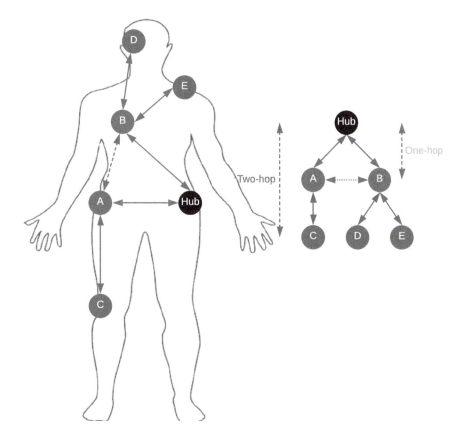

Figure 15.1 WBAN topology [24].

of communication although few researchers suggested adding a new tier of communication between nano and micro nodes [85]. However, generally, three tiers of communication are recognized in the standard of WBAN as follows [48, 53, 58]:

- *Tier-1 intra-BAN communication*: All the communication within the BAN itself is regarded as a tier-1 communication, including the communication between the sensor nodes and the sink, and the communication between the sensor nodes themselves. The used frequency and data rates vary depending on the used physical layer.

- *Tier-2 inter-BAN communication*: The communication between BANs and the access points (APs), and between BANs themselves are regarded as a tier-2 of communication.

- *Tier-3 beyond BAN communication*: The communication beyond the WBAN ecosystem is regarded as a tier-3 communication, which includes all the communication between the APs and the remote medical server.

The three tiers of communication are illustrated in Figure 15.2. The figure shows a set of in-body and on-body sensor nodes forming two BANs. All sensor nodes can

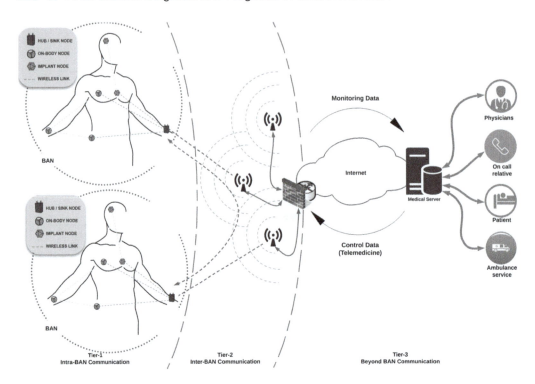

Figure 15.2 WBAN architecture.

communicate with the hub directly if they are in the direct communication range or via relaying nodes.

15.3.3 Security in WBAN

Ensuring a high level of security and privacy in WBAN plays a pivotal role in adopting this technology on a large scale. The sensitive information must be protected during all phases from sensing bio-signals to storage. Any compromise could disclose a patient's health records, and even endanger the patient's life. WBAN inherits many security vulnerabilities and concerns from WSN. These inherited security concerns, in addition to the strict resource constraints impose unprecedented security requirements, and open the way to further scientific research to meet these requirements. The basic security requirements of WBAN are outlined as follows:

- *Confidentiality*: Data must be protected from being disclosed to any unauthorized party [10]. Adopting a proper encryption algorithm could protect data during transmission and storage phases. Unprotected data can be readily disclosed during transmission in open channels by eavesdropping attacks, or when stored in plain format when medical servers or nodes got compromised.

- *Integrity*: The attacker can intercept data during the transmission phase and delete, inject, or modify the transmitted packet. Confidentiality alone cannot protect the data from alteration. Therefore, the receiver must be able to ensure that the received data is original and has not been modified on its way [66].

- *Availability*: WBAN provides critical services. The adversary may launch an attack to disrupt the communication between the caregivers and the sensor nodes [3]. Disrupting the network's operation may endanger the patient's life. Therefore, maintaining the ability to access the required data under any circumstances is a crucial requirement.

- *Data Authentication*: When the integrity requirement is fulfilled, the receiver can ensure that the received data is intact and has not been modified during the transmission phase. However, the receiver cannot verify that the received message was sent by the original sender, which it is believed to be [14]. Therefore, to achieve data authentication, IEEE 802.15.6 standard defines the message authentication code (MAC) in order to ensure that the received message is sent by the original sender.

- *Data Freshness*: Data freshness is an essential requirement to ensure that no adversary can capture messages and replay them later [44]. A replay attack may cause instability and confusion in the network and could make the physicians take wrong decisions based on inaccurate information. Two levels of freshness can be achieved, strong and weak freshness. In strong freshness, the received message must be in order and on time; however, there is no guaranteed delay in the weak freshness.

- *Secure Management*: Cryptographic security countermeasures, such as authentication, encryption, and integrity validation require security keys. Therefore, secure management is essential to ensure that the security keys are distributed in a secure manner [43].

There are three levels of security defined in the IEEE 802.15.6 standard. All nodes, including the sink, have to choose one of these levels in accordance with their application requirements.

- *Level-0 unsecured communication*: In the first security level, there is no security countermeasure applied. Unsecured frames are transmitted without authentication, encryption, integrity, or even replay defence.

- *Level-1 authentication*: In the second security level, authenticated frames are exchanged but not encrypted. Moreover, integrity validation and replay defence are provided; however, no confidentiality and privacy protection.

- *Level-2 authentication and encryption*: The third security level provides the highest security. Authenticated and secured frames are exchanged at this level. Moreover, confidentiality, privacy protection, replay defence, and integrity validation are all provided at this level.

During the association process, each node and the hub need jointly to select one of the aforementioned levels based on their respective requirements. The security key generation is shown in Figure 15.3. A pre-shared master key (MK) is activated or established during the association process for secure unicast communication. Afterward,

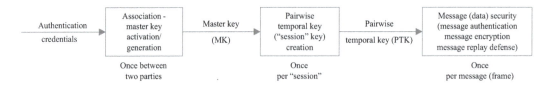

Figure 15.3 Security structure.

pairwise temporal key (PTK) is created by calling the PTK creation procedure, which is used for a communication session between two nodes. Moreover, the hub generates a group temporal key (GTK) and shares it with the corresponding multicast group nodes in a unicast manner in order to establish multicast secured communication.

On the other hand, recent research shows that despite all the incorporated security measures in the IEEE 802.15.6 standard, it still has some vulnerabilities. For example, analyzing the key agreement protocols to establish the pre-shared MK shows that four protocols are vulnerable to key compromise impersonation (KCI) and do not fulfill the forward secrecy requirement. Furthermore, one protocol is prone to offline dictionary attack [81, 82].

15.4 WBAN THREATS

WBAN is vulnerable to different kinds of attacks and malicious activities. There are different classifications for attacks. Based on the attack origin, they can be classified into internal and external as follows [1]:

- *Internal attacks*: These attacks are launched by internal intruders. For instance, when a node got compromised. Internal attacks are more complicated and challenging to defeat as compromised nodes have already passed the traditional security countermeasure and could have a copy of the keys. Therefore, more advanced security measures are required to protect the network [23].

- *External attacks*: These attacks are sourced from outside the network by external adversaries.

Attacks on WBAN can also be classified into passive and active as follows [1]:

- *Passive attacks*: Information gathering is the main goal of passive attacks. Although it is less harmful than active attacks, it violates the patient's privacy. Attackers can take advantage of the gathered data to launch more advanced attacks.

- *Active attacks*: A vast range of attacks are regarded as active attacks, such as DoS attacks, packet alteration attacks, and route poisoning attacks. Attackers can target the network operation to deplete the resources and degrade the overall performance.

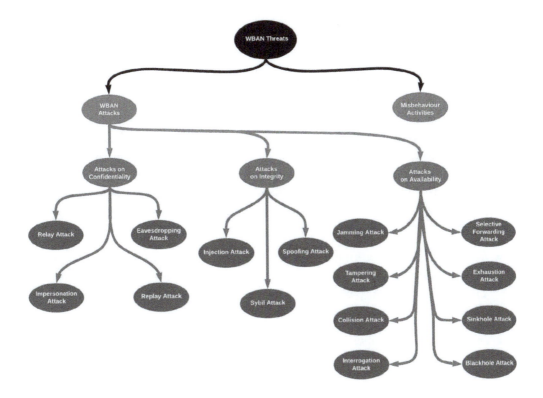

Figure 15.4 WBAN threat taxonomy.

Moreover, WBAN is also vulnerable to different kinds of malicious activities launched by WBAN nodes. Different reasons are behind these malicious activities. Nodes could launch internal attacks when they got compromised. Moreover, even benign nodes could act selfishly in order to save resources. For example, when a relay node stops relaying packets to save power, it could disrupt the network operation. Therefore, misbehaving activities launched even by legitimate nodes are very deleterious and dangerous because cryptographic security measures are not able to defeat them. However, some advanced countermeasures, such as AI-based TMS [22, 23], which will be discussed later in this chapter, can detect malicious activities like dropping attacks.

In what follows, a wide range of attacks is discussed and grouped based on CIA (confidentiality, integrity, and availability) security requirements they violate. It is worth mentioning that this list of attacks is not exhaustive.

15.4.1 Attacks on Confidentiality

There are many attacks that can compromise confidentiality as listed below:

- *Eavesdropping Attack*: It is a passive attack where the adversary can sniff on the transmission media to capture the exchanged packets with a view to getting access to sensitive information [2]. According to the IEEE 802.15.6 security levels, no encryption service is provided in security level-0 and level-1, which allows the adversary to readily capture and analyze the exchanged plain frames. Moreover, even at the third security level, intelligent adversaries can capture the secret keys in the key exchange phase.

- *Replay Attack*: It is an active attack where the adversary capture and store the exchanged frames to be replayed later into the network [54]. Replay attack is a severe attack in WBAN because the replayed frames are still valid, which may cause serious consequences when a decision is made based on these messages. Therefore, a replay defence mechanism has been incorporated in security level-1 and security level-2 in the IEEE 802.15.6 standard. The first octet of the MAC frame body, namely, "Low-Order Security Sequence Number" is used to verify message freshness and detect replay attacks [34].

- *Relay Attack*: It is a kind of man-in-the-middle (MITM) attack. The attacker tries to intercept the communication between the sender and the receiver [44]. Thus, while the two parties think they are in direct communication with each other, the attacker can intercept all the exchanged messages.

- *Impersonation Attack*: The attacker can exploit the sniffed messages to impersonate a legal node. Successful impersonation attacks can cause deleterious consequences [71].

15.4.2 Attacks on Integrity

Integrity attacks are regarded as active attacks as the attacker tries to delete, inject, or modify the exchanged frames.

- *Spoofing Attack*: Spoofing attacks can be launched in different ways to disrupt the network's operation. Attackers try to alter messages to get legitimate access to resources [54]. The IEEE 802.15.6 standard supports exchanging authenticated frames. The message integrity code (MIC) field is set to the message authentication code (MAC) with a view to verify the authenticity and the integrity of the received frames.

- *Modification/Injection Attack*: Modification and injection attacks are kinds of MITM attacks. The attacker tries to inject new frames or alter the exchanged frames before replaying them [2]. When no integrity validation service is running, such a simple attack could cause severe consequences and affect the patient's life. For instance, the altered message could contain an emergency bio-signal sent to the physicians, or it could be a command from a remote healthcare center to an insulin pump to release an insulin dose. In both scenarios, frame modification could endanger the patient's life.

- *Sybil Attack*: The attacker in this kind of attack impersonates fake identities. Then, the adversary uses the illegitimate identity to launch attacks until they are detected. Once detected, the attacker generates a new fake identity to appear as a benign node [52]. Meanwhile, the attacker has the opportunity to intercept the exchanged messages and continue the malicious activities.

15.4.3 Attacks on Service Availability

The most common attacks that impact service availability are DoS attacks, in which the adversary disrupts the network's operation and deprives other entities from accessing the required resources. DoS attacks can target different stack layers. The most common DoS attacks are discussed below:

- *Jamming Attack*: Jamming attacks target the physical layer. It was first discussed in the literature in 1982 [86]. Since then, wireless networks have always been vulnerable to jamming attacks. The adversary blocks legitimate communication within the network by intentionally interfering with the used frequency, which notably decreases the signal-to-interference-plus-noise ratio (SINR). In WBAN, jammers can easily degrade network performance, which could delay urgent and critical bio-signal messages or medication commands. Moreover, it could deplete the batteries of the sensor nodes because of the re-transmission attempts.

- *Tampering Attack*: A tampering attack can occur when the adversary can access WBAN nodes physically. The attacker can cause hardware damage or get access to critical data, such as encryption keys [54]. Although sensor nodes are usually close to the human body, raising awareness among patients about who is authorized to handle these sensor nodes can help defeat tampering attacks.

- *Collision Attack*: A collision attack is a data link attack that occurs when more than one node transmits at the same time [38]. When an overlapped transmission happens, senders enter the re-transmission phase, which depletes the node's resources and degrades the network's performance. The attacker overlaps the other's transmission intentionally, which results in a collided frame. This causes the cyclic redundancy check (CRC) process on the receiver side to fail verifying the received frame and discarding it.

- *Selective Forwarding Attack*: It is one of the dropping attacks in which malicious or selfish nodes drop received frames instead of forwarding them. In a selective forwarding attack, the malicious or the selfish node selectively forward some packets and drops others [37]. For example, in extended two-hop star topology in WBAN, relay nodes can drop all the frames sourced from a particular node.

- *Exhaustion Attack*: It is a sort of DoS attack in which the attacker tries to deplete the resources of the victimized node. For instance, in the denial of sleep attack, the victim's battery becomes depleted significantly [70].

- *Sinkhole Attack*: It is a packet dropping attack in which the adversary tries to attract all the traffic within the WBAN and drop it [22]. It can be launched by sending fake routing updates showing the attacker as the best route.

- *Blackhole Attack*: It is similar to selective forwarding attacks; however, in a blackhole attack, the malicious node drop all the received messages [61].

15.5 WBAN THREAT COUNTERMEASURES

The WBAN nature and its critical health applications impose achieving a high level of security and privacy. Data must be protected during collection, transfer, processing, and storing. Any kind of vulnerabilities could be exploited by adversaries to launch a fatal attack. On the other hand, any proposed security measures must take into account the security requirements of WBAN and the strict resource limitations. In this section, various security countermeasures will be discussed at different levels. It is worth mentioning that WBAN security is still an open area of research and there is a limited number of WBAN-specific schemes proposed in the literature. Therefore, some security schemes discussed throughout this section are generally proposed for WSN, which have a high potential to fit WBAN requirements; however, further investigation before the adoption is highly recommended. Figure 15.5 shows a high-level perspective of the security countermeasures discussed throughout this section.

15.5.1 Secure Communication

Secure communication is regarded as the cornerstone to guarantee the confidentiality, privacy, and integrity of the WBAN. Therefore, in this section, the security requirements, authentication and key establishment, integrity validation, and encryption are discussed.

15.5.1.1 Security Requirements

To ensure an end-to-end secure communication, there are some security requirements to be met by any proposed scheme that provides authentication, encryption, and integrity validation.

- *Lightweight*: As WBAN has limited resources, any security scheme must be computationally lightweight [67].

- *Anonymity*: To ensure that privacy is guaranteed, outsiders should not be able to identify the involved parties in the authentication process [72].

- *Mutual authentication*: In mutual authentication, the two parties can authenticate each other with a view to protecting from impersonation attacks [91].

- *Unlinkability*: Unlinkability is a critical requirement to prevent an attacker from tracing the identity of the nodes. Moreover, the unlinkability must still be maintained even if the attacker is able to capture two frames belonging to the

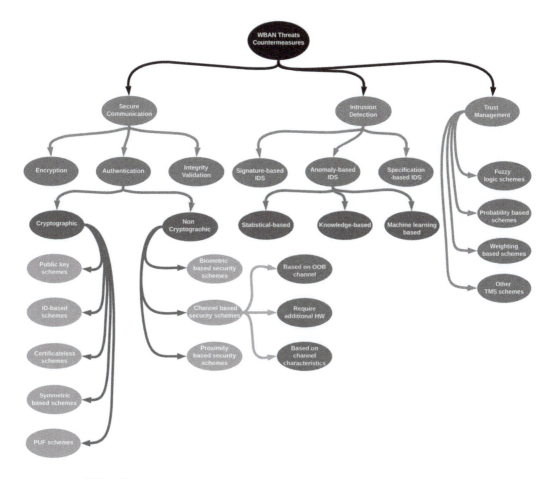

Figure 15.5 WBAN security countermeasure taxonomy.

same sender, which means it should not be any link or association between the captured frames and the sender [50].

- *Session key establishment*: Once the authentication process is achieved successfully, a session key should be created and exchanged in a secure manner for subsequent communication between the nodes [91].

- *Forward secrecy*: This implies that the session key must still be secured in case of one or both of the communicating parties are compromised. Moreover, it should still be secure even when an attacker gets one or both private keys [28]

- *Revocability*: Revocability requirements allow revoking any malicious node effectively to keep the network secure [91].

- *Non-repudiation*: By fulfilling a non-repudiation requirement, senders cannot deny their messages [72].

- *Resilience to well-known attacks*: Security schemes must be resilient to well-known attacks, such as replay and impersonation attacks [14].

15.5.1.2 Authentication and Key Establishment

Authentication is the cornerstone to secure the network on all tiers of communication in WBAN. Considerable research has been put forward to fill this research gap in the literature. However, the majority of the proposed schemes are vulnerable to specific attacks or do not fulfill all the security requirements [64, 90], which makes proposing a practical and secure authentication scheme for WBAN an open area for further research [50]. In what follows, the potential authentication schemes will be presented.

Non-cryptographic security schemes. The non-cryptographic authentication schemes take advantage of some physical characteristics of the targeting network, such as human physiological signals. They have been classified as non-cryptographic because of the used technique; however, many proposed schemes are able to create secret keys to encrypt the exchanged messages.

1. *Biometric-based security schemes*: Many lightweight authentication schemes are proposed in the literature based on the body bio-signals because there is a thought that these signals are difficult to be forged. A novel security scheme is proposed for WBAN in [69]. The proposed scheme is able to generate and distribute symmetric keys by sensing the ECG signal. The authors used time synchronization to avoid broadcasting the ECG signal. Moreover, the study proves the robustness of the proposed method by running informal and formal security analyses.

2. *Channel-based security schemes*: Another approach to use the physical characteristics to authenticate nodes is built on the assumption that the communication channel qualities between two nodes are the same. The proposed schemes in the literature can be categorized into:

   - *Security schemes based on an out-of-band (OOB) communication channel*: Some schemes introduce the use of an auxiliary channel to authenticate nodes assuming that this out-of-band channel is not prone to eavesdropping attacks. For example, a visual OOB channel is introduced in [49] to help the patient authenticates sensor nodes by comparing LED blinking patterns.

   - *Security schemes that require additional hardware*: Few studies introduced adding additional hardware to facilitate the authentication process, such as the good neighbor scheme where the authors used multiple antennas at the receiver side [12].

   - *Security schemes based on channel characteristic measurements*: Many schemes considered the channel characteristic measurements, such as received signal strength (RSS) measurements. Authors in BANA [73] and MASK-BAN [74] proposed lightweight authentication schemes based on RSS. In such schemes, there are some nodes distribution restrictions where the nodes are usually distributed within the half-wavelength range.

3. *Proximity-based security schemes*: In proximity schemes, the secret key can be extracted by taking advantage of the small-scale fading variations and a third-party radio-frequency (RF) source [55].

Cryptographic security schemes. These vary depending on the key types, and can be classified into the following categories.

1. *Public key signature schemes*: The public key cryptography (PKC) is a robust security approach to provide authentication. It could be mathematically implemented as an integer factorization problem like RSA or a discrete logarithmic problem like in elliptic curve cryptography (ECC). However, both approaches are still regarded as greedy in consuming the limited resources of WBAN. Authors [87] proposed a hybrid multiplication method to reduce the memory access rate, which results in speeding up the process by around seven times. Moreover, an ECC-PKC library has been introduced to be used for WSN [87].

2. *ID-based signature schemes*: It is a kind of PKC where the public key includes identity information, and the private key is generated by a trusted third party (TTP), namely, private key generator (PKG). ID-PKC schemes do not meet all the security requirements of WBAN [13, 36, 93]. They are vulnerable to key escrow problems because of the existing TTP. Moreover, as PKG has all the private keys, it can easily decrypt all the exchanged messages and forge any signature.

3. *Certificateless signature schemes*: This is a kind of PKC that has been introduced to reduce the resource consumption of PKC and the key escrow problem of ID-PKC [4]. The inborn key escrow problem in ID-PKC has been resolved by introducing a key generator center (KGC), which holds a master key instead of the private keys. The KGC is responsible for sending a partial private key (D_A) to nodes, which in turn can create their private keys. Many remote authentication schemes between the hub and the application providers have been introduced in the literature [39, 71, 76, 91, 95]. However, further research is still going on to produce a security scheme that meets all the security requirements and is not vulnerable to attacks.

4. *Symmetric-based schemes*: One group [50] proposed a symmetric-based authentication scheme using a pre-shared key and unique IDs to achieve mutual authentication, unlinkability, and forward secrecy security requirements. However, the adversary can take advantage of the unmasked value (γ) to link two sessions to the same node. Therefore, a modification has been suggested [41] to fulfill the unlinkability and forward secrecy. However, this security scheme still has key escrow problem because the hub still has all the IDs in addition to the master key.

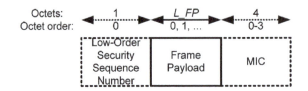

Figure 15.6 MAC frame body format [34].

5. *Physical unclonable function (PUF) schemes*: Some security schemes consider the unavoidable unique differences between nodes, which naturally appear during semiconductor manufacturing. Authors in [88] proposed a mutual authentication scheme between two WBAN sensor nodes with the help of the coordinator. Another security scheme is introduced in [78] for multi-hop BAN. A hierarchical authentication method has been used for nodes that are not in the communication range of the hub. The challenge-response pairs (CRPs) are stored on the cloud in order to minimize storage consumption.

15.5.1.3 Integrity Validation

Integrity validation allows the receiver to ensure that the received message is intact and has not been manipulated during the transmission. A vast range of manipulation can occur during the transmission, which changes the message content, whether it is caused by malicious activities or transmission errors. Changing content may include adding, removing, or transposing fragments. Deploying an integrity validation module can be feasibly achieved after generating cryptographic keys [15]. However, this task is still challenging for WBAN because of its resource scarcity [53].

As discussed earlier, IEEE 802.15.6 has three levels of security where integrity validation is provided in the second and third levels [34]. Figure 15.6 shows the WBAN MAC frame body. The length is variable and can expand to a maximum of 255 octets. WBAN entities can exchange secured and unsecured frames. Sensor nodes are to choose a security level that fulfills their security requirements during the association process. Two additional fields are used in the secured frames, MIC and "Low-Order Security Sequence Number." The latter is used to verify message freshness to detect replay attacks, while the MIC field is used to validate message integrity by setting it as the MAC [34].

15.5.1.4 Encryption

WBAN data plays a critical role in disease diagnosis and treatment. Confidentiality is essential to protect this sensitive information either during transmission or storage. Abundant encryption algorithms have been introduced in the literature, such as 3-DES [59] and RSA [65]. However, not all of them can fit the rigid resource constraints of WBAN. For example, encryption algorithms with long keys, countless number of rounds, and huge block sizes are inapplicable to WBAN. Therefore, in 2015, the National Institute of Standards and Technology (NIST) started the process to standardize a lightweight encryption algorithm for constrained devices such as WBAN sensor nodes [56]. To design a lightweight encryption algorithm that is resource and energy efficient, the following aspects must be taken into account [75]:

- *Key size*: Constrained devices have minimal storage capacity. MICAz, for example, has only 4 KB storage [31]. Therefore, the small key size is an essential factor for such devices. Authors [92] have proposed SIMECK, an encryption algorithm with encryption key size that could be 64, 96, or 128 bits. It is a hardware-oriented block encryption algorithm inspired by SIMON's algorithm design [8].

- *Block size*: Smaller block size is another essential factor in building a lightweight and resource-efficient encryption algorithm. Authors in [8] proposed two lightweight block based encryption algorithms SPECK and SIMON. SPECK is built on Addition-Rotation-XOR (ARX) structure and supports different block sizes ranging between 32 and 128 bits. On the other hand, SIMON belongs to the same family as SPECK, but it is software-oriented.

- *Number of rounds*: Lightweight encryption algorithms tend to use simple operations in order to meet the stringent resource limitations. However, using simple operations like in the ARX structure requires increasing the number of rounds. Therefore, choosing an algorithm that needs a fewer number of rounds is desirable for constrained devices. One group of authors [83] introduced LWE, a 3-round block cipher encryption algorithm. it uses 64 bits for key and block size with a view to fit the resource constraints of the medical sensors. Furthermore, it has been contrasted with well-known lightweight algorithms such as Rectangle [96] and TWINE [77]. The performance results show that LWE performed better in encryption/decryption rates without creating a heavy load on the infrastructure.

15.5.2 Intrusion Detection Systems

IDS is a vital cyber-security tool to monitor the network and detect any malicious activities. It is able to resist inside attacks launched by nodes that have passed the traditional cryptographic security measures. It is regarded as an additional layer of protection to detect and defeat internal and external abuses. Different methods are proposed in the literature to achieve an effective IDS scheme which could be classified based on the detection method.

15.5.2.1 Signature-Based IDS

It is also called rule-based IDS. In this kind of IDS, a profile (signature) is created for each previously known attack. This signature is then used to detect any malicious activity that matches the pre-defined attack pattern. The main disadvantage of this kind of detection method is the inability to detect unknown attacks. This requires updating the signature database periodically in order to detect new attacks. Authors [5] have proposed a lightweight IDS inspired by the human immune system for resource constrained networks. It adopted the properties of the immune cells with a signature database. For example, the detection nodes represent the dendritic cells and similarly, T-cells and B-cells, are mapped to appointed nodes in the detection process. The detection node is able to stimulate other members in the detection process.

15.5.2.2 Anomaly-Based IDS

In this kind of IDS, the normal network operation and node behavior are profiled. The detection engine is then able to report anomalies when there is a certain amount of variation. Researchers [11] have suggested a classification of anomaly-based IDS as follows.

Statistical-Based. The statistical-based anomaly IDS builds a reference profile for the normal network operation without malicious activities. Afterward, the IDS monitors the network, periodically generates a profile and compares it with the reference profile to compute the anomaly score. If the anomaly score exceeds a certain threshold, then an anomaly is detected. One group [30] proposed a sink assistant statistical IDS for WBAN. The proposed scheme is successfully able to detect replay attacks, jamming attacks, data forging attacks, exhausting DoS attacks, and selective forwarding attacks. The detection performance shows high true-positive and low false-positive rates. Another statistical IDS is a proposed scheme using a variety of statistics, such as forward percentage (FP), maximum sequence counter (MSC), malicious flooding on a specific target, global forward percentage (GFP) and local forward percentage (LFP), to detect abnormal behavior [32]. Moreover, the proposed scheme is able to provide more details about the attack, such as the attack type and source. The performance results show high accuracy in detecting selfish activities and less accuracy in detecting blackhole and spoofing attacks.

Knowledge Based. A knowledge-based anomaly IDS depends on having prior knowledge about the network conditions in both cases normal operation and under certain attacks. Different techniques could be used in this kind of IDS, such as expert systems, description languages, finite state machines, data clustering, and outlier detection [5, 11].

Machine Learning Based. Machine learning-based IDS is an intelligent approach to detect abnormal activities in the network. A detection model is built using a machine learning algorithm and trained using example patterns from a real network. The advantage of this data-driven detection engine is the ability to detect even unknown attacks. Many machine learning algorithms have been used to build a detection model, such as SVM [9] and random forest [51]. One group of authors [79] proposed iDetect, a distributed intelligent IDS to detect WBAN attacks. The model is built using a multi-objective genetic algorithm to make a trade-off between high-detection performance and resource consumption. The performance results show a good detection accuracy against jamming attacks, random jamming attacks, and selective forwarding attacks. A distributed IDS framework with a mobile agent has been introduced [80] for WBAN. The detection process migrates from one sensor node to another in the WBAN, which allows all sensor nodes to share the detection overhead. The results show that the proposed framework was able to reduce power consumption. These authors [62] also used the mobile agent technique. They proposed a hierarchical and distributed IDS with autonomous mobile agents. The proposed framework has been evaluated for the following machine learning algorithms decision tree (DT), SVM, RF, Naïve Bayes classifier (NBC), and K-nearest neighbor (KNN). The reported performance results showed a rise of 6% in the consumed power. HEKA IDS has been proposed [60], which is a passive IDS that can monitor and detect anomalies. The authors first launched several attacks on medical devices to find vulnerabilities. They ran eavesdropping attacks, MITM attacks, replay attacks, and DoS attacks. The proposed model is evaluated for four machine learning algorithms, SVM, RF, DT, and KNN. The performance results show an accuracy around 98% for the aforementioned attacks. Furthermore, they evaluated the proposed scheme for the composite attacks, false data injection attack with MITM attack, and replay attack with MITM attack. The results show an accuracy of around 95%. One group [21] built a healthcare monitoring testbed. They built a dataset of 16,000 records of normal and malicious conditions. A combination of network and biometrics features are used to train and test four machine learning algorithms KNN, ANN, SVM, and RF. The performance results show that using combined features can improve the accuracy between 7% and 25% in some cases.

15.5.2.3 *Specification-Based IDS*

It is somehow located between the anomaly-based IDS and the signature-based IDS. First, the specifications and the constraints which describe a program or a protocol are defined. Then the system will monitor the running program or protocol with respect to the defined specifications and constraints [11].

15.5.3 Trust Management Systems

Cryptographic security measures are imperative to protect WBAN from security breaches and privacy violations. However, internal attacks, malicious activities, and selfish behavior could not be detected and defeated using this kind of security countermeasures. For instance, a sensor node, which passed all cryptographic measures and is regarded a legitimate node, could stop relaying frames for others and consequently disrupt the network operation. Therefore, a trust relationship between WBAN nodes could enhance the overall security and protect the network from malicious activities. Therefore, it is necessary to deploy an effective TMS that can continuously monitor the behavior of sensor nodes with a view to differentiate between trustworthy nodes and untrustworthy ones. The trust relationship can be defined as follows *Node X trusts node Y if and only if X has adequate confidence in Y's behavior and performance in the future* [24]. As with other security schemes, deploying a trust management scheme in WBAN is a challenging task [35]. Thus, more consideration must be given to the WBAN architecture, resource limitation, communication overhead, and

TMS attacks. In addition to its desirable security protection, trust management has been introduced in different applications, such as trust-based routing protocols [94], access control and role assignment [57].

The trust relationship is usually evaluated using two components based on the source of information: direct trust and indirect trust. In direct trust, the trustor directly monitors the trustee and records the successful and unsuccessful interactions. The indirect trust component is evaluated based on the received recommendations from other nodes in the network. The trustor could always consider this second-hand information, or may only consider them when no sufficient observation history is available to assess the trustee. Unfortunately, the indirect trust process is resource consuming and prone to dishonest recommendation attacks [25].

Although TMS is introduced to enhance the overall security, it can be vulnerable to some internal attacks, which makes designing a robust TMS a challenging task.

- *On-off attack*: In an on-off attack, the smart adversaries change their behavior alternately to keep themselves undetected and their trust values above the trust threshold. This kind of attack can manipulate the TMS and disrupt the network operation [46].

- *Bad-mouthing attack*: One of the dishonest recommendations for attacks. In this kind of attack, the recommender tends to give negative recommendations about a trustworthy node to destroy its reputation [27].

- *Ballot-stuffing attack*: Another type of dishonest recommendation for attacks, in which the recommender gives positive recommendations about malicious nodes to promote them [42].

- *Collusion attack*: Unlike bad-mouthing and ballot-stuffing attacks where just one malicious node provides dishonest recommendations, in collusion attacks, a group of malicious nodes colludes to provide dishonest recommendations. Collusion attacks are challenging to detect and can mislead the system to make unfair decisions [26].

- *Selective forwarding attack*: Discussed in Section 15.4.

The trust management schemes can be divided into four main groups based on their trust evaluation method.

15.5.3.1 Fuzzy Logic-Based TMS

The trust relationship is evaluated in this kind of TMS using fuzzy logic and pre-defined criteria that have a fuzzy nature. The authors [29] proposed DTMS, a distributed fuzzy logic-based trust management scheme. Each node inside the network monitors the others behavior and forecasts their trustworthiness with a view to remove malicious, compromised and selfish nodes. DTMS used two weighting techniques. The first is used to estimate the current trust based on direct observations and second-hand information from neighbor nodes, while the second is used to evaluate overall trust based on the current trust and the trust history. DTMS shows superior performance compared with benchmark schemes. However, using trust matrices and tables in addition to considering many factors to estimate the trust produce a significant network overhead [40]. Fuzzy-Based Trust Management–The Internet of Medical Things (FTM-IoMT) [6] is another fuzzy-based TMS. It has been proposed for the IoMT to protect from Sybil attacks. The authors adopted a centralized approach that uses features like integrity, receptivity, and compatibility to evaluate the trust value. The performance results show a good detection accuracy. However, it also shows a noticeable processing overhead. Therefore, further investigation is required to reduce server-side overhead and enhance packet delivery delays.

15.5.3.2 Probability-Based TMS

The probability distribution theory is widely used in the literature to infer the trust value from former estimations. The beta probability distribution is approximately the most used one. However, few schemes are built on different probability distributions such as exponential probability distribution and binomial probability distribution [18, 97]. A reputation-based framework for sensor networks (RFSNs) [20] is regarded as the first beta trust management scheme in the literature [24]. The authors used the watchdog technique to monitor the behavior of adjacent sensor nodes and collect observations which were then used to update the posterior reputation value. Authors [23] have proposed a lightweight trust management scheme for Wireless Medical Sensor Network (WMSNS). LTMS (Lightweight Trust Management Systems) provided two algorithms to evaluate the trust relationship. The first is proposed to fit the strict resource limitations of in-body sensor nodes, while the second provides further protection from on-off attacks and proposed to fit on-body and off-body sensor nodes. The two proposed algorithms have contrasted with benchmark trust schemes and showed superior performance in terms of attack detection and overhead processing. An ETRES (exponential-based trust and reputation evaluation system) [97] is another probability-based trust scheme. The authors used exponential distribution to evaluate the trust relationship, assuming that the future behavior should have the same mode as the old one. The authors suggest using the entropy theory to evaluate the uncertainty to reduce the overhead caused by considering an indirect trust module. Moreover, the authors used a weighting technique to emphasize the most reputable recommenders and recent observations. The performance results of ETRES showed a slight improvement compared to RFSN [19] and BTMS [17].

15.5.3.3 Weighting-Based TMS

In this kind of TMS, the trust relationship is evaluated by weighting the performance of other sensor nodes over time. It is easy to implement and deploy; however, it does not have a solid statistical or mathematical foundation [35]. The authors in [45] proposed a weighting-based trust scheme with a risk assessment to ensure a quick reaction to malicious activities. The risk assessment module makes destroying the trust easier than building it, which enhances the reliability of the proposed scheme. The overall trust value is evaluated using direct trust, received recommendations, risk factor, and the previous trust value. Another weighting-based TMS is RaRTrust [47]. The authors used both reputation and risk to evaluate the trust relationship with a view to defeat on-off attacks. Moreover, they adopted a timing window for ratings to reduce network congestion and delay. As a result, RaRTrust shows resiliency to on-off attacks and bad-mouthing attacks, while TMR is just able to defeat on-off attacks.

15.5.3.4 Other TMSs

Some trust management schemes do not fall within the previous classification. For instance, one group [33] proposed a cluster based with a 3-tier architecture trust management scheme. Tier-1 only considers nodes registration, while tier-2 defines five levels of misbehavior activities to secure the data communication between nodes, and the third tier is to monitor the energy consumption and migrate the cluster head process to another node. The proposed scheme evaluates the trust relationship using previous information and second-hand information to discover malicious nodes.

15.6 CONCLUSION

In this chapter, the current research on WBAN has been comprehensively investigated with a particular focus on security. A brief introduction about WBAN and its architecture and topology was presented first to provide the reader with the required information to comprehend the chapter's content readily. Next, attacks on WBAN and security requirements have been widely discussed. Finally, security countermeasures at different levels have been investigated, including secure communication, intrusion detection, and trust management.

Due to the sophisticated nature of modern attacks on WBAN, traditional signature-based methods would not be sufficient to mitigate them effectively. Instead, more advanced methods supported by recent developments in AI should be implemented. AI systems are trained to identify malware, recognize network traffic patterns, and detect APT attacks before reaching the target. Therefore, integrating AI into WBAN security would be the greatest method to detect and respond to WBAN attacks in real time and provide authenticity protection.

Bibliography

1. Farhan Abdel-Fattah, Khalid A Farhan, Feras H Al-Tarawneh, and Fadel Al-Tamimi. Security challenges and attacks in dynamic mobile ad hoc networks manets. In *2019 IEEE Jordan International Joint Conference on Electrical Engineering and Information Technology (JEEIT)*, pages 28–33. IEEE, 2019.

2. Moshaddique Al Ameen, Jingwei Liu, and Kyungsup Kwak. Security and privacy issues in wireless sensor networks for healthcare applications. *Journal of Medical Systems*, 36(1):93–101, 2012.

3. Samaher Al-Janabi, Ibrahim Al-Shourbaji, Mohammad Shojafar, and Shahaboddin Shamshirband. Survey of main challenges (security and privacy) in wireless

body area networks for healthcare applications. *Egyptian Informatics Journal*, 18(2):113–122, 2017.

4. Sattam S Al-Riyami and Kenneth G Paterson. Certificateless public key cryptography. In *International Conference on the Theory and Application of Cryptology and Information Security*, pages 452–473. Springer, 2003.

5. Vishwa Teja Alaparthy and Salvatore Domenic Morgera. A multi-level intrusion detection system for wireless sensor networks based on immune theory. *IEEE Access*, 6:47364–47373, 2018.

6. Ahmad Almogren, Irfan Mohiuddin, Ikram Ud Din, Hisham Al Majed, and Nadra Guizani. FTM-IOMT: FTM-IOMT trust management for preventing Sybil attacks in Internet of Medical Things. *IEEE Internet of Things Journal*, 2020.

7. Deena M Barakah and Muhammad Ammad-uddin. A survey of challenges and applications of wireless body area network (WBAN) and role of a virtual doctor server in existing architecture. In *2012 Third International Conference on Intelligent Systems Modelling and Simulation*, pages 214–219. IEEE, 2012.

8. Ray Beaulieu, Stefan Treatman-Clark, Douglas Shors, Bryan Weeks, Jason Smith, and Louis Wingers. The SIMON and SPECK lightweight block ciphers. In *2015 52nd ACM/EDAC/IEEE Design Automation Conference (DAC)*, pages 1–6. IEEE, 2015.

9. Mohammadreza Begli and Farnaz Derakhshan. A multiagent based framework secured with layered SVM-based IDS for remote healthcare systems. *arXiv preprint arXiv:2104.06498*, 2021.

10. KR Siva Bharathi and R Venkateswari. Security challenges and solutions for wireless body area networks. In *Computing, Communication and Signal Processing*, pages 275–283. Springer, 2019.

11. Ismail Butun, Salvatore D Morgera, and Ravi Sankar. A survey of intrusion detection systems in wireless sensor networks. *IEEE Communications Surveys & Tutorials*, 16(1):266–282, 2014.

12. Liang Cai, Kai Zeng, Hao Chen, and Prasant Mohapatra. Good neighbor: Ad hoc pairing of nearby wireless devices by multiple antennas. In *NDSS*, 2011.

13. Xuefei Cao, Xingwen Zeng, Weidong Kou, and Liangbing Hu. Identity-based anonymous remote authentication for value-added services in mobile networks. *IEEE Transactions on Vehicular Technology*, 58(7):3508–3517, 2009.

14. Sravani Challa, Mohammad Wazid, Ashok Kumar Das, and Muhammad Khurram Khan. Authentication protocols for implantable medical devices: Taxonomy, analysis and future directions. *IEEE Consumer Electronics Magazine*, 7(1):57–65, 2017.

15. Djamel Djenouri, Lyes Khelladi, and Nadjib Badache. Security issues of mobile ad hoc and sensor networks. In *IEEE Communications Surveys Tutorials*, volume 7, pages 2–28. IEEE Communications Society, 2005.

16. Andre Esteva, Brett Kuprel, Roberto A Novoa, Justin Ko, Susan M Swetter, Helen M Blau, and Sebastian Thrun. Dermatologist-level classification of skin cancer with deep neural networks. *Nature*, 542(7639):115–118, 2017.

17. W Fang, X Zhang, Z Shi, Y Sun, and L Shan. Binomial-based trust management system in wireless sensor networks. *Chin J Sens Actuat*, 28(5):703–708, 2015.

18. Weidong Fang, Chunsheng Zhu, Wei Chen, Wuxiong Zhang, and Joel JPC Rodrigues. BDTMS: Binomial distribution-based trust management scheme for healthcare-oriented wireless sensor network. In *2018 14th International Wireless Communications & Mobile Computing Conference (IWCMC)*, pages 382–387. IEEE, 2018.

19. Saurabh Ganeriwal, Laura K Balzano, and Mani B Srivastava. Reputation-based framework for high integrity sensor networks. *ACM Transactions on Sensor Networks (TOSN)*, 4(3):15, 2008.

20. Saurabh Ganeriwal and Mani B. Srivastava. Reputation-based framework for high integrity sensor networks. In *Proceedings of the 2Nd ACM Workshop on Security of Ad Hoc and Sensor Networks*, pages 66–77. ACM, 2004.

21. Anar A Hady, Ali Ghubaish, Tara Salman, Devrim Unal, and Raj Jain. Intrusion detection system for healthcare systems using medical and network data: A comparison study. *IEEE Access*, 8:106576–106584, 2020.

22. Muhammad Shadi Hajar, M Omar Al-Kadri, and Harsha Kalutarage. ETAREE: An effective trend-aware reputation evaluation engine for wireless medical sensor networks. In *2020 IEEE Conference on Communications and Network Security (CNS)*, pages 1–9. IEEE, 2020.

23. Muhammad Shadi Hajar, M Omar Al-Kadri, and Harsha Kalutarage. LTMS: A lightweight trust management system for wireless medical sensor networks. In *2020 IEEE 19th International Conference on Trust, Security and Privacy in Computing and Communications (TrustCom)*, pages 1783–1790. IEEE, 2020.

24. Muhammad Shadi Hajar, M Omar Al-Kadri, and Harsha Kumara Kalutarage. A survey on wireless body area networks: architecture, security challenges and research opportunities. *Computers & Security*, page 102211, 2021.

25. Muhammad Shadi Hajar, Harsha Kalutarage, and M Omar Al-Kadri. Trustmod: A trust management module for NS-3 simulator. In *2021 IEEE 20th International Conference on Trust, Security and Privacy in Computing and Communications (TrustCom)*. IEEE, 2021.

26. Guangjie Han, Jinfang Jiang, Lei Shu, Jianwei Niu, and Han-Chieh Chao. Management and applications of trust in wireless sensor networks: A survey. *Journal of Computer and System Sciences*, 80(3):602–617, 2014.

27. Daojing He, Chun Chen, Sammy Chan, Jiajun Bu, and Athanasios V Vasilakos. Retrust: Attack-resistant and lightweight trust management for medical sensor networks. *IEEE Transactions on Information Technology in Biomedicine*, 16(4):623–632, 2012.

28. Debiao He and Sherali Zeadally. Authentication protocol for an ambient assisted living system. *IEEE Communications Magazine*, 53(1):71–77, 2015.

29. J Hossein, RA Mohammad, et al. A fuzzy fully distributed trust management system in wireless sensor networks. *International Journal of Electronics and Communications*, 9(17):1–10, 2016.

30. Xuyang Hou, Jingjing Wang, Chunxiao Jiang, Sanghai Guan, and Yong Ren. A sink node assisted lightweight intrusion detection mechanism for WBAN. In *2018 IEEE International Conference on Communications (ICC)*, pages 1–6. IEEE, 2018.

31. CMT. http://courses.ece.ubc.ca/494/files/MICAz_Datasheet.pdf. Micaz. Accessed: 07-11-2019.

32. Yi-an Huang and Wenke Lee. A cooperative intrusion detection system for ad hoc networks. In *Proceedings of the 1st ACM Workshop on Security of Ad Hoc and Sensor Networks*, pages 135–147. ACM, 2003.

33. Syed Asad Hussain, Imran Raza, and Muhammad Mohsin Mehdi. A cluster based energy efficient trust management mechanism for medical wireless sensor networks (MWSNS). In *2018 5th International Conference on Electrical and Electronic Engineering (ICEEE)*, pages 433–439. IEEE, 2018.

34. IEEE. IEEE standard for local and metropolitan area networks—part 15.6: Wireless body area networks. *IEEE Std 802.15.6-2012*, pages 1–271, Feb 2012.

35. Farruh Ishmanov, Aamir Saeed Malik, Sung Won Kim, and Bahodir Begalov. Trust management system in wireless sensor networks: design considerations and research challenges. *Transactions on Emerging Telecommunications Technologies*, 26(2):107–130, 2015.

36. Sk Hafizul Islam and GP Biswas. A more efficient and secure ID-based remote mutual authentication with key agreement scheme for mobile devices on elliptic curve cryptosystem. *Journal of Systems and Software*, 84(11):1892–1898, 2011.

37. Saeideh Sadat Javadi and MA Razzaque. Security and privacy in wireless body area networks for health care applications. In *Wireless Networks and Security*, pages 165–187. Springer, 2013.

38. Minho Jo, Longzhe Han, Nguyen Duy Tan, and Hoh Peter In. A survey: Energy exhausting attacks in MAC protocols in WBANs. *Telecommunication Systems*, 58(2):153–164, 2015.

39. Philemon Kasyoka, Michael Kimwele, and Shem Mbandu Angolo. Certificateless pairing-free authentication scheme for wireless body area network in healthcare management system. *Journal of Medical Engineering & Technology*, 44(1):12–19, 2020.

40. Farwa Kazmi, Muazzam A Khan, Ayesha Saeed, Nazar Abbas Saqib, and Muhammad Abbas. Evaluation of trust management approaches in wireless sensor networks. In *2018 15th International Bhurban Conference on Applied Sciences and Technology (IBCAST)*, pages 870–875. IEEE, 2018.

41. Haibat Khan, Benjamin Dowling, and Keith M Martin. Highly efficient privacy-preserving key agreement for wireless body area networks. In *2018 17th IEEE International Conference on Trust, Security and Privacy In Computing and Communications/12th IEEE International Conference on Big Data Science and Engineering (TrustCom/BigDataSE)*, pages 1064–1069. IEEE, 2018.

42. Tayyab Khan, Karan Singh, Mohamed Abdel-Basset, Hoang Viet Long, Satya P Singh, Manisha Manjul, et al. A novel and comprehensive trust estimation clustering based approach for large-scale wireless sensor networks. *IEEE Access*, 7:58221–58240, 2019.

43. Nesrine Khernane, Maria Potop-Butucaru, and Claude Chaudet. BANZKP: A secure authentication scheme using zero knowledge proof for WBANs. In *2016 IEEE 13th International Conference on Mobile Ad Hoc and Sensor Systems (MASS)*, pages 307–315. IEEE, 2016.

44. Marko Kompara and Marko Hölbl. Survey on security in intra-body area network communication. *Ad Hoc Networks*, 70:23–43, 2018.

45. Nabila Labraoui. A reliable trust management scheme in wireless sensor networks. In *2015 12th International Symposium on Programming and Systems (ISPS)*, pages 1–6. IEEE, 2015.

46. Nabila Labraoui, Mourad Gueroui, and Larbi Sekhri. On-off attacks mitigation against trust systems in wireless sensor networks. In *IFIP International Conference on Computer Science and Its Applications*, pages 406–415. Springer, 2015.

47. Nabila Labraoui, Mourad Gueroui, and Larbi Sekhri. A risk-aware reputation-based trust management in wireless sensor networks. *Wireless Personal Communications*, 87(3):1037–1055, 2016.

48. Benoît Latré, Bart Braem, Ingrid Moerman, Chris Blondia, and Piet Demeester. A survey on wireless body area networks. *Wireless Networks*, 17(1):1–18, 2011.

49. Ming Li, Shucheng Yu, Joshua D Guttman, Wenjing Lou, and Kui Ren. Secure ad hoc trust initialization and key management in wireless body area networks. *ACM Transactions on Sensor Networks (TOSN)*, 9(2):18, 2013.

50. Xiong Li, Maged Hamada Ibrahim, Saru Kumari, Arun Kumar Sangaiah, Vidushi Gupta, and Kim-Kwang Raymond Choo. Anonymous mutual authentication and key agreement scheme for wearable sensors in wireless body area networks. *Computer Networks*, 129:429–443, 2017.

51. Chao Liu, Zhaojun Gu, and Jialiang Wang. A hybrid intrusion detection system based on scalable K-means+ random forest and deep learning. *IEEE Access*, 9:75729–75740, 2021.

52. Ruixia Liu and Yinglong Wang. A new Sybil attack detection for wireless body sensor network. In *2014 Tenth International Conference on Computational Intelligence and Security*, pages 367–370. IEEE, 2014.

53. Vikash Mainanwal, Mansi Gupta, and Shravan Kumar Upadhayay. A survey on wireless body area network: Security technology and its design methodology issue. In *2015 International Conference on Innovations in Information, Embedded and Communication Systems (ICIIECS)*, pages 1–5. IEEE, 2015.

54. Mohammad Masdari and Safiyeh Ahmadzadeh. Comprehensive analysis of the authentication methods in wireless body area networks. *Security and Communication Networks*, 9(17):4777–4803, 2016.

55. Suhas Mathur, Robert Miller, Alexander Varshavsky, Wade Trappe, and Narayan Mandayam. Proximate: Proximity-based secure pairing using ambient wireless signals. In *Proceedings of the 9th International Conference on Mobile Systems, Applications, and Services*, pages 211–224. ACM, 2011.

56. Kerry McKay, Lawrence Bassham, Meltem Sönmez Turan, and Nicky Mouha. Report on lightweight cryptography. Technical report, National Institute of Standards and Technology, 2016.

57. Sudip Misra and Ankur Vaish. Reputation-based role assignment for role-based access control in wireless sensor networks. *Computer Communications*, 34(3):281–294, 2011.

58. Samaneh Movassaghi, Mehran Abolhasan, Justin Lipman, David Smith, and Abbas Jamalipour. Wireless body area networks: A survey. *IEEE Communications Surveys & Tutorials*, 16(3):1658–1686, 2014.

59. Aamer Nadeem and M Younus Javed. A performance comparison of data encryption algorithms. In *2005 International Conference on Information and Communication Technologies*, pages 84–89. IEEE, 2005.

60. AKM Iqtidar Newaz, Amit Kumar Sikder, Leonardo Babun, and A Selcuk Uluagac. HEKA: A novel intrusion detection system for attacks to personal medical

devices. In *2020 IEEE Conference on Communications and Network Security (CNS)*, pages 1–9. IEEE, 2020.

61. Pejman Niksaz and Mashhad Branch. Wireless body area networks: Attacks and countermeasures. *International Journal of Scientific and Engineering Research*, 6(19):565–568, 2015.

62. Adedayo Odesile and Geethapriya Thamilarasu. Distributed intrusion detection using mobile agents in wireless body area networks. In *2017 Seventh International Conference on Emerging Security Technologies (EST)*, pages 144–149. IEEE, 2017.

63. Office of National Statistics. National population projections: 2016-Based statistical bulletin, 2016. Accessed: 14-05-2019.

64. Anyembe Andrew Omala, Kittur P Kibiwott, and Fagen Li. An efficient remote authentication scheme for wireless body area network. *Journal of Medical Systems*, 41(2):25, 2017.

65. B Padmavathi and S Ranjitha Kumari. A survey on performance analysis of DES, ACS and RSA algorithm along with LSB substitution. *IJSR, India*, 2:2319–7064, 2013.

66. Pangkaj Chandra Paul, John Loane, Gilbert Regan, and Fergal McCaffery. Analysis of attacks and security requirements for wireless body area networks—A systematic literature review. In *European Conference on Software Process Improvement*, pages 439–452. Springer, 2019.

67. Monalisha Polai, Sujata Mohanty, and Shreeya Swagatika Sahoo. A lightweight mutual authentication protocol for wireless body area network. In *2019 6th International Conference on Signal Processing and Integrated Networks (SPIN)*, pages 760–765. IEEE, 2019.

68. Ribers, M. A. and Ullrich, H. (2019). Battling antibiotic resistance: Can machine learning improve prescribing? *arXiv preprint arXiv:1906.03044*.

69. Amal Sammoud, Mohamed Aymen Chalouf, Omessaad Hamdi, Nicolas Montavont, and Ammar Bouallegue. A new biometrics-based key establishment protocol in WBAN: Energy efficiency and security robustness analysis. *Computers & Security*, page 101838, 2020.

70. Vladimir V Shakhov. Protecting wireless sensor networks from energy exhausting attacks. In *International Conference on Computational Science and Its Applications*, pages 184–193. Springer, 2013.

71. Jian Shen, Shaohua Chang, Jun Shen, Qi Liu, and Xingming Sun. A lightweight multi-layer authentication protocol for wireless body area networks. *Future Generation Computer Systems*, 78:956–963, 2018.

72. Jian Shen, Ziyuan Gui, Sai Ji, Jun Shen, Haowen Tan, and Yi Tang. Cloud-aided lightweight certificateless authentication protocol with anonymity for wireless body area networks. *Journal of Network and Computer Applications*, 106:117–123, 2018.

73. Lu Shi, Ming Li, Shucheng Yu, and Jiawei Yuan. BANA: Body area network authentication exploiting channel characteristics. *IEEE Journal on Selected Areas in Communications*, 31(9):1803–1816, 2013.

74. Lu Shi, Jiawei Yuan, Shucheng Yu, and Ming Li. MASK-BAN: Movement-aided authenticated secret key extraction utilizing channel characteristics in body area networks. *IEEE Internet of Things Journal*, 2(1):52–62, 2015.

75. Saurabh Singh, Pradip Kumar Sharma, Seo Yeon Moon, and Jong Hyuk Park. Advanced lightweight encryption algorithms for IoT devices: Survey, challenges and solutions. *Journal of Ambient Intelligence and Humanized Computing*, pages 1–18, 2017.

76. Upasna Singh and Bhawna Narwal. A novel authentication scheme for wireless body area networks with anonymity. In *Progress in Advanced Computing and Intelligent Engineering*, pages 295–305. Springer, 2021.

77. Tomoyasu Suzaki, Kazuhiko Minematsu, Sumio Morioka, and Eita Kobayashi. Twine: A lightweight, versatile block cipher. In *ECRYPT Workshop on Lightweight Cryptography*, volume 2011, 2011.

78. Xiao Tan, Jiliang Zhang, Yuanjing Zhang, Zheng Qin, Yong Ding, and Xingwei Wang. A PUF-based and cloud-assisted lightweight authentication for multi-hop body area network. *Tsinghua Science and Technology*, 26(1):36–47, 2020.

79. Geethapriya Thamilarasu. iDetect: An intelligent intrusion detection system for wireless body area networks. *International Journal of Security and Networks*, 11(1-2):82–93, 2016.

80. Geethapriya Thamilarasu and Zhiyuan Ma. Autonomous mobile agent based intrusion detection framework in wireless body area networks. In *2015 IEEE 16th International Symposium on a World of Wireless, Mobile and Multimedia Networks (WoWMoM)*, pages 1–3. IEEE, 2015.

81. Mohsen Toorani. On vulnerabilities of the security association in the IEEE 802.15. 6 standard. In *International Conference on Financial Cryptography and Data Security*, pages 245–260. Springer, 2015.

82. Mohsen Toorani. Security analysis of the IEEE 802.15. 6 standard. *International Journal of Communication Systems*, 29(17):2471–2489, 2016.

83. Sezer Toprak, Akhan Akbulut, Muhammet Ali Aydın, and Abdül Haim Zaim. LWE: An energy-efficient lightweight encryption algorithm for medical sensors and iot devices. *Electrica*, 20(1):71–81, 2020.

84. United Nations, Department of Economic and Social Affairs. World population ageing 2019, 2019. Accessed: 07-11-2021.

85. Muhammad Usman, Muhammad Rizwan Asghar, Imran Shafique Ansari, and Marwa Qaraqe. Security in wireless body area networks: From in-body to off-body communications. *IEEE Access*, 6:58064–58074, 2018.

86. Satish Vadlamani, Burak Eksioglu, Hugh Medal, and Apurba Nandi. Jamming attacks on wireless networks: A taxonomic survey. *International Journal of Production Economics*, 172:76–94, 2016.

87. Haodong Wang and Qun Li. Efficient implementation of public key cryptosystems on mote sensors (short paper). In *International Conference on Information and Communications Security*, pages 519–528. Springer, 2006.

88. Weichao Wang, Xinghua Shi, and Tuanfa Qin. Encryption-free authentication and integrity protection in body area networks through physical unclonable functions. *Smart Health*, 2018.

89. World Health Organization. Global status report, 2010. Accessed: 14-05-2019.

90. Hu Xiong. Cost-effective scalable and anonymous certificateless remote authentication protocol. *IEEE Transactions on Information Forensics and Security*, 9(12):2327–2339, 2014.

91. Hu Xiong and Zhiguang Qin. Revocable and scalable certificateless remote authentication protocol with anonymity for wireless body area networks. *IEEE Transactions on Information Forensics and Security*, 10(7):1442–1455, 2015.

92. Gangqiang Yang, Bo Zhu, Valentin Suder, Mark D Aagaard, and Guang Gong. The Simeck family of lightweight block ciphers. In *International Workshop on Cryptographic Hardware and Embedded Systems*, pages 307–329. Springer, 2015.

93. Jen-Ho Yang and Chin-Chen Chang. An ID-based remote mutual authentication with key agreement scheme for mobile devices on elliptic curve cryptosystem. *Computers & Security*, 28(3-4):138–143, 2009.

94. Guoxing Zhan, Weisong Shi, and Julia Deng. Design and implementation of TARF: A trust-aware routing framework for WSMs. *IEEE Transactions on Dependable and Secure Computing*, 9(2):184–197, 2012.

95. Lu Zhang, Jingwei Liu, and Rong Sun. An efficient and lightweight certificateless authentication protocol for wireless body area networks. In *2013 5th International Conference on Intelligent Networking and Collaborative Systems*, pages 637–639. IEEE, 2013.

96. Wentao Zhang, Zhenzhen Bao, Dongdai Lin, Vincent Rijmen, Bohan Yang, and Ingrid Verbauwhede. Rectangle: A bit-slice lightweight block cipher suitable for multiple platforms. *Science China Information Sciences*, 58(12):1–15, 2015.

97. Jin Zhao, Jifeng Huang, and Naixue Xiong. An effective exponential-based trust and reputation evaluation system in wireless sensor networks. *IEEE Access*, 7:33859–33869, 2019.

Machine Learning-Based Security and Privacy Protection Approach to Handle Physiological Data

M. Humayun Kabir

Department of Electrical and Electronic Engineering, Islamic University, Kushtia, Bangladesh

CONTENTS

ARTIFICIAL INTELLIGENCE (AI) is one of the emerging fields in the modern computing domain. Using AI techniques, machines can obtain knowledge through data analysis, which helps them to act rationally. AI-empowered healthcare has proven to be efficient and affordable with personalized features. The power of AI can accelerate diagnostics, image analysis, resource optimization, and medical diagnosis. Therefore, it is essential to aggregate AI technologies to healthcare. Recent advancements in AI technology demonstrate remarkable performance in implementing AI and machine learning in healthcare. Healthcare systems are capturing various

DOI: 10.1201/9781003251903-16

types of physiological data to provide required services to users. It requires a high level of accountability and transparency. Most wireless protocols support authentication to ensure data security even though this data can be vulnerable to traditional and zero-day attacks. So, security and privacy are two main issues that are the barriers to accommodating technology in healthcare. This chapter presents an analysis of security and privacy solutions utilizing an AI-based method in physiological data protection. In addition, we discuss the possible drawbacks which affect the security and privacy in machine learning (ML)-based approaches. Moreover, a motivation to overcome these flaws is also presented.

16.1 INTRODUCTION

Nowadays, the span of application of wearable technology is increasing day by day. Consumers use wearable devices to capture physiological data for personal and clinical use. Different types of portable devices (watches, glasses, wristbands, jewellery) from several companies are available; most use Bluetooth and Wi-Fi as the communication protocols. Wearable devices record physiological (heartbeat, respiration, blood pressure), personal data (profile), and geolocation which can interact through the internet and cloud servers. They become an entity on the Internet of Things (IoT). Wearable devices use embedded technology which provides low computational power and storage capacity [1]. As a result, they are prone to face security and privacy challenges compared to the traditional approaches. Since it is part of the IoT, unauthorized persons can get into these devices due to weak security. Wearable devices are vulnerable to user data security due to poor security firmware systems [2].

Healthcare deals with analysis, processing of personal health-related physiological data and creating a bridge between users and healthcare providers utilizing various types of communication technology. Any misuse or misinterpretation can cause a life threat even death for its user. The term security is used to handle data integrity, confidentiality, and availability. Whereas privacy maintains the authentication and proper handling of individual's data. The scope and area of both privacy and security is different. The measure of security cannot guarantee the protection of user privacy. So, measures should be taken individually to preserve the rights of users' security and privacy in the healthcare domain.

Nowadays different types of security and privacy issues arise. Among them the lack of authentication due to insecure PINs, intercepting messages (spoofing) between the cloud server and the healthcare devices, third-party access to sensitive information from devices including location information, insecure HTTPS protocols, image and video recordings, without users' consent are most noticeable. These security and privacy issues create various types of threats i.e. data injection, denial of service (DOS), Wi-Fi signal hijacking, eavesdropping, spyware, and malware. Different types of mitigation methods are recommended: (i) Firmware updates in regular intervals, (ii) Usage of fixed and private identity resolving are key while pairing with mobile devices, (iii) Generation of MAC address randomly, at regular interval. As the application area of physiological data usage has expanded it's become complex to handle the healthcare domain.

AI technology can be used in diagnosis and prediction of disease, help in assisted living, as well as detection and protection of security and privacy. This chapter focus on usages of AI in security and privacy of physiological data protection. As the scope and area of AI in the healthcare domain increases day by day, it becomes increasingly difficult to give an overview of the usage of AI in healthcare. Different types of technology involvement makes the healthcare domain very complex. We try to differentiate the application of AI in the healthcare domain into four groups (wearable devices, cloud platform, big data, telemedicine and e-Healthcare) based on the similarity of application. ML is a subset of AI used to analyze data and produce meaningful information. ML can learn from the available amount of data which can help to determine security and privacy threats. The three flavors of ML: supervised, unsupervised, and semi-supervised can be used in privacy and security solutions of physiological data protection. Existing solutions are not aware of time complexity, unable to share domain knowledge with others, or support black-box model which has lack of trust to users, while sharing the data it cannot guarantee the users' security as well as privacy. In order to raise the security, and trust the traditional black-box method, there is a need to be replace with explainable features which can express the cause and facts behind the reasoning. In addition, distributed machine learning features can be adopted which can reduce the time complexity and at the same time reduce security and privacy threats. Hence, explainable AI and federated learning could be a better solution in this regard.

The organization of this chapter is as follows: Section 16.2 presents an overview of the security and privacy issues that exist in physiological data handling in the healthcare domain. The potential security and privacy issues handling the physiological data are categorized based on the similarity of the application area. Section 16.3 describes the machine learning-based solutions available in physiological data protection. Mitigation methods are explained based on the category mentioned in Section 16.2. Limitation of existing AI-based approaches is presented in Section 16.4. Motivation to overcome the limitations of existing ML methods are described in Section 16.5. Finally, we conclude in Section 16.6 by mentioning the scope and future of this domain.

16.2 OVERVIEW OF SECURITY AND PRIVACY PROTECTION IN PHYSIO-LOGICAL DATA

Security and privacy are burning questionz in information and communication technology (ICT)-based services. As the healthcare domain is related to processing and distributing personal health-related physiological data, it is not an exception. Security deals with data confidentially, integrity as well as availability. It concentrates on keeping data from harmful attacks and misuse for the sake of benefit. On the other hand, privacy in the healthcare domain ensures the safety of personal healthcare-related physiological data. Privacy policies should be taken to ensure the proper use of individuals' data. It also regulates authorization to collect, share, and utilize physiological data in appropriate ways. The measures of security cannot guarantee the protection of user privacy. Figure 16.1 illustrates an overview of data security and

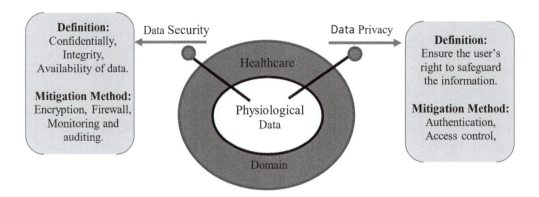

Figure 16.1 Overview of data security and privacy of physiological data in the healthcare domain.

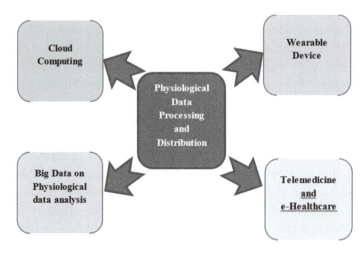

Figure 16.2 The sub-domain of physiological data handling.

privacy of physiological data in the healthcare domain. In this chapter, we differentiate the uses of physiological data in the four sub-domains: wearable devices, cloud computing, big data, telemedicine and e-healthcare. A brief description of each sub-domain is illustrated in the following sections. The sub-domain of physiological data handling is shown in Figure 16.2.

16.2.1 Wearable Devices

In this section, we have investigated wearable devices' security and privacy features and their strength to handle threats. In addition, a detailed scenario is depicted on the present state-of-the-art technique. Moreover, an indication of future research has also been illustrated in wearable devices. One group [3] has presented an analysis on how unauthorized persons can access user data through current wearable devices. They also discovered challenges to securing the data on wearable devices. The study was based on three different mobile health devices Fitbit, Jawbone, and Google Glass. A summary of the study is tabulated in Table 16.1.

TABLE 16.1 Summary of the Potential Security Threat and Mitigation Methods in Wearable Devices

Device name	Security and privacy concern	Common threat type	General mitigation method
Fitbit Tracker [4]	• Lack in authentication • Drainage of Bluetooth Low Energy(BLE) technology privacy [5] • Intercepting messages (spoofing) between the cloud server and fitness tracker [7] • Third-party access to the sensitive information from the devices, including location information [8]	• Data injection • Denial of service (DoS) • Battery drain hacks	• Regular Firmware update [6] • Firmware needs fixed and private
Jawbone [9]	• Lacks privacy features • User's location information from devices • Unsecure HTTPS protocols	• Denial of service (DoS) • Generated false fitness data for their individual account	• Identity Resolving Key (IRK) while pairing with mobile phones
Google Glass [10]	• Lack of authentication due to insecure PIN [11] • Image and video recording without user's consent [12] • Insecure network • Subjected to the hostile environment	• Wi-Fi signal hijacking • Eavesdropping • Installation of Spyware • Easy recording system by people nearby due to gesture base authentication scheme • Malware such as QR Photobombing	• Random generation of MAC address at a regular interval

Fitbit tracker is a wearable wrist device that tracks different types of user activity and discrete events such as the number of steps walked, sleep patterns and quality. It also can measure various health measurements like body temperature, blood pressure, pulse rate, food habits, and body weight. Moreover, it can automatically recognize users' exercises and record the data through a smartphone app. Users need to wear the Fitbit device throughout the day for continuous service. The data is captured through a mobile application. It supports Android, iOS, and Windows applications. Data is forwarded to the Fitbit cloud server using an internet connection during data synchronization. Another potential health activity monitoring device is Jawbone. It's a wrist-like wearable device that tracks users' daily activity and recommends helpful tips related to lifestyle. Jawbone periodically transmits geolocation information when users log-in to apps or synchronize their wearable devices. It supports Android and iOS. Google Glass is the pioneer of wearable technology. Google Glass utilizes an eyewear computing device interfaced with a pair of glasses. Google Glass is usable for enterprise versions only and not available for individual's usage.

TABLE 16.2   Summary of the Potential Security Threat and Mitigation Methods in Cloud Platforms

Device name	Security and privacy concern	Common threat type	General mitigation method
Cloud platform	• Password-based authentication has generic limitations, risks • Heterogeneity and diversity of services demand fine-grained access control policies • Cloud service providers use the Web Services Description Language (WSDL), which can't fully meet the cloud computing services description requirements	• Data injection • Distributed Denial of Service (DDoS) • Eavesdropping • Flooding attacks by viruses or malware, Spyware	• A user-centric Identity and Access Management (IAM), a mechanism should be introduced in the authentication • Security Audit to protect the data from a data breach • Role-based access policy to secure the data from unauthorized handling • Introducing the digital signature in identify the right sources

TABLE 16.3   Summary of the Potential Security Threat and Mitigation Methods in Big Data Domains

Device name	Security and privacy concern	Common threat type	General mitigation method
Big physiological data analytics	• Due to a large amount of data, it is quite hard to maintain the user's privacy • Protecting data sources from authorized access is a challenging task • Vulnerable to data mining-based attack	• Network layer related Packet Sniffers [16] • Spoofing • Defacing • Application-Layer related Denial of Service (DoS) [17] • DDoS Spam • Viruses • Worms • HTTP • Exploits	• Sniffing detection platform, cleared the cache at a specific interval • Firewall, Intruder detection system

16.2.2   Cloud Platform for Physiological Data Storage

Cloud computing is a blessing of current technology, which has enabled low resource devices to overcome their limitations by sharing from the repository. Users can utilize the processing power and memory storage ubiquitously. It provides a remote access facility. Due to the enormous benefits of cloud platforms, concern in security and privacy arises [13]. Table 16.2 represents a summary of security threats and mitigation methods in cloud platforms for physiological data storage.

16.2.3   Big Physiological Data Analytics

Healthcare systems handle physiological data to provide services to the end user. A massive amount of data needs to be processed and distributed among the networks. So, measures should be taken to protect the data from unauthorized handling [14,15]. A summary of potential security threats and practicing mitigation methods in this sub-domain are presented in Table 16.3.

TABLE 16.4 Summary of the Potential Security Threat and Mitigation Methods in Telemedicine and e-Health

Device name	Security and privacy concern	Common threat type	General mitigation method
Telemedicine and e-health	• Requirement of user identification • Extension of network boundary • Variation of standard security monitoring	• Hackers use flooding attacks by virus or malware programs • Distributed Denial of Service (DDoS)	• Deploying management tools in order to secure networks • Next-Generation • Firewalls • Intrusion Detection Systems • Advanced Malware Protection

16.2.4 Telemedicine and E-Healthcare

Nowadays, network-based applications have emerged in the digital world. Healthcare devices have been adopted within communication facilities to converge with networks. Healthcare devices network telecommunication closes the distance between the patient and expert in the medical domain. As a result, telemedicine and e-healthcare services are evolving. Telemedicine and e-healthcare services open the door to provide remote patient care and monitoring facilities. Beyond its large scope of benefits, it suffers some security and privacy concerns tabulated in Table 16.4.

16.3 MACHINE LEARNING-BASED SOLUTIONS IN PHYSIOLOGICAL DATA PROTECTION

The branch of AI which can mimic human learning from data is known as ML [18]. ML can be used to analyze data and produce meaningful information. Traditional data security and privacy solutions are not an application for low-resource embedded devices. The ML technique can learn from the vast amounts of data, which can help to find security and privacy threats [19]. ML can be classified into three types: supervised, unsupervised and semi-supervised. The first category is supervised learning, handling label data of known classes. For malware detection, supervised learning techniques, i.e. support vector machine (SVM), decision tree, neural network (NN), deep learning (DL), are mainly used. DL is an advanced form of NN that can process the data layer-wise like the human brain and extract automatic features from the data. There are several flavours of DL: Convolution neural network (CNN), recurrent neural network (RNN), long-short term memory (LSTM), autoencoders (AEs), Boltzman neural network (BM), generative adversarial network (GAN), feedforward deep networks (FDNs), and deep belief networks (DBNs). For anomaly detection, AE is used.

On the other hand, unsupervised ML deals with unlabeled data in an indeterministic approach. For anomaly detection, unsupervised learning, i.e. K-nearest neighbour (KNN), self-organizing map (SOM), latent Dirichlet allocation (LDA), is used. A semi-supervised approach could be followed when some data are labeled among the wide number of learning data. For attack detection, semi-supervised learning is

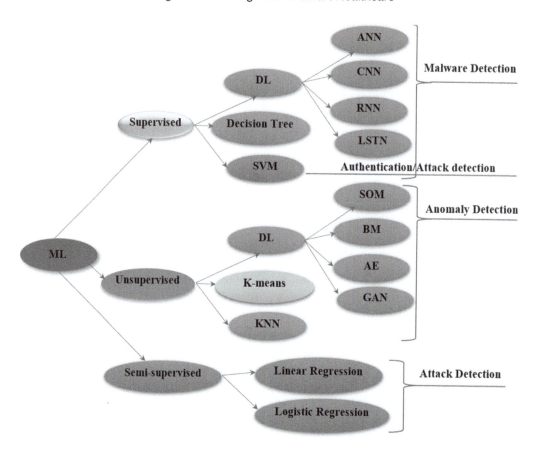

Figure 16.3 Machine learning approach used in physiological data protection.

utilized. Several ML techniques have been used in attack detection and mitigation methods. This section presents the existing ML techniques proposed in the healthcare domain, especially in physiological data protection. We try to differentiate the application into four sub-domains based on the similarity of technology usage as described in the previous section. Figure 16.3 shows an overview of ML methods used for security and privacy of physiological data protection.

16.3.1 Wearable Device Domain

Wearable devices and biosensors are utilized to capture the physiological (temperature, blood glucose, hearbeat, respiration, brain) signals from the human body [20]. The captured raw data is then processed and stored on a local or remote server for further application. Different types of wireless communication protocols (Bluetooth, Bluetooth Low Energy [BLE], Near Field Communication [NFC], Radio Frequency Identification [RFID], Zigbee, Z-wave, Wi-Fi, etc.) are used for sending and receiving data. The wearable devices assist the user in leading an easier and healthier life. The security and privacy of this domain are essential because they can threaten the patient's life. Wearable devices are tiny, and their computational resource and power cannot handle existing security and privacy solutions. Hence, vulnerable to traditional and zero-day attacks. Faulty measurements create a false-positive result, which has a negative impact on users. So, the initiative should be taken to differentiate between the real situation with the observed state. One group of authors [21] present a sensor anomaly method to detect the sensor anomaly through proper analysis of collected physiological data from the sensor of medical devices. This method can efficiently differentiate the false report from the true report considering the historical data with the collected data. Sequential minimal optimization (SMO) regression is utilized to detect the anomaly in physiological data. To evaluate the proposed approach, they have used physiological data from the multiple intelligent monitoring in intensive care (MIMIC) dataset of PhysioNet [22]. The detection result is compared with Mahalanobis (MD), linear SVM, and J48 Decision tree approaches for performance analysis. The experimental results show that the proposed system achieves a high detection rate of 100% for all three medical datasets.

The anomaly can be created by damaged sensor nodes or malicious third parties, which creates misdiagnosis and may hamper life. O Salem et al. [23] proposed a framework which converged data mining and ML algorithms with sensor fusion techniques to anomaly detection in wireless body area network (WBAN). A SVM is used to classify the irregularity in sensor data. They have utilized a physiological dataset [24] from PhysioNet, which contain seven attributes: Blood pressure (BP) mean, systolic blood pressure (BPsys), diastolic blood pressure (BPdias), heart rate (HR), pulse, respiration (RESP), oxygen saturation (SpO2). The authors reported that their approach able to identify anomaly quickly than other state-of-art algorithms: Linear regression, additive regression, K-NN, decision tree J48, Naïve Bayes, decision Table.

Owe to the rapid development of sensing technology, the scope of developing the healthcare device for elderly care is emerging. Threats related to medical devices have been studied [25], and a novel intrusion detection technique is proposed. This technique consists of three steps: Data pre-processing, feature extraction, and determination of intrusion behavior. For feature extraction, a stacked autoencoder (SAE) is utilized. On the other hand, XGBoost is used to detect intrusions by the learned feature from SAE. They have considered a connected healthcare system for evaluation of the proposed method. Authors have claimed that SAE can extract more fine-grained features as well as reduce the feature dimensions.

Confidentiality of physiological data plays a vital role in healthcare device development. The limitation of computing power and memory is a drawback of implementing conventional robust cryptography. Alassaf et al. [26] present a performance comparison of three lightweight cryptography (LWC) algorithms: AES, SPECK, SIMON, based on execution time, power consumption, memory occupation, and speed. The result shows that SPECK outperforms the other two cyphers algorithms considering all the performance metrics. SPECK is 27% faster than AES and 45% than the SIMON. One author [27] presents an SVM-based authentication approach and shows that due to the hardware resource limitations, it will be a good practice to include other domains in the authentication process, such as physiological biometrics and behavioral biometrics. In order to make a cheap and light device, most wireless medical devices have not adopted with authentication techniques. Hence, they are vulnerable to various security and privacy threats. Consequently, biometric-based lightweight mutual authentication and key agreement protocol is proposed for a real-time wearable medical sensor network [28]. This approach used fuzzy extractors, elliptic curve cryptography (ECC), and dynamic identity for mutual authentication. They focus on developing a lightweight protocol for low resources wearable devices. This approach consists of five steps: Firstly, medical personnel and wearable device registration, secondly log-in by medical staff, thirdly is the authentication and key agreement step, fourthly monitoring, and finally encryption key changing steps.

A medical device that is implanted within the patient's body is called an implantable medical device. Implantable medical devices (IMDs) are used for patients suffering from chronic diseases like diabetes, cardiac arrhythmia, Parkinson's. Nowadays, IMDs are mostly connected with an outside program for control and monitoring. Various types of IMDs are available recently: Oximeters [29], defibrillators [30], pacemakers, etc. The market of IMDs is emerging. So, security and privacy issues arise in this domain for secure and smooth application. IMDs contains limited resources in terms of computation, storage, and energy. These authors [31] addressed resource depletion (RD) attacks which deplete IMDs resources drastically. RD include denial of service (DoS), and authentication attacks. A novel SVM based approach is proposed to detect the RD attacks. They have utilized the linear and nonlinear SVM classifier and made a performance comparison. Experimental results showed that the proposed approach could detect 90% of attacks using linear SVMs and 97% for nonlinear SVMs. A novel detection and prevention system, CardiWall has proposed to protect implantable cardioverter defibrillators (ICDs) from cyber attacks [32]. Moreover, it can design, detect, and implement the bugs in the software to prevent human errors. The proposed system achieved a high true positive rate (TPR) with an accuracy of 91.4% and 94.7%, respectively.

The security and privacy of implantable devices can be affected by wireless communication, which can allow an unwanted person to get into the network in case of any shortage of security and privacy features. The software and hardware of implantable systems can be affected by the security and privacy attack. Rathore et al. [33] proposed a multilayer perception model-based authentication method that provides security for data, network, and application. The ECG-based biometric authentication approach uses Legendre polynomial extraction and an Multi-Layer Perceptron (MLP) classifier for identification and authorization. The proposed model reports 99.99% accuracy. Researchers [34] have evaluated the adaptability of usage of the ML algorithm to detect security threats in implantable medical devices. A developed feature set was used to experiment with various ML algorithms, including decision tree, SVM, and K-means algorithms. Experimental results show that the decision tree-based algorithm outperforms the other ML algorithm's accuracy and computational complexity.

16.3.2 Cloud Computing Domain

Security key and biometric authentication suffer some drawbacks, which raises questions on the reliability and efficiency of the system. Some authors [35] present a framework based on a highly secure cancelable biometric authentication [36] system using a cloud platform. They have shown theoretically and experimentally that the error rate of this system is minimum when compared with the state-of-the-art techniques. The minimal time complexity led to its suitability for IoT environments. Nowadays, electronic health records (EHRs) can use cloud storage, which enhances the scope of sharing the patient's physiological data with remote medical practitioners and healthcare providers. These new paradigms open the door of low cost, flexible and ubiquitous EHRs. As the operating area of EHRs expands, the patient physiological data security and privacy concern arise. One group of authors [36] utilized blockchain and decentralized interplanetary file system (IPFS) to present a framework on a mobile cloud platform for secure EHRs. The framework is implemented on a mobile App with an Amazon cloud platform. The evaluation results show that the framework could guarantee reliable data exchange on mobile clouds against potential threats. The performance comparison results claim that this framework can take minimum network latency with high data security and privacy level than the state-of-art techniques. Medical cyber-physical systems (MCPS) are a convergence network of systems with sensing and computing devices in the healthcare domain. This network is established to provide high-quality care by remote monitoring and caring. MCPS handles personal physiological data; hence lack of security and privacy can cause severe threats to patients' health and life.

16.3.3 Big Data Domain

Big data is a recent development in information and communication technology (ICT) domain which provides the scope of ubiquitous computing and storage. The combination of big data and smart healthcare devices can expand the application in the healthcare domain. Due to the blessing of ICT, huge amounts of data are generated and handled electronically in the healthcare domain. Data mining using the ML approach is necessary to exploit the information and knowledge in the healthcare domain. The concept of big data is to handle a large volume of data, so it is quite challenging for the conventional method to manipulate it. DL-based ML solutions can be utilized to handle security and privacy features. Technology related to the prevention or management of chronic illness is a requirement for the digital world to save costs and provide a better life. This technology should handle the devices used for remote monitoring physiological data and assist users on demand. Nowadays, patients are utilizing their mobile phones to manage their health-related needs using mobile applications.

Mobile medical devices are now equipped with telemedicine and telehealth facilities using IoT and big data. Big data analysis can indicate the digital disruption of the healthcare world and can accelerate business processes and real-time decision-making [37]. People around the world are now using wearable biosensors, and it is emerging. Hence a reliable, adaptive, safe, and cost-effective solution is needed. One group [38] introduced a big data-based smart healthcare system framework (BSHSF) that enables a new business model in the healthcare context.

16.3.4   Telemedicine and E-Healthcare Domain

An MCPS paved a new path for monitoring, diagnosing and treating a patient in the clinical integrated environment (ICE), providing automatic, efficient, and reliable services. Current MCPS systems in the ICE domain do not consider security and privacy issues. Hence it is prone to cyber-security threats. Ransomware attacks cause 85% [39] of all malicious software attacks in the healthcare domain. F. Maimo et al. [40] introduce an ML-based ransomware detection solution in ICE environment. Moreover, network function virtualization (NFV) and software-defined networking (SDN) is utilized to mitigate the spreading of a ransomware attack. The experimental reports show that the proposed solution can accurately classify 99.9% of a ransomware attack. The Internet of Medical Things (IoMT) is a special paradigm of IoT in which medical devices communicate with each other for sharing physiological data. Hence patients can get better care from healthcare providers using the ICT.

As the application area and capacity increase, so security and privacy increase. Various types of security and privacy-related threats are nowadays drawing attention to researchers in this domain like replay, man-in-the-middle, impersonation, privileged-insider, remote hijacking, password guessing, DoS and malware attacks. One group of authors [41] presents a framework to determine and avoid the risk of insecure IoMT. This security assessment framework aims to overcome the gap between the IoMT solution provider and consumer irrespective of security, privacy, and trust. This framework builds awareness in IoMT users (patient and medical professional) and makes them conscious of the existing low-level security solution in the IoMT domain. Mobile healthcare systems utilize the sensor to collect physiological data for providing various types of health-related services to their users. Inaccurate or erroneous physiological data can produce wrong decisions that can hamper the service and sometimes create a life threat to users. AUDIT [42], anomalous data detection and isolation approach, is proposed for mobile healthcare systems. It can distinguish faults or errors in real-time measurements. The proposed method followed two approaches; firstly, a pre-processing is done after a real-time detection is utilized based on the principal component analysis (PCA) technique. To validate the performance in terms of computational complexity and performance, experiments have been done on real medical datasets. The experimental results show that it can detect and isolate anomalous physiological measurements with high recall and low false-positive rates. HealthGuard [43], an ML-based security framework, proposes the ability to detect malicious activities in smart healthcare systems (SHS). This framework can observe the vital signs of different devices in SHS to detect malicious activities. It has used ANN, decision tree, random forest, and K-nearest neighbour. HealthGuard reported a high accuracy and F1-Score of 91% and 90%, respectively. One author [44] proposes an effective and efficient intrusion detection system (IDS) in the IoMT domain based on deep learning (DL) techniques to classify unforeseen cyber attacks. Through rigorous testing, they claim that the proposed DNN model outperforms the existing machine learning approaches in terms of accuracy and computational complexity by 15% and 32%, respectively. P. M. Shakeel et al. [45] proposed a deep Q-network to reduce malware attacks in healthcare-related applications. It follows a layer-wise approach to handle data with less computational complexity. The experimental results show its potentiality against malware threat protection.

16.4   LIMITATION OF EXISTING AI-BASED APPROACHES

Introducing AI in healthcare systems makes it more efficient, affordable, and personalized. ML as a subset of AI plays a huge role in security and privacy of physiological data protection. Physiological data handling requires a high level of security, privacy, and trust platform. Hence, these issues should be carefully taken into account to expand the usage of physiological data in different domains. Recently, a number of ML-based solutions have been involved with handling the security and privacy features. These solutions are not fully able to maintain the users' security and privacy. As the application area of physiological data usage is expanded it become complex to handle the domain. Existing solutions are not aware of time complexity, unable to share domain knowledge with others, support the black-box model which has a lack of trust to users, while sharing data.

The traditional ML approach followed a black-box model, which hid all the processes and only published the reasoning outcome. So the expert or practitioner could not guess which factors were considered for reasoning, and could not justify knowledge behind the belief. While performing diagnosis or producing prescriptions, an expert system should explain the justification behind the decision. Conventional ML approaches use central servers/processors to process data that may cause harm to data security and privacy during transmission. The applications in healthcare domains are safety critical and contain personal sensitive information. Hence, attention should be taken while training the model and doing prediction or classifications.

16.5   MOTIVATION TO OVERCOME THE LIMITATIONS

To decrease the computational complexity in the ML approach, a distributed approach should be followed. The distributed approach such as the federated learning (FL) [46] method may be helpful in this regard. The distributed ML approach saves computational cost as well as maintains the security and privacy of the client. It does not allow the local data for processing in a central server, rather process it locally. This decentralization keeps the data in source and maintains security and privacy.

In order to raise the security, and trust the traditional black-box method need to be replace with explainable features which can express the cause and facts behind the reasoning. So, explanation is necessary to ensure security and to build trust on the platform. Explainable AI (XAI) can be successfully used in this regard [47]. A number of XAI commercial platforms are engaged in providing this service. Among them IBM, Google, Darwin AI, Flowcast, Imandra, Kybdi, Factmata are most noticeable [48].

XAI- and FL-supported ML models are crucial in the healthcare domain for security and privacy protection. XAI and FL are mainly focused on prediction and classification in diagnosis, health records, and prediction of diseases. Another group [49] utilized FL to minimize the communication and computation costs compared to the traditional ML-based approach. This approach is evaluated with real patient data and against security DoS, data modification, and data injection attacks. The evaluation result shows that it can get average accuracy of 99%. There is more scope in detection of anomaly, authentication, attack detection, and malware using XAI and FL methods.

16.6 CONCLUSION

Physiological data is the personal property of a person hence confidential. So, security and privacy are significant concerns in the study of the healthcare domain. This current chapter illustrates a scenario of security and privacy in physiological data based on AI. A brief explanation of the step-by-step physiological data flow from the signal source to the user's end is presented. The role of AI handling the security and privacy issues in each step is also briefly described, as well as the gap between the current ML approaches. In addition, the motivation of the mitigation method in this domain is also depicted.

Bibliography

1. J. Al-Muhtadi, D. Mickunas, and R. Campbell, "Wearable security services," in *Proceedings 21st International Conference on Distributed Computing Systems Workshops*, pp. 266–271, IEEE, 2001.

2. H. Fortify, "Internet of Things security study: Smartwatches," Technical Report, 2015.

3. C. Zhang, H. Shahriar, and A. K. Riad, "Security and privacy analysis of wearable health device," in *2020 IEEE 44th Annual Computers, Software, and Applications Conference (COMPSAC)*, pp. 1767–1772, IEEE, 2020.

4. Fitbit, "`https://www.fitbit.com/global/us/technology`" (accessed: 05.11. 2021).

5. M. L. Hale, K. Lotfy, R. F. Gamble, C. Walter, and J. Lin, "Developing a platform to evaluate and assess the security of wearable devices," *Digital Communications and Networks*, vol. 5, no. 3, pp. 147–159, 2019.

6. J. A. Martin, "10 security risks of wearables, `https://www.csoonline.com/article/3186164/10-security-risks-of-wearables.html`" (accessed: 05.11.2021).

7. T. Seals, "Fitbit vulnerabilities expose wearer data, `https://www.infosecurity-magazine.com/news/fitbit-vulnerabilities-expose/`" (accessed: 05.11.2021).

8. "How fitbits, other bluetooth devices make us vulnerable to tracking, `https://www.bu.edu/articles/2019/fitbit-bluetooth-vulnerability/`" (accessed: 05.11.2021).

9. Jawbone, "`https://www.wearablezone.com/companies/jawbone/`" (accessed: 05.11.2021).

10. Google Glass, "`https://www.google.com/glass/start/`" (accessed: 05.11.2021).

11. K. W. Ching and M. M. Singh, "Wearable technology devices security and privacy vulnerability analysis," *International Journal of Network Security & Its Applications*, vol. 8, no. 3, pp. 19–30, 2016.

12. S. Safavi and Z. Shukur, "Improving Google Glass security and privacy by changing the physical and software structure," *Life Science Journal*, vol. 11, no. 5, pp. 109–117, 2014.

13. M. Mamun-Ibn-Abdullah and M. H. Kabir, "A multilayer security framework for cloud computing in Internet of Things (IoT) domain," *Journal of Computer and Communications*, vol. 9, no. 7, pp. 31–42, 2021.

14. F. Lombardi and R. Di Pietro, "Heterogeneous architectures: Malware and countermeasures," in *Secure System Design and Trustable Computing*, pp. 421–438, Springer, 2016.

15. K. Abouelmehdi, A. Beni-Hessane, and H. Khaloufi, "Big healthcare data: Preserving security and privacy," *Journal of Big Data*, vol. 5, no. 1, pp. 1–18, 2018.

16. X. Chen, S. Chen, X. Zeng, X. Zheng, Y. Zhang, and C. Rong, "Framework for context-aware computation offloading in mobile cloud computing," *Journal of Cloud Computing*, vol. 6, no. 1, pp. 1–17, 2017.

17. Y. Chen, X. Li, and F. Chen, "Overview and analysis of cloud computing research and application," in *2011 International Conference on E-Business and E-Government (ICEE)*, pp. 1–4, IEEE, 2011.

18. M. Kubat, "Induction in multi-label domains," in *An Introduction to Machine Learning*, pp. 251–271, Springer, 2017.

19. F. Hussain, R. Hussain, S. A. Hassan, and E. Hossain, "Machine learning in iot security: Current solutions and future challenges," *IEEE Communications Surveys & Tutorials*, vol. 22, no. 3, pp. 1686–1721, 2020.

20. Valadkhani, Z., Lotfi, F., and Rodić, B. (2021). A vision of the Internet of Things: A review of critical challenges. *International Journal of Computer and Information Technology* (2279-0764), 10(4).

21. S. A. Haque, M. Rahman, and S. M. Aziz, "Sensor anomaly detection in wireless sensor networks for healthcare," *Sensors*, vol. 15, no. 4, pp. 8764–8786, 2015.

22. PhysioNet, "`https://www.physionet.org/content/mimicdb/1.0.0/`" (accessed: 05.11.2021).

23. O. Salem, A. Guerassimov, A. Mehaoua, A. Marcus, and B. Furht, "Anomaly detection in medical wireless sensor networks using SVM and linear regression models," *International Journal of E-Health and Medical Communications (IJEHMC)*, vol. 5, no. 1, pp. 20–45, 2014.

24. "`https://physionet.org/about/database/`," (accessed: 05.11.2021).

25. D. He, Q. Qiao, Y. Gao, J. Zheng, S. Chan, J. Li, and N. Guizani, "Intrusion detection based on stacked autoencoder for connected healthcare systems," *IEEE Network*, vol. 33, no. 6, pp. 64–69, 2019.

26. N. Alassaf and A. Gutub, "Simulating light-weight-cryptography implementation for iot healthcare data security applications," *International Journal of E-Health and Medical Communications (IJEHMC)*, vol. 10, no. 4, pp. 1–15, 2019.

27. A. A. Mawgoud, A. I. Karadawy, and B. S. Tawfik, "A secure authentication technique in internet of medical things through machine learning," *arXiv preprint arXiv:1912.12143*, 2019.

28. N. R. Mohsen, B. Ying, and A. Nayak, "Authentication protocol for real-time wearable medical sensor networks using biometrics and continuous monitoring," in *2019 International Conference on Internet of Things (iThings) and IEEE Green Computing and Communications (GreenCom) and IEEE Cyber, Physical and Social Computing (CPSCom) and IEEE Smart Data (SmartData)*, pp. 1199–1206, IEEE, 2019.

29. Nonin Medical, Inc. "`https://www.nonin.com/resources/`" (accessed: 05.11.2021).

30. Medtronics, Inc. "`https://www.medtronic.com/me-en/index.html`" (accessed: 05.11.2021).

31. X. Hei, X. Du, J. Wu, and F. Hu, "Defending resource depletion attacks on implantable medical devices," in *2010 IEEE Global Telecommunications Conference GLOBECOM 2010*, pp. 1–5, IEEE, 2010.

32. M. Kintzlinger, A. Cohen, N. Nissim, M. Rav-Acha, V. Khalameizer, Y. Elovici, Y. Shahar, and A. Katz, "Cardiwall: A trusted firewall for the detection of malicious clinical programming of cardiac implantable electronic devices," *IEEE Access*, vol. 8, pp. 48123–48140, 2020.

33. H. Rathore, C. Fu, A. Mohamed, A. Al-Ali, X. Du, M. Guizani, and Z. Yu, "Multi-layer security scheme for implantable medical devices," *Neural Computing and Applications*, vol. 32, no. 9, pp. 4347–4360, 2020.

34. S. Gao and G. Thamilarasu, "Machine-learning classifiers for security in connected medical devices," in *2017 26th International Conference on Computer Communication and Networks (ICCCN)*, pp. 1–5, IEEE, 2017.

35. M. R. Abdmeziem and D. Tandjaoui, "An end-to-end secure key management protocol for e-Health applications," *Computers & Electrical Engineering*, vol. 44, pp. 184–197, 2015.

36. N. K. Ratha, S. Chikkerur, J. H. Connell, and R. M. Bolle, "Generating cancelable fingerprint templates," *IEEE Transactions on Pattern Analysis and Machine Intelligence*, vol. 29, no. 4, pp. 561–572, 2007.

37. C. L. Ross, T. Teli, and B. S. Harrison, "Electromagnetic field devices and their effects on nociception and peripheral inflammatory pain mechanisms," *Alternative Therapies in Health & Medicine*, vol. 22, no. 3, 2016.

38. M. I. Pramanik, R. Y. Lau, H. Demirkan, and M. A. K. Azad, "Smart health: Big data enabled health paradigm within smart cities," *Expert Systems with Applications*, vol. 87, pp. 370–383, 2017.

39. S. Widup, M. Spitler, D. Hylender, and G. Bassett, "Verizon data breach investigations report," *Retrieved from https://www. researchgate. net/publication/324455350_2018_Verizon_Data_Breach_Investigations*, 2018.

40. L. Fernandez Maimo, A. Huertas Celdran, A. L. Perales Gomez, F. J. Garcia Clemente, J. Weimer, and I. Lee, "Intelligent and dynamic ransomware spread detection and mitigation in integrated clinical environments," *Sensors*, vol. 19, no. 5, p. 1114, 2019.

41. F. Alsubaei, A. Abuhussein, and S. Shiva, "A framework for ranking iomt solutions based on measuring security and privacy," in *Proceedings of the Future Technologies Conference*, pp. 205–224, Springer, 2018.

42. L. Ben Amor, I. Lahyani, and M. Jmaiel, "Audit: Anomalous data detection and isolation approach for mobile healthcare systems," *Expert Systems*, vol. 37, no. 1, p. e12390, 2020.

43. A. I. Newaz, A. K. Sikder, M. A. Rahman, and A. S. Uluagac, "Healthguard: A machine learning-based security framework for smart healthcare systems," in *2019 Sixth International Conference on Social Networks Analysis, Management and Security (SNAMS)*, pp. 389–396, IEEE, 2019.

44. Swarna Priya R.M., Praveen Kumar Reddy Maddikunta, Parimala M., Srinivas Koppu, Thippa Reddy Gadekallu, Chiranji Lal Chowdhary, and Mamoun Alazab, "An effective feature engineering for DNN using hybrid PCA-GWO for intrusion detection in IoMT architecture," *Computer Communications*, vol. 160, pp. 139–149, 2020. ISSN 0140-3664, https://doi.org/10.1016/j.comcom.2020.05.048.

45. P. M. Shakeel, S. Baskar, V. S. Dhulipala, S. Mishra, and M. M. Jaber, "Maintaining security and privacy in health care system using learning based deep-Q-networks," *Journal of Medical Systems*, vol. 42, no. 10, pp. 1–10, 2018.

46. T. S. Brisimi, R. Chen, T. Mela, A. Olshevsky, I. C. Paschalidis, and W. Shi, "Federated learning of predictive models from federated electronic health records," *International Journal of Medical Informatics*, vol. 112, pp. 59–67, 2018.

47. Y. Xie, G. Gao, and X. Chen, "Outlining the design space of explainable intelligent systems for medical diagnosis," *arXiv preprint arXiv:1902.06019*, 2019.

48. M. Humayn Kabir, K. Fida Hasan, M. Kamrul Hasan, and K. Ansari, "Explainable artificial intelligence for smart city application: A secure and trusted platform," *arXiv e-prints*, pp. arXiv–2111, 2021.

49. W. Schneble and G. Thamilarasu, "Attack detection using federated learning in medical cyber-physical systems," in *28th International Conference on Computer Communications and Networks (ICCCN)*, pp. 1–8, 2019.

## CHAPTER 17

Future Challenges in Artificial Intelligence for Smart Healthcare

Faisal Tariq

James Watt School of Engineering, University of Glasgow, UK

S. M. Riazul Islam

Department of Computer Science, University of Huddersfield, Huddersfield, UK

Ghita Kouadri Mostefaoui

Department of Computer Science, University College London, UK

CONTENTS

ARTIFICIAL INTELLIGENCE (AI) techniques in general have proliferated in all areas of healthcare including disease detection, electronic health records, patient management, clinical decision making, image analysis, disease prediction and treatment. Another rapidly evolving area is personalized medicine research and design. With more powerful hardware processing, software agility, networking and communication capabilities, AI is becoming a dominant player in smart healthcare systems. Despite rapid progress made over the last decade, there exist significant challenges in various aspects of technology and biological factors that need to be addressed in order to ensure successful use of AI in smart healthcare. In this chapter, we will briefly outline the major challenges for AI adoption in various aspects of a smart healthcare system.

DOI: 10.1201/9781003251903-17

17.1 INTRODUCTION

Artificial Intelligence (AI) research for smart healthcare has been increasing rapidly but the actual clinical implementation is still lagging far behind the expected level of adoption due to various challenges and limitations [7]. The rest of the chapter identifies some key challenges for AI adoption in smart healthcare and steps required to potentially transfer these research outcomes into real-life clinical practices.

17.2 OPTIMIZING AI ARCHITECTURE FOR SMART HEALTHCARE

The AI system is not optimally or customized for healthcare. A number of challenges remain from a technical perspective. For example, most power chips are not specifically designed for AI algorithms and thus often lack efficiency. Also, they are very expensive which hinders widespread adoption. So, designing low-cost solutions is of paramount importance.

Massive proliferation of the Internet of Things (IoT) in healthcare requires AI systems to work in distributed fashion. However, communication and networking technology to support the distributed nature of AI algorithms are still not up to the mark. Particularly latency, reliability, and ubiquitous availability of networks seriously hamper development of distributed AI solutions for IoT-enabled healthcare systems [8]. Therefore, a disruptive design is needed that can support extreme latency and reliability which is crucial for remote surgery and remote patient monitoring.

17.3 AI-ENABLED DISEASE DIAGNOSIS USING IMAGING

AI has been used widely for predicting, detecting, and diagnosing diseases. For example, various cancers can be diagnosed from magnetic resonance imaging (MRI), computed tomography (CT), and positron emission tomography (PET) scans [4, 7]. AI algorithms are also used for detecting chest problems from radiographs, Alzheimer disease, cardiac disease, and so on.

While the performance of deep learning algorithms often exceeds the performance of expert radiologists and consultant oncologists, they are mostly evaluated in the controlled and historical image datasets. Thus the lack of so-called gold standard benchmarking via randomized control trials (RCTs) remains a major challenge. So, more RCTs need to be carried out to make AI solutions clinically viable. Also, the metrics used for AI systems are from traditional AI performance parameters (such as specificity and selectivity), which may not translate to effective clinical performance. Thus more clinically suitable metrics need to be defined to ensure appropriate evaluation of clinical AI systems.

Generalisation is also a major challenge for imaging. For example, skin cancer detection from images is very effective but performance varies widely among various ethnic groups due to different skin tones. Thus, a mechanism needs to designed to avoid this bias in AI system training datasets.

17.4 AI IN BIOINFORMATICS AND MULTIOMICS

In multiomics analysis (see Chapter 8), multiple datasets are analyzed in an integrated manner to uncover the pattern associated with a particular disease. AI-enabled multiomics analysis helps us to identify new biomarkers and design targeted therapy. To date, researchers have developed many such tools to make multitier datasets more effective [6]. Nevertheless, we need to develop more precise and efficient tools, especially for clinical practice applications. In this context, effective collaborations between molecular biologists, medical practitioners, and computer scientists are essential to developing biological knowledge-guided holistic frameworks.

17.5 AI-ENABLED ELECTRONIC HEALTH RECORDS

The Electronic Health Record (EHR) is a widely adopted mechanism for managing patient information and administering treatment/medicine. Traditionally, labor intensive and expert dependent phenotyping has been used to make them useful for further exploitation as it is really challenging to produce workable analytic models for EHR data for further processing. Deep learning has been used for automating the process and reducing human expert dependency but has thus far gained limited success due to a lack of generalizability of datasets in various institutions and settings [9]. Also, finding a true pattern from long-term data is very challenging due to the complexity and diversity of various clinical events along with fluctuation of data density and irregular sampling problems. Multimodality of data is another challenge for automation and makes the task of labeling very complicated. Most machine learning (ML) approaches consider neural networks as black box and hence lack interpretability, explainability, and transparency of their inner workings, which makes it difficult for efficiency benchmarking as well as reusing the system in different settings. Transfer learning [2] can be exploited to improve the performance of AI-based EHR systems, which can prevent the above-mentioned pitfalls.

17.6 EXPLAINABLE AI IN HEALTHCARE

AI-driven systems are undoubtedly well-performing. Yet the lack of explainability continues to trigger criticism, especially when it comes to its applications in healthcare. Explainability "can assist clinicians in evaluating the recommendations provided by a system based on their experience and clinical judgment. This allows them to make an informed decision whether or not to rely on the system's recommendations and can, consequently, strengthen their trust in the system" [1]. The paradigm of explainable AI (XAI) in healthcare is enormous because of its multidimensional perspectives. It involves not only the technical work but also medical, ethical, legal, and social aspects that indeed need in-depth investigation [1]. Currently, some studies on disease diagnosis from explainability context based on generic XAI tools such as SHAP and LIME exist [3]. However, healthcare-focused domain-specific XAI frameworks need to be developed.

17.7   SECURITY AND PRIVACY CHALLENGES

While AI techniques have shown promising improvement in disease diagnosis and prognosis, security remains a major challenge. As identified in Chapters 15 and 16, there are significant vulnerabilities that need addressing. Particularly, many algorithms have been found to have loopholes that could be compromised to adversarial attacks. This can be life-threatening for some cases. For example, images of benign moles can be misdaignosed as malignant by deliberatly adding adversarial noise or even just by introducing a simple rotation [5]. Appropriate steps need to be taken to ensure such attacks can be detected and remedied.

Bibliography

1. Julia Amann, Alessandro Blasimme, Effy Vayena, Dietmar Frey, and Vince I Madai. Explainability for artificial intelligence in healthcare: A multidisciplinary perspective. *BMC Medical Informatics and Decision Making*, 20(1):1–9, 2020.

2. Edward Choi, Mohammad Taha Bahadori, Andy Schuetz, Walter F Stewart, and Jimeng Sun. Doctor AI: Predicting clinical events via recurrent neural networks. In *Machine Learning for Healthcare Conference*, pages 301–318. PMLR, 2016.

3. Shaker El-Sappagh, Jose M Alonso, S. M. Riazul Islam, Ahmad M Sultan, and Kyung Sup Kwak. A multilayer multimodal detection and prediction model based on explainable artificial intelligence for Alzheimer's disease. *Scientific Reports*, 11(1):1–26, 2021.

4. Andre Esteva, Brett Kuprel, Roberto A Novoa, Justin Ko, Susan M Swetter, Helen M Blau, and Sebastian Thrun. Dermatologist-level classification of skin cancer with deep neural networks. *Nature*, 542(7639):115–118, 2017.

5. Samuel G Finlayson, John D Bowers, Joichi Ito, Jonathan L Zittrain, Andrew L Beam, and Isaac S Kohane. Adversarial attacks on medical machine learning. *Science*, 363(6433):1287–1289, 2019.

6. Sijia Huang, Kumardeep Chaudhary, and Lana X Garmire. More is better: recent progress in multi-omics data integration methods. *Frontiers in Genetics*, 8:84, 2017.

7. Christopher J Kelly, Alan Karthikesalingam, Mustafa Suleyman, Greg Corrado, and Dominic King. Key challenges for delivering clinical impact with artificial intelligence. *BMC Medicine*, 17(1):1–9, 2019.

8. Faisal Tariq, Muhammad RA Khandaker, Kai-Kit Wong, Muhammad A Imran, Mehdi Bennis, and Merouane Debbah. A speculative study on 6g. *IEEE Wireless Communications*, 27(4):118–125, 2020.

9. Cao Xiao, Edward Choi, and Jimeng Sun. Opportunities and challenges in developing deep learning models using electronic health records data: A systematic review. *Journal of the American Medical Informatics Association*, 25(10):1419–1428, 2018.

Index